Areas of barren rock and scree around the edge of Antarctica provide a breeding ground for two of the continent's most well-known species of bird: the south polar skua and the Adélie penguin. This book considers the relationship between these two species, taking as its study site Ross Island. Through detailed observations of the foraging ecology of the skua, the traditional view that skuas are totally dependant on penguin eggs and chicks for food is challenged. In addition, studies of the impact of skuas on penguin breeding and the extent to which the skua breeding cycle is functionally related to that of the penguin provide further evidence to suggest that the two species occur together independently as a consequence of limited breeding space, rather than as a result of a distinct predator–prey relationship.

T0155732

Skua and penguin: predator and prey

Studies in Polar Research

This interdisciplinary series, aimed at all scientists with an interest in the world's cold regions, reflects the growth of research activity in the polar lands and oceans and provides a means of synthesizing the results. Each book in the series covers the present state of knowledge in a given subject area, resulting in a series which provides polar scientists with an invaluable, broad-ranging library.

Skua and penguin:
predator and prey

EUAN YOUNG

School of Biological Sciences,
University of Auckland,
Auckland,
New Zealand

CAMBRIDGE
UNIVERSITY PRESS

CAMBRIDGE UNIVERSITY PRESS
Cambridge, New York, Melbourne, Madrid, Cape Town, Singapore, São Paulo

Cambridge University Press
The Edinburgh Building, Cambridge CB2 2RU, UK

Published in the United States of America by Cambridge University Press, New York

www.cambridge.org
Information on this title: www.cambridge.org/9780521322515

First published 1994
This digitally printed first paperback version 2005

A catalogue record for this publication is available from the British Library

ISBN-13 978-0-521-32251-5 hardback
ISBN-10 0-521-32251-0 hardback

ISBN-13 978-0-521-01813-5 paperback
ISBN-10 0-521-01813-7 paperback

Contents

Preface

The drama of antarctic bird life is not without its villain. Theft and pillage, murder, cannibalism and infanticide, these crimes are all in the repertory of the South Polar Skua

Siple and Lindsey, 1937.

On 16 November 1959 I arrived by helicopter at Cape Royds on Ross Island to make up the second half of a New Zealand team studying penguins and skuas. For the rest of that quite beautiful summer we worked together to unravel as much as we could of the biology of these two extraordinary species. We were living in Shackleton's 1908 hut and working about the local area made so familiar from the photographs and accounts of early expeditions.

Much of the skua's biology seemed pretty ordinary and they acted as one would have expected of any large gull, with a scavenging–predatory life-style. However, there were two surprises.

First, for some seemingly inexplicable reason the parents allowed the older of the two chicks to harass the younger and chase it from the nest area. Almost none of the younger chicks survived. This problem has since engaged a great deal of research, with a fine study of its causation being undertaken later at Cape Bird by one of the students there. It also set me off on a comparative study of chick behaviour and chick survival of the brown skua among the grasses and shrubs of the benign Chatham Islands environment. The consequences of this decision are still being played out with a long-term study of communal breeding among the skuas there. But that is a separate story.

Second, most skuas were found to feed not at the penguin colony, as expected from the lurid accounts of penguins and skuas by earlier writers, but foraged at sea – plunge diving to take small fish. The penguin colony was defended by a few pairs and these kept all the other skuas in the area away. But even those resident on the colony also fed at sea at times during the season. This re-appraisal of the skua's feeding ecology was not universally accepted for two reasons. First, because the Royds colony was so small it had little food anyway compared with that to be found in a really big colony – like Crozier or Hallett. And second, because at other (bigger) colonies there was bound to be so many penguin chicks dying during summer they would provide a 'larder' of food for the skuas when the penguins had departed for the year.

There did not seem to be any way to answer these criticisms except by looking at skuas at a larger penguin colony. In early November 1965 we began working at Cape Bird, selecting the Northern Colony of the three there from aerial photographs as the best place to work. It was a good choice. It was much more sheltered than the others, much more scenic and an excellent natural laboratory with a wide range of penguin breeding groups in different situations and with lots of skuas. One of its greatest advantages was that there were several marvellous observation points overlooking groups of skuas and penguins. We got to know H Beach and EF Block very well from these high points over the coming years.

The first season was spent in tents. For the following year a hut was in place, which has since been expanded and refurbished. It has now been a summer home to numerous research people.

This then is an account of the first five years' work on the penguins and skuas at Cape Bird. It should have been written up years ago. However, the long delay has been of value. The study of birds has advanced wonderfully in the meantime; there is now a much better understanding of population processes and of the behavioural motivation of animals; there have been numerous studies of both penguins and skuas at many different places around the continent and its islands, and there have been studies also of related species; and much more is known also of the glacial history of Antarctica and of the history of life there. There has also been the opportunity to return twice to the research area and to neighbouring colonies, bringing memories up to date and providing new insights into the association between the two species. All of this information has changed the way the original results have been analysed and initial conclusions re-interpreted. Even so, the long delay in publication is inexcusable.

The research in Antarctica was carried out while the author was a member of the Zoology Department at Canterbury University. The writing up has been done since appointment to the University of Auckland.

Acknowledgements

First, thanks must go to the University of Canterbury and the Zoology Department for supporting this long-term research programme and to the Antarctic Division, Department of Scientific and Industrial Research for support and transport while in Antarctica. It was the wintering-over staff at Scott Base in 1966 that prefabricated the hut and transported and erected it at Cape Bird in time for the next summer's research. The programme each year was funded by the Research Committee of the University Grants Committee. I am grateful for the invariably sympathetic and interested involvement of this committee – and mourn its replacement by a faceless Wellington bureaucracy.

Each year several students and colleagues from the Zoology Department assisted with the study, either directly as research assistants or indirectly by carrying out complementary research programmes for their postgraduate courses. In the first year, however, Antarctic Division provided a field assistant cartographer. Reg Blezard, over two seasons, produced the maps of all three penguin colonies at Cape Bird that have underpinned research there ever since. These people are listed below for each year of study with notes of their research involvement.

1965–66 Reg Blezard.
1966–67 John Darby, Jim Peterson and Dennis Procter.
Many of the photographs used in the book were taken by John. Dennis Procter carried out research on sibling aggression and survival in skua chicks while Jim Peterson researched hormone balance and change in incubating Adélie penguins – an ambitious project for the time.
1967–68 John Darby, Jim Peterson, Eric Spurr and Morgan Williams.
Eric began a four-year study of penguin breeding behaviour for his doctoral thesis, most of which appeared to be done from the top of a step-ladder overlooking his study groups, and Morgan Williams began a two-year masters degree study of skua territorial behaviour.
1968–69 Eric Spurr and Morgan Williams with Trevor Crosby and Tony Harrison as research assistants in the general research programme on skua–penguin interactions.
1969–70 Eric Spurr with Peter Kettle, David Smith and Pamela Young as research assistants.

It has been most gratifying over the years since this Antarctic experience to see the universal success of these students and colleagues to senior positions within the New Zealand science community. Pamela Young, the first New Zealand woman to work in a science programme in Antarctica, has since this time raised to independence the family shown as small children in her book of the Antarctic while teaching at Epsom Girls Grammar School in Auckland.

The work of producing this account has been encouraged and helped by colleagues at Auckland and overseas. Bob Furness of Glasgow University and John Croxall at the British Antarctic Survey, Cambridge have been especially helpful and tolerant of the many questions put to them, on providing unpublished information and in reading sections of the text. At Auckland statistical advice has come from Brian McArdle and Diane Brunton. The art work heading each chapter and most of the figures in the text were produced by Vivian Ward and all of the photographs were produced by Iain MacDonald.

Some photographs of skua–penguin interaction have appeared previously in Young, P. M. (1971).

E. C. Young

1

Introduction

PENGUIN
(pe.ngwin, pe.ngwin) 1578. [Of unkn. origin.]
1. The Great Auk or Garefowl (*Alca impennis*) – 1792.
2. Any bird of the family Spheniscidae, including several genera of sea-fowl inhabiting the southern hemisphere, distinguished by having the wings represented by scaly 'flippers' or paddles with which they swim under water 1588.

SKUA
(skiu.a). 1678. [- mod. L. – Faeroese skugvur = ON. skufr, of unkn. origin]
A predatory gull belonging to the genus *Stercorarius*, esp. the largest European species, *S. catarrhactes*, which breeds in Shetland, the Faeroes, and Iceland. Also s.-gull.

The Shorter Oxford English Dictionary. Third Edition, 1978.
Oxford, Clarendon Press.

Prey, in natural populations are generally not eradicated by their predators. Indeed, if this were not so, complete extinction of both prey and predators would soon ensue, the elimination of the prey species leaving the predator to die by starvation

Klopfer, P. H., 1962.

Many cases are on record showing how complex and unexpected are the checks and relations between organic beings, which have to struggle together in the same country

Charles Darwin, 1859.

1.1   *Introduction*

The overwhelming impression of Antarctica is of snow and ice, blinding sunlight or mist, and fierce cold. But there are also substantial areas of barren rock and scree dotted around the continent's edge allowing a limited terrestrial biota a precarious existence. In general it is thought that these biota are all recently arrived, colonising the increasingly ice-free ground as the glacial maxima waned.

This book is about the two most visible and accessible birds on the continent – the Adélie penguin and the south polar skua – familiar both to Antarctic visitors and through photography and film among the most easily recognisable of the world's bird species. There is a long history of research on both, with reports dating from the very first days of Antarctic scientific expeditions. No other Antarctic birds are as accessible or as well known – the Emperor penguins breed at the wrong time of the year for

Fig. 1.1. The two most visible and accessible birds on the continent.

easy observation and tourism and the several petrels are either tucked away on mountains or spend their time in crevices and holes. They are, anyway, not nearly so interesting for the casual visitor, nor so evocative of Antarctic life. However, all share the common feature of being sea birds dependent entirely on the food web of the oceans for life; the land is barren.

It is also about a special place: the warm and sheltered slopes lying along the western margin of Mt Bird on Ross Island looking out over McMurdo Sound to the Royal Society Mountains on southern Victoria Land. The three Adélie penguin colonies spaced out along the shoreline of these slopes, known prosaically as the Northern, Middle and Southern Cape Bird colonies, provided the study populations for this research.

The penguins and skuas begin to arrive on the breeding areas on the continent from early October and because snow-free places with good access to open water are restricted they settle often in the same places – Adélies tightly aggregated into breeding groups, the skuas more widely dispersed in loose-knit colonies. Where the two species breed together at

Fig. 1.2. The Northern Penguin Colony at Cape Bird. The slopes of the upper colony look to the north and the sea and the bounding ice cliffs of the Mt Bird glacier.

the same site they come into specially close contact, with the skuas foraging among penguins for carrion and live eggs and chicks.

The early view of this association between the two species was that the skua was entirely dependent on the penguins for food. This was an eminently reasonable supposition for the time. The science of ethology, the systematic observation of animals in the field, was in its infancy and even such basic techniques as banding to allow recognition of individual birds was not yet in vogue. In terms of public relations the skua is its own worst enemy and their high visibility when taking penguin eggs and struggling to kill penguin chicks have for decades dismayed and disgusted visitors. Not surprisingly, these casual observations led almost universally it seemed to a biased view of their habits and ecology. They were the hyenas of the Antarctic paradise – that is, before Hans Kruuk (1972) rehabilitated hyenas by demonstrating their importance as predators.

Although specifically about the skuas and penguins, this account has wider generality about predators and prey. The advantage of doing such research in Antarctica is that the association is a relatively simple one – one in which the general rules might become more obvious than in more complex systems with numerous predators and prey interacting in a very complex way. At Cape Bird the penguins could only feed at sea; on krill, other zooplankters and small fish. The predator could feed either at the penguin colony or at sea. Moreover, at the colony the predator and prey were fully exposed to direct observation in all parts of their active ranges and throughout the whole 24-hour day. The way the two species interact, the shifts and changes in behaviour and the apparent decisions should all become pretty obvious, given patience, time and good fortune. R. J. Taylor (1984) has defined predation as occurring:

1. 'when one organism kills another for food' which is narrowly defined if restricted to animal prey but broadly defined if including herbivory. This is the most generally accepted definition by behavioural ecologists.
2. 'When individuals of one species eat living individuals of another' This definition is significantly broader than the first embracing herbivory and parasitism as it does not require that prey must die during the process of predation; or
3. as a 'process by which one population benefits at the expense of another' a definition that focusses only on the ecological result; or
4. 'any ecological process in which energy and matter flow from one species to another', which is the broadest definition of all and really

has no exceptions as it would embrace both heterotrophs and decomposers.

All these definitions are applicable to the relation between skua and penguin but in this study, focussing largely on behavioural interactions, the first is the most appropriate.

The key constraint for this study in the definition of predation is not, however, in the selection of the most appropriate mode, but in distinguishing between scavenging (for carrion – abandoned or dead eggs and chicks) and predation (attacks on living prey). This difficulty will be considered in detail later in the text so that an example in illustration is all that is required at this point. When skuas take penguin eggs that have been lost from the nest are they scavenging or preying? The embryo is probably still alive but functionally it must be considered already dead. And to what extent did previous skua attacks or disturbance of the colony contribute to the loss or abandonment of the egg in the first place?

There are a number of special attributes of this prey–predator association. Both species are birds; they come into contact only during the breeding season for a few months each year; both are carnivorous marine species; the predator is a powerful flying species whereas the prey species is flightless; the prey are the eggs and chicks of the penguins and for most of the breeding season must be defended at a fixed nest position so that fleeing is not an option for attacked penguins. The situation here where the prey are protected by parents, is not conceptually different from any other interaction where the prey are the offspring of animals too big or dangerous to be attacked themselves. Such interactions are common between carnivores and ungulates; between wolves and muskox or fox and sheep, for example. Only in exceptional circumstances are the adults themselves attacked as prey, and the attacks on them, as in the attacks of skuas on penguin adults, are mock attacks to distract, harass and intimidate, so that the real prey, the offspring, can be killed.

The interaction is not entirely one-sided, however, and penguins can inflict severe injury on attacking skuas and cause the loss of their eggs and chicks. But at all times it is an asymmetrical relationship for even when the penguins inadvertently or intentionally break skua eggs or trample young chicks, or even severely injure the adults in defence against predatory or scavenging attacks – it is not as prey. Penguins do not eat skuas.

For skuas nesting away from penguin colonies the decisions on foraging are the straightforward ones of when to forage, which area of sea to search and which fish or plankton species to take.

For skuas breeding within the penguin colonies there is the complicating factor of food availability at the colony, and with this a remarkably clear separation of choices and decisions – whether to forage at the colony and in what way (i.e. scavenging or preying), or whether to forage at sea. There is the added complexity also of the timing of the switch from one foraging choice to the other – of when to give up an unsuccessful foraging bout on the colony after initially choosing to forage there.

These are the sort of decisions that are considered by some as best understood through application of optimal foraging theory (OFT) in which decision-making rules are favoured through natural selection to maximise foraging gain or efficiency (Stephens and Krebs, 1986). Although this study of penguins and skuas was not based on such models (it was undertaken before they had gained their current scientific standing) it is nevertheless difficult not to be intrigued by the way different skuas acted in choosing where to forage and how an individual bird appeared to reach decisions on how long to attack each penguin, when to shift to another breeding group, when to abandon the foraging bout entirely – and when to leave the area and forage at sea. To a large degree the day to day observations made of the skuas seemed to be taken up with just these questions. For this reason, as it figures so importantly in the way the study was undertaken, a summary of foraging choices and points of decision by the skuas is illustrated in Fig. 1.3. In the absence of any way to listen in to the thoughts or conversations of these birds one can only record and interpret the behaviour observed. Did they actually 'decide' consciously and after weighing up the situation and needs, or were they merely operating automatically, governed by rules established through evolution-ary history? It is impossible to tell. But often when watching a foraging skua over the long afternoons and identifying closely, perhaps too closely, with its difficulties and frustrations in not being able to get an egg or a chick it was all to easy to translate the pauses and breaks in the attack bouts, while the bird seemed to be re-assessing and gathering resolution for another bout, as consciously weighing up the possible risks and gains before deciding on what to do next. How else to interpret, for example, the routine quick searching flight over the penguins after exhausting and fruitless hours in attack bouts at the colony before going off to sea as anything except, 'Well, I will give it one more chance and if there is nothing really encouraging in view I will have to head for sea'. This story suggests that skuas prefer to forage at the colony rather than at sea. For a fair proportion of birds this was certainly the case. As will become apparent in the text, however, this attitude was not universal, and some

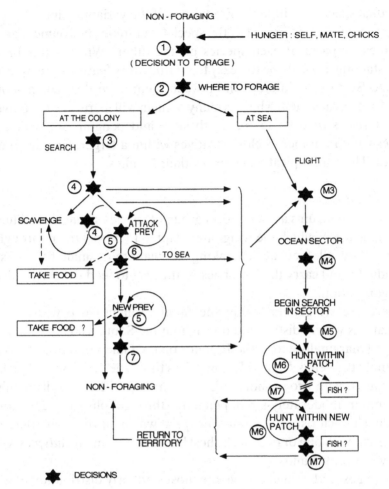

Fig. 1.3. Schematic outline of the options and decision points for skuas foraging on penguin colonies and at sea. 1–7 indicate decision points for skuas on land, M3–M7 indicate decision points for skuas foraging at sea.

skuas holding territories containing penguins routinely foraged at sea as their first preference. Of course, before penguin eggs appeared in spring and after the last penguin chicks had departed in late summer all skuas must forage at sea, as they do for the whole time they are away from the breeding area.

1.2   *The two species in taxonomic and biogeographic context*
Understanding the behaviour and feeding biology of the two species of this association requires some background about each bird's systematic position to appreciate what can be expected of them as

individual species – within the limitations of the variability inherent in the group as a whole, the impact of the special Antarctic environment and the reciprocal impacts of each species on the other. Without this broader understanding it would be too easy to ascribe every feature of their biology at Cape Bird to the interaction between them or to the demands of the Cape Bird environment whereas many of them will merely be the universal characteristics of each taxon. Both skua and penguin are widespread species with a number of close relatives within a larger, similar group of species. They are typical members of their families.

1.2.1 *Penguins*

The penguins are a uniform group of species strongly modified for an aquatic life in which the wings are adapted to propel them through the water. They all feed at sea taking zooplankton, small fish or squid individually and carry the food back to the nest to feed the chicks directly by regurgitation.

There are six genera within the family Spheniscidae with breeding populations widely distributed throughout the Southern Hemisphere and reaching marginally across the Equator on the Galapagos Islands. Within the family the genus *Pygoscelis* comprises three species of squat black and white birds, with broad short bills and long tails. They are distinguished from other moderately-sized penguins through plumage and colour-pattern characteristics – from *Eudyptes* which have a red bill, from *Spheniscus* with one or two black chest bands and from *Megadyptes* with its yellow eye-brow bands.

The pygoscelid penguins have the most southerly distribution of all the penguins, with the exception of the Emperor penguin *Aptenodytes forsteri*, with the largest populations for the genus occurring on the Antarctic continent itself. But only the Adélie penguin breeds on the continental mainland. There are three species.

> *Pygoscelis papua* Gentoo penguin. This species has the most temperate distribution of the three (with over 40% of breeding birds in the cold temperate zone) and is widely distributed throughout the Southern Ocean islands reaching, as a breeding bird, as far north as Macquarie Island. It is also found, however, with the other species on the Scotia Arc Islands and on the Antarctic Peninsula.
>
> *Pygoscelis antarctica* Chinstrap penguin. This is the dominant pygoscelid of the Scotia Arc Islands, making up over 85% of the

breeding population of the three species there. These islands are the centre of its population range with over 90% of the species in this sector of the Antarctic zone. But there are also surprisingly distant outlying populations; on Isla Hornos at the southern tip of South America, and even more extraordinarily, a few birds on the opposite side of the continent on the Balleny Islands south of New Zealand.

Pygoscelis adeliae Adélie penguin. This species has the most southerly distribution of the three, being confined to the Scotia Arc Islands, the Antarctic Peninsula and the continent. It is manifestly a species of the Extreme Antarctic Zone, with over 75% of its numbers breeding on the Antarctic mainland and close islands. By contrast, fewer than 5% breed on the Scotia Arc Islands and none breed as far north as South Georgia.

Only on the Scotia Arc Islands and the Antarctic Peninsula do the breeding ranges of all three species overlap. Recent counts of chinstrap, gentoo and Adélie penguins at three selected locations (G. J. Wilson, 1983) reflect the increasing dominance of the Adélie with proximity to the continent. At Signy Island (South Orkney Islands) there were 370, 66 000, and 31 000 breeding pairs of each species respectively, on King George Island in the South Shetland group 6000, 57 000 and 47 000 and on Anvers Island, midway down the Antarctic Peninsula, 2000, 6000 and 17 000 pairs of each.

The three species are much the same size. None show colour pattern dimorphism but all three have moderate, though scarcely discernible in the field, sexual size dimorphism. The gentoo is both the tallest and heaviest species of the three, with the chinstrap being the shortest and lightest. Their body sizes, as given by Volkman *et al.* (1980) are shown in Table 1.1.

The body lengths given are the standard ones of ornithologists, with the birds stretched out and length recorded from the tip of the beak to the tip of the tail. A standing bird in its normal posture would appear much shorter.

This very general survey of the penguins, with rather more detail for the genus *Pygoscelis*, is nevertheless sufficient to demonstrate how small the variation within the family is. One should not expect, therefore, extraordinary behaviour among the Adélies at Cape Bird. They are but one population of a widely dispersed species, itself typical enough of an even more widely dispersed genus. The penguins are moreover a conservative

Table 1.1. *Size and weight of pygoscelid penguins*

	Body weight (kg)	Body length (mm)
P. papua		
male	5.5	652
female	5.1	622
P.adeliae		
male	4.5	580
female	4.2	563
P. antarctica		
male	4.3	579
female	3.7	548

group, with few taxonomically distinguishable forms within individual species. As far as can be told at present the Adélies are identical throughout their Antarctic range, as are the Emperors,. They form, apparently, a single panmictic population. Moreover, the Adélies similarly confront the same skua species, the south polar skua, at all mainland breeding sites and only in the Antarctic Peninsula and on the Scotia Arc Islands do they encounter the brown skua as well. This taxonomic and ecological uniformity allows ready comparison with other research on these species.

1.2.2   Skuas

In contrast to the penguins, which seem a well ordered and well recognised group with explicitly defined species and genera, there is still uncertainty and debate about the systematic ordering of the skuas. Three questions have long taxed systematists: 1. the relationship of the skuas to the (other) gulls; 2. the relationship between the jaegers (the smaller species) and the large skuas; and 3. the classification of the numerous and widespread populations of skuas. It is the last of these questions that is most germane to the present account because it impacts on the validity of comparative accounts of their biology.

The most recent general review (Furness, 1987) of the systematics of the group was based largely on the descriptions of plumage and measurements of body size, essentially a 'museum collections' approach but supported by an increasing knowledge of biology from field studies. Furness distinguished the skuas from the gulls, placing them in the family Stercorariidae, separated the smaller jaegers within a separate genus from the larger skuas and recognised three species of skuas, one with four subspecies. This is the most detailed and comprehensive account so far. His classification is as follows:

Family Laridae (gulls)
Family Stercorariidae (jaegers and skuas)
Genus *Stercorarius* (jaegers or small skuas)
1. *Stercorarius parasiticus*   arctic skua
2. *Stercorarius pomarinus*   pomarine skua
3. *Stercorarius longicaudus*   long-tailed skua

Genus *Catharacta* (skuas)
1. *Catharacta skua skua*   great skua (bonxie)
   Catharacta skua hamiltoni   Tristan skua
   Catharacta skua lonnbergi   brown skua (southern or subantarctic skua)
   Catharacta skua antarctica   Falkland skua
2. *Catharacta maccormicki*   south polar skua (McCormick's or Antarctic skua)
3. *Catharacta chilensis*   Chilean skua

Since this work there has been a comprehensive revision of avian systematics by Sibley and Monroe (1990) on the basis of DNA–DNA hybridisation techniques for assessing affinity. In their classification of the skuas and jaegers the group is considered a tribe (Stercorariini) of the Laridae and the different skuas as allospecies of a superspecies taxon of which the senior name is *skua*. They were persuaded to this view by observing that '*C. antarctica* and *C. lonnbergi* are usually considered conspecific with *C. skua*, [but] the degree of difference between these allopatrically breeding populations is comparable to that between *C. lonnbergi* and *C. maccormicki*, which breed (with some hybridisation) in the South Shetland Islands (Parmalee [sic] 1988)'. They include the populations on Tristan da Cunha and Inaccessible Island within *C. antarctica*.

The classification used for the skuas in the present account follows their revision. These scientific and common names are:

Catharacta skua   great skua
C. antarctica   Falkland or southern skua

C. lonnbergi	brown skua
C. chilensis	Chilean skua
C. maccormicki	south polar skua

Debate will nevertheless continue. A key question is the significance to be given to the hybridisation between *C. lonnbergi* and *C. maccormicki* throughout the Antarctic Peninsula and as far north as Signy Island (Devillers, 1978; Parmelee, 1988). Hybridisation is recorded also between *C. chilensis* and *C. antarctica* on the Patagonia coast of South America (Devillers, 1978). Both these hybridisations are of considerable taxonomic interest as each is taking place between the most distinctive, and most surely regarded species on one side (*chilensis* and *maccormicki*) with members of the much more variable and less certain taxa on the other.

The map of the breeding distribution of the large skuas in the Southern Hemisphere (Fig. 1.4) emphasises the enormous geographic range of *C. lonnbergi*, distributed around the continent among the widely dispersed Southern Ocean islands (but because of the distribution of these islands in fact appearing in three loci, the New Zealand and Australian islands, islands in the southern Indian Ocean and those in southern Atlantic and Scotia Arc) and the essentially Antarctic range of the south polar skua. The broad extent of range overlap between these two species is also apparent. Hybridisation occurs in the Atlantic sector and would no doubt take place on the Balleny Islands south of New Zealand if *C. lonnbergi* does indeed occur there. By contrast *C. chilensis* and *C. antarctica* on South America and the Falklands and Patagonia respectively have restricted ranges and range overlap.

All these skuas have a rather similar breeding biology. They occupy and defend large territories, which may be breeding territories or breeding and feeding territories (Hinde, 1956). In at least two taxa, *C. lonnbergi* and *C. maccormicki*, both sorts may occur within the same local population. On the Chatham Islands, for example, all *C. lonnbergi* pairs have similarly large territories but some 20% lack food. Over most of its geographic range *C. maccormicki* defend breeding territories but in places where they nest near penguins or petrels territories may also contain food. All skuas are ground nesters and lay two eggs. During incubation the eggs are held within symmetrically-placed brood pouches beneath the wings. Young chicks are brooded there also during the first few days from hatching and later during severe weather. In many populations both chicks may be raised to fledging, with a resulting overall high breeding success. South polar skuas at high latitude sites are exceptional because of their generally poor and variable success. In part this is a direct consequence of a more

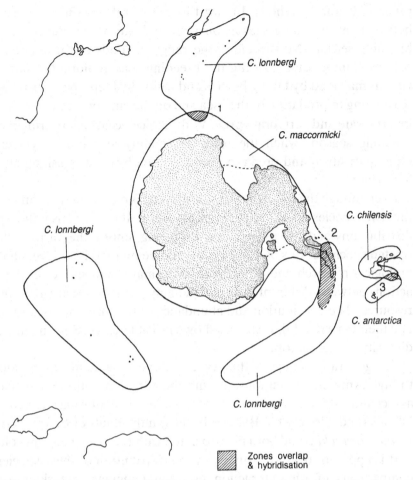

Fig. 1.4. The breeding distribution of skua species on the Antarctic, Southern Ocean islands and South America. Sympatry and hybridisation occurs (hatched areas) between *C. maccormicki* and *C. lonnbergi* on the Antarctic Peninsula and the southern islands of the Scotia Arc and between *C. chilensis* and *C. antarctica* in Patagonia. 1. Balleny Islands; 2. Antarctic Peninsula and Scotia Arc; 3. Patagonia coast.

rigorous breeding environment but the invariable loss of the younger chick of pairs hatching contributes significantly (Young, 1963a; Procter, 1975).

Flexibility and opportunism are the hallmarks of skua feeding biology. This is more exaggerated for populations on the richer Southern Ocean islands. Birds on neighbouring islands may feed differently and pairs on the same island may specialise on different prey. In these habitats each population must be studied individually. Tradition and learning are clearly important in skua feeding ecology. But skuas are remarkable also for their range of feeding methods. In part this simply reflects their divergent life between the breeding season on land and life at sea during the non-breeding season. Nevertheless, the range of foods taken by great skua species is impressive. During the breeding season alone the same individuals may feed by taking berries and fruits, by kleptoparasitism, and by scavenging or predation both at sea and on the land on a range of different invertebrate and vertebrate prey. Wide choice occurs also during the non-breeding season, when the birds are mainly at sea with scavenging, kleptoparasitism and predation on other sea birds, and fishing all being important.

Even though the focus of the present study is on the interaction between the two species in Antarctica, it is useful to summarise the species distributions and the associations occurring among the various closely-related species of this pair. The comparative method is a powerful tool in biological research and differences, both major and subtle, in the interactions between different skua and penguin populations are an important resource. The distribution and abundance of the three pygoscelid species and their associated skua species shown in Table 1.2 disclose an impressive diversity of interaction.

The geographic range of the pygoscelid species extends from temperate to high Antarctic conditions. Within this range the different species are associated with four skua taxa. On the Antarctic Peninsula and Scotia Arc Islands (and possibly the Balleny Islands) individuals of *C. lonnbergi* and *C. maccormicki* could both be associated with the same penguin colonies.

It is apparent from this survey of the distribution of these species that comparisons of the interaction between penguins and skuas on the Antarctic mainland will be of the same species but that comparisons with studies elsewhere will need to take into account that different species may be involved living in quite different environments.

A full assessment of this interaction between south polar skuas and Adélie penguins can only be made in the context of the Antarctic environment and of their history together. An earlier attempt (Young, 1981) to relate the history of colonisation of Antarctic birds to the amelioration of climate since the last glacial maxima suggested that both penguins and skuas could have first begun to re-colonise Antarctica more than 10 000 years before present (BP), and to have reached as far as McMurdo Sound by 5–6000 years BP as the ice steadily contracted. This evidence of habitability from glaciological research has not yet been complemented by a detailed record of the history of bird or seal occupation. Although penguin remains dated from 5340 ± 70 years BP have

Table 1.2. *Distribution and approximate numbers of breeding pairs of pygoscelid penguins in each biogeographic zone and associated skua species*

	P. papua	P. antarctica	P. adeliae	Skua species
Mild Antarctic Zone				
S. American coast		1000		*chilensis*
Falkland Islands	100 000			*antarctica*
Cool Antarctic Zone				
Southern Ocean islands[a]	7500			*lonnbergi*
South Georgia	100 000	2000		*lonnbergi*
Kerguelen	7500			*lonnbergi*
Heard Island	10 000			*lonnbergi*
Cold Antarctic Zone				
Scotia Arc Islands[b]	22 000	810 000	105 000	*lonnbergi and maccormicki*[c]
Antarctic Peninsula	22 000	70 000	390 000	*lonnbergi and maccormicki*
Extreme Antarctic Zone (continent and near islands)				
Balleny and Scott islands		10	6000	*lonnbergi & maccormicki*[d]
Bouvetoya Island		9000	60	?[e]
Peter Oy Island		10	?	?
Continental mainland			1 730 000	*maccormicki*

[a] Marion and Prince Edward Islands, Crozet Island, the New Zealand 'Subantarctic' islands, Macquarie Island.
[b] South Sandwich, South Orkney and South Shetland islands. Clarence and Elephant Islands.
[c] *C. maccormicki* is not on the South Sandwich Islands.
[d] Occurrence of *lonnbergi* to be confirmed.
[e] No published information.
Data from G. J. Wilson (1983), Poncet and Poncet (1987), Taylor *et al.* (1990); Antarctic climate zones from Longton (1988).

been found at Franklin Island, 125 km north of Ross Island, there are very few radiometric datings of subfossil penguin remains from Ross Island. Dated remains, until very recently, were all found to be of recent dates, no more than a few hundred years BP (Spellerberg, 1970a; Stonehouse, 1970). This earlier picture of recent colonisation of Ross Island has now been put in doubt by the publication of the dating of penguin remains found in a beach profile at the Middle Colony at Cape Bird (Heine and Speir, 1989). This material has a conventional radiocarbon age of 8080 ± 160 years BP and a corrected age of 6870 ± 170 years BP (correcting for the Antarctic Standard). It is therefore almost contemporaneous with the final decay of the McMurdo glacier identified by the transition from grounded ice to shelf ice in the southern McMurdo Sound, which is dated at 7750 years ^{14}C BP (Kellogg *et al.*, 1990). The difficulty with this research in establishing colonisation date is not so much in the techniques of dating as in the discovery of material from the earliest colonisation stage. A systematic search for material from the very earliest stages has not yet been carried out. All the dates so far are of penguins. Skuas, because of their ability to fly long distances for feeding, are perhaps even better colonisers than penguins and could have been present in the area even before the first penguin colonies were established. The association between the two species at Cape Bird is, therefore, a long-standing one, measured in some thousands of years, and hundreds of generations. Elsewhere on the continent and fringing islands it is of course of even longer duration, measured in many thousands of years. These conclusions derive from a model of biogeographic range movement in which all species were forced from the continent during Pleistocene glaciation maxima, with subsequent reoccupation during warmer interglacial periods. This simple model has been challenged recently by the discovery that petrels were able to breed on the continent during the peak of the last major glaciation (Hiller *et al.*, 1988). Petrels are a special case, however, nesting in protected crevices and able to range widely across the oceans when foraging. Skuas and penguins are far more constrained by their local environment, by the need for open ground for nesting and by near access to open water. In any case McMurdo Sound was still glaciated 10 000 years ago, precluding any possible colonisation of penguins there until those glaciers had largely decayed.

1.3   *Research philosophy and definitions*

1.3.1   *Research philosophy*

The most important results from this study have come from the direct observation of groups of skuas and penguins apparently going quite normally about their lives. In the light of recent studies of the sensitivity of the Adélies to human disturbance (Culik *et al.*, 1990), this normality may have been more apparent than real. In hindsight, the siting of the observation points for this work at a distance from the penguins so that the behaviour of a fair number of skua pairs among them could be recorded simultaneously has probably quite fortuitously protected the general conclusions from the criticism that they are invalidated because of disturbance. However, there was undoubtedly a significant impact on penguin breeding from the regular counts of breeding groups, from weighing chicks and from working near them when catching and handling skuas, and this should be borne in mind when assessing and comparing breeding statistics. Of equal concern was the impact of the study on the breeding skuas. There was some reassurance of normal behaviour in the frequency birds left to forage at sea, in the way they fed mates and chicks on return and in the long incubation spells undertaken by both birds of the pair. In disturbed pairs all these behaviours are disrupted: the birds are uneasy on the nest, they are reluctant to leave the territory and they will not feed mates or chicks. Fortunately, these pointers were all identified in the earlier research at Cape Royds and recognised again in an ancillary study on naive skuas quite early in the work at Cape Bird.

Antarctic skuas are notoriously poor breeders, and this study records a dismally low breeding success. Although the statistics of little-disturbed control groups provide valuable reassurance that perhaps working among them does not affect their breeding all that much, readers should bear the possibility in mind that the skua statistics are also flawed.

Even when this field work was being undertaken, long before the current great interest in Antarctic conservation and protection, quite firmly policed regulations and ethical rules applied. From my own inclinations and in the terms of my permits for working at Cape Bird, the study was carried out with minimal experimentation and without killing or maiming any birds. Indeed, the only birds killed during the five years at Cape Bird were two penguins and two skuas demanded urgently by the New Zealand Government during the height of the great 'DDT in Antarctica' scare. To the best of our abilities and within the knowledge of the time this was a study of undisturbed birds interacting when breeding together. It is an account of a large-scale natural experiment in which there was a wide range of different predator–prey circumstances, positively encouraging comparison.

A major shortcoming of the study is that it was constrained by the difficulties of access and travel more or less to the mid summer months, so that large parts of the breeding cycles of both birds were missed out. But the gaps in the cycle differed between the two species. By the time research began in mid November the penguins were already in full occupation and most were incubating but the skuas were just beginning to occupy their territories consistently. The events of the end of summer were also unrecorded. When the station was closed for the year in late January penguin chicks were already leaving for the sea, completing their cycle, but the skua chicks were still weeks away from fledging. These are the shortcomings of working from a summer field station. The span of records here are the usual ones for this area. No one seems to have seen the skuas leave or followed the moulting of Adélie penguins since Brian Reid was at Cape Hallett and Rowley Taylor and I were at Cape Royds in 1960.

1.3.2   *Definitions*

With three important exceptions the definitions of terms in penguin biology used in this study are those of Ainley *et al.* (1983), most of which have a long history of acceptance since being originally defined by Sladen (1958) and further refined by Penny (1968). The exceptions are in the use of the term colony for their rookery and breeding group for colony, and the use of post-guard stage instead of crèche stage for chicks. These are significant departures from established usage and have not been taken lightly. The established terms are, however, so misleading or ambiguous that they should not be used.

Throughout the text the term *colony* is used to describe an aggregation of breeding groups within a discrete geographic area. On Ross Island this term defines the six (or seven if the new one at Cape Barne is included) penguin colonies of Cape Crozier, Cape Bird and Cape Royds. Within each colony the breeding penguins are aggregated into *breeding groups*. These definitions reflect the force of the arguments of Oelke (1975) for a more rational terminology for these breeding aggregations.

The use of the term crèche or crèche stage (spelled variously crèche, créche, crêche) to describe both the stage in the chick's life when it is left unguarded on the nest and the behaviour of tight grouping, bunching or huddling is very misleading. For example, what is to be made of the Yeates (1968) statement (p. 483) 'I was somewhat surprised that creching did not occur in 1964–65 but attributed it to the warm conditions'. Does this mean that the chicks were not left unguarded in this year, or that they never came together in groups after being left?

The need is to recognise that two distinct processes occur at this stage. The first is the behavioural change of the adults to leaving the chicks alone on the nest for all or part of each day. The second is the behaviour of the chicks after being left on their own in the breeding group. Richdale (1951) concluded that the chick stage of all penguin species was characterised by a guard stage in which the chick was brooded or guarded continuously and a post-guard stage in which it was left on the nest or breeding area while both parents foraged. In some species chicks left on the breeding area might collect together in crèches; in species nesting in burrows or under shelter most would remain separately on the nest area. In the present account the chick stages are, therefore, termed the *guard* and *post-guard* stages. When the Adélie chicks collect closely together, as defined by Davis (1982a) for example, these are termed *crèches*. As Yeates (1968) has noted, although the chicks are left on the breeding area by the parents they might not necessarily form crèches at any time. The utility of distinguishing between the parent's behaviour of leaving the chicks and the subsequent behaviour of the chicks is apparent when describing the very earliest phases of the post-guard stage. During the first few days at Cape Bird there were abandoned chicks on nests or about the edges in each breeding group but none was in crèches as usually defined.

The terms used in the account of skua breeding and behaviour are those of Young (1970), Spellerberg (1971a) and Furness (1987).

The key questions that motivated this study and determined its direction were: 'To what degree are the skuas dependent on penguins?' and 'In what ways has the association between the two species impacted on each – in their morphology, behaviour, breeding location and seasonality?' These seemed simple enough goals at the outset, but as this work shows they were extremely demanding in research execution. The present account is little more than an introductory skirmish, indicating contentious issues and pointing towards larger topics and difficulties.

2

The study area: Ross Island and the Cape Bird penguin colonies

2.1 Introduction

Ross Island lies in the southwestern corner of the Ross Sea (Fig. 2.1). During summer, open water of the Ross Sea sweeps around the northern coast and McMurdo Sound on the western side separates the island from the rising slopes of continental Antarctica, which stretch across the western horizon to the Transantarctic Mountains and the polar ice cap. In contrast, the whole of the southern coast is invested permanently, summer and winter, by the Ross Ice Shelf. The island is of recent volcanic origin, and is dominated by Mt Erebus (3749 m). Three substantial but lower volcanic centres are spaced about Erebus to give the island overall a triangular form. Mt Bird (*not* Byrd!) (1800 m) lies to the north, Mt Terror (2150 m) is on the eastern flank and the Hut Point Peninsula runs away to the south.

As expected for an Antarctic locality much of the island is surmounted by ice cap, with centres on each of the three mountains, but in the current climate this does not cover the island everywhere to sea level and ice-free areas of rock, scree and silt exist on the northeastern point (at Cape

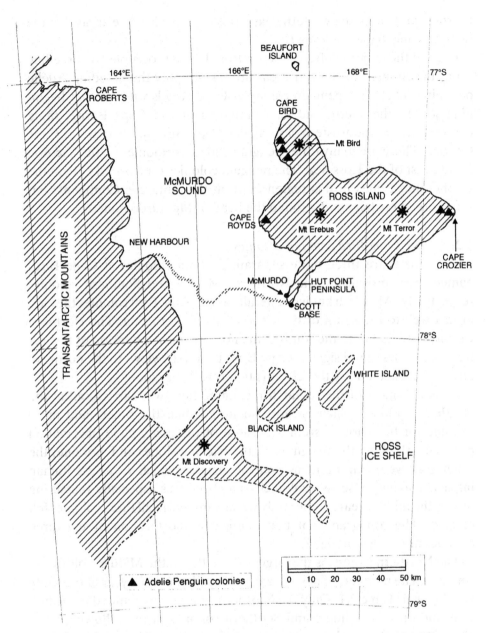

Fig. 2.1. Ross Island and McMurdo Sound showing the location of Adélie penguin colonies and their proximity to the McMurdo (US) and Scott Base (NZ) research stations.

Crozier) and along the western coast. It is on these ice-free areas that the penguins and skuas can breed.

Each of the major ice-free areas on Ross Island is occupied by breeding skuas in summer and smaller areas are taken up by the breeding Adélie penguins. There is a penguin colony at Cape Royds with another beginning just to the south, and an abandoned one at Cape Barne, three colonies are on the western slopes near Cape Bird and two are at Cape Crozier. There are many other penguin and skua colonies on the Victoria Land coast of the Ross Sea. The penguin colonies on Ross Island are the southernmost ones in this region (and on the Antarctic Continent) but skuas breed further south around the head of McMurdo Sound.

2.2   *The Cape Bird penguin colonies*

There are three quite separate penguin colonies at Cape Bird, named most prosaically the Northern, Middle and Southern colonies respectively. Although known generally as the Cape Bird colonies they are in fact well to the south of the Cape on its western side. The Northern Colony is c. 7.5 km to the south at the northernmost extension of the ice-free land and is out of sight of Cape Bird itself. Any views to the north are anyway blocked out by the ice cliffs that invest the colony.

Although most of the research recounted here was carried out at the Northern Colony, which is more sheltered than the others and more suitable for behavioural studies because of the many good observation points overlooking the penguins on the backing slopes and moraine, the other two were important as controls against which to assess human impact. Having these two colonies as a check was especially reassuring during the middle years of the study when numbers in all three colonies fell equally. The topography of the region and the location of the three colonies are shown in Fig. 2.2.

The Northern Colony is the largest of the three, the Middle Colony the smallest. The numbers of nests in early December in each year of the study are shown in Table 2.1. Only the Northern Colony was counted in the first year and only the Middle and Southern colonies in the following year. Thereafter all three colonies were counted each year. The numbers for the second year (1966–67) have been estimated by proportionality from the counts of the study breeding groups in each year. Counts were of occupied nests. In large breeding groups it was impossible to discriminate between nests with and without eggs, but from separate studies it is known that virtually all nests at this date contain eggs. For example, in the 1969 count

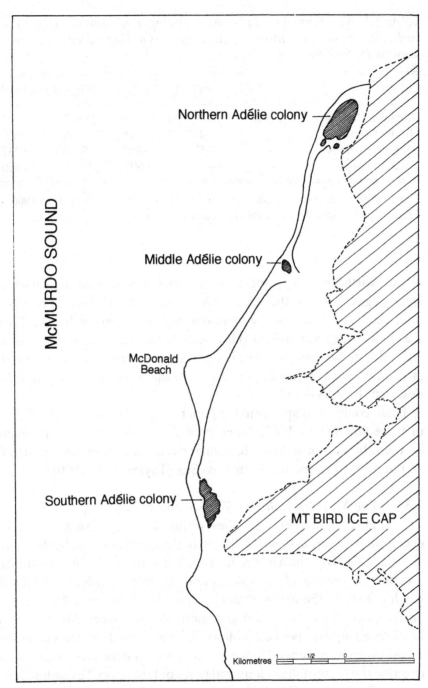

Fig. 2.2. The location of the three Adélie penguin colonies at Cape Bird.

Table 2.1. *Numbers of occupied nests during minimum occupation in early December in the three penguin colonies at Cape Bird in the five seasons 1965–1969*

Colony	1965–66	1966–67	1967–68	1968–69	1969–70
Northern Colony	25 065	25 727[a]	21 560	15 067	19 803
Middle Colony	—	1419	1183	835	1260
Southern Colony	—	10 577	8865[b]	5046	8237
Total numbers	—	—	31 608	20 948	29 300

[a] Estimated from counts made in study breeding groups, 647 nests in 1966–67.
[b] Estimated from a 70% partial count of the colony.

of the Northern Colony there were 19 803 occupied nests of which 597 lacked eggs, 3.0% of the total nests at that time. The errors in counting total occupied nests are less than one per cent. Except for the 1968–69 season few eggs were lost at this colony before the December count. For example, in the 1966–67 season 4.1% were lost to this date in the study groups, of which fewer than one half of the losses led to the nest being deserted. In 1968–69 13.6% were lost.

These counts also appear in the paper by K-J. Wilson (1990) of changes in these colonies to 1987. From 1969–70 to 1987 the total numbers of breeding birds almost doubled. Similar overall increases have occurred for all the colonies in the southern Ross Sea (Taylor *et al.*, 1990).

2.3   *The Northern Colony (77°14′ S, 166°28′ E)*

This colony (Fig. 2.3) lies within a shallow basin on the coast enclosed to the south by low hills and to the north and east by the moraine and glacial cliffs of the Mt Bird ice cap. The topography broadly limits the breeding dispersion of both skuas and penguins, neither of which have much colonised the exposed slopes that stretch away to the south. The local topography of beach flat and gentle slopes dissected by river valleys has broken up the penguin colony into more or less discrete breeding areas, which were defined in the first mapping and have been used since to describe the parts of the colony and to set out the special experimental and observation areas (Fig. 2.4). This favourable pattern of well defined breeding groups of penguins of course reflects this topography as well. Five water courses dissect the basin. In the first years they were given rather exotic names, but are best referred to as they are numbered from

Fig. 2.3. Aerial photograph looking south over the Northern Colony. The ice-free area is backed by the Mt Bird ice cap. The individual breeding groups of penguins show up as lighter coloured patches on the dark volcanic rock slopes. G Block is to the left, B and H Blocks to the right.

the north of the colony, with First Stream lying hard against the ice cliffs beyond G Block. The southernmost of the series flows down onto the beach beside the field station and retains its original name of Lab Stream. In winter and spring they are filled with snow, but in mid and late summer they carry a surprising volume of broken water flowing from the snow and ice faces around the colony basin. On warm afternoons in December and January the water levels are so high in these rivers that they cannot be easily crossed on foot. The river marking the boundary between A and G blocks in the north of the colony floods each year across penguin breeding areas washing out eggs and young chicks from the nests (Fig. 2.5). In spite of this regular event the penguins nevertheless persisted in occupying the same hazardous site year after year.

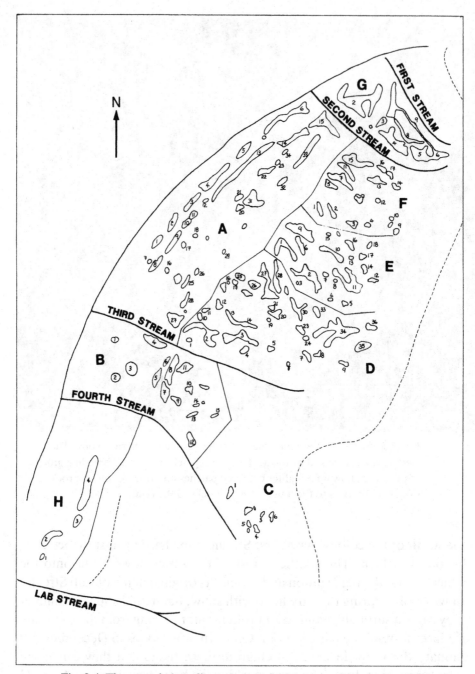

Fig. 2.4. The penguin breeding groups at the Northern Colony, Cape Bird. The different blocks on the colony used to identify the individual breeding groups and define research areas are labelled A–H. The field station is located inland of H Block.

Fig. 2.5. Summer flooding of Second Stream between F and G blocks.

To a lesser degree the topography has defined the distribution of breeding skuas. They are concentrated within the same basin as the penguins, but with a linking line of territories to the south along the narrow beach, so that it is appropriate to refer also to a northern skua colony.

2.3.1 *The penguin breeding groups*

The colony (Fig. 2.4) is made up of two parts containing similar numbers of breeding birds, a lower almost flat beach area containing a series of shallow beach ridges (areas H, A and parts of B and G) and a more steeply sloping (but still gentle) area rising to the foot of the enclosing moraine (areas D, E, F and parts of B and G). On the upper area the ridges, on which the penguins nest, run down the slopes, i.e., perpendicular to those on the beach. These upper breeding groups are less clearly defined and more irregular in outline than those on the lower beach, most of which conform to the regular shape of the underlying beach ridges. The two parts of the colony are separated by a steep step of some 5–7 m, behind A and the lower part of G. In addition to these major areas there is a small outlier of the colony on the moraine (area C).

The colony was first mapped in November, 1965, and individual breeding groups identified and demarcated at natural breaks in the breeding aggregations. Over the subsequent seasons with the fall-off in nest numbers these breaks in the larger groups became even more pronounced and new gaps occurred in some. This had the effect of increasing the numbers of small to medium sized breeding groups at the expense of the larger ones. These changes resulted naturally in exposing a higher proportion of penguins to skuas foraging along the edges of breeding groups. During peak occupation and reoccupation periods when groups were invested by non-breeders the opposite trend occurred and groups became less discrete and with fewer natural breaks in the pattern of nests. In 1965–66 there were 29 groups with less than 40 nests, 49 with between 41 and 100, 38 with between 101 and 400 and 20 with more than 401 nests. But not too much should be made of this distribution; about half are of artificial groups only weakly demarcated from neighbours which in another classification might well be run together.

2.3.2   *Skua territories*

Skua territories cover the entire basin area and extend out from it over the enclosing moraine. There are few, however, on the open slopes above and to the south of the colony area. The highest concentrations of birds, and so conversely the smallest territories, are on the lower slopes of the moraine backing the penguin colony, in the basin between B and D penguin areas and on H beach. Densities of breeding birds varied widely throughout the colony area. On the lower beach the territories of 47 pairs averaged 1740 m^2 (0.57/1000 m^2), almost identical to those of 23 pairs on the upper colony areas of BDEFG blocks, which averaged 1840 m^2 (0.54/1000 m^2) but both areas were less densely occupied than those on the colony perimeter where 166 pairs there without penguins had territories averaging 407 m^2 (2.45/1000 m^2).

There were wide differences also in the numbers of penguin nests within the skua territories. Because of the rigorous defence of territories by skuas this number can be estimated very precisely from the maps of territorial boundaries as they range across the penguin colony. In contrast to the skuas on the Cape Hallett colony described by Trillmich (1978) there was very little overlap in territory areas and where this occurred it was confined to space over large penguin breeding groups that could only be demarcated in flight. Most boundaries were determined by ground display and attack. Figure 2.6, showing skuas disputing a boundary and settling close

Fig. 2.6. Pairs of skuas on territories on the colony. In the upper photograph they are disputing a boundary, with all birds in display, the long-call with oblique body position. In the lower photograph two pairs sit closely together within a metre or so of a well established and recognised boundary.

together near the boundary, demonstrates just how finely demarcated these territories were.

In 1965–66 the numbers of breeding pairs of penguins per skua territory were as follows:

> 1–50 pairs, 17;
> 51–150 pairs, 21;
> 151–500 pairs, 24;
> 501–1000 pairs, 10;
> 1001–1500 pairs, 4;
> >1501 pairs, 2.

The smallest skua territory had 14 nests (at 10 December) and the largest 1715. The figures can be proportionally reduced for subsequent years as numbers of breeding penguins declined. The overall pattern of territories on colony skuas changed very little in these years (Young, 1990b). The three pairs with the largest numbers of breeding penguins within the territory all had territories sited entirely within the penguin colony area. Those pairs with small numbers of penguins nested about the edge of the colony with the territory touching on a single or a few breeding groups.

2.4   *The Cape Bird environment*

The surroundings present a drab picture of dark grey and black rocks and gravels, the local volcanic moraine, enlivened here and there with exotic rocks and pebbles carried here long ago by the McMurdo Glacier from the continental mainland. Overall, however, the background is of dark rock and scree. On the colony area this has, however, been overlain with weathered guano, a much lighter coloured material, that is again lightened by white and pink (due to krill) penguin excreta during the breeding season. Overwinter rafts of ice are thrown up onto the shore and encase it with high broken ridges well into the summer. Only from about late December is there an open beach of black sand along the colony edge, and even at the end of summer, when the penguin chicks are leaving the colony, eroded heaps of ice still exist in places. Offshore the colony is invested with sea-ice, at first by fast-ice, the northern edge of the McMurdo Ice sheet, and later in summer with the ebb and flow of pack-ice on the currents and winds. Only in late summer, or after a lengthy spell of strong southerly winds, is the sea free of any ice cover in this area.

2.4.1 *The local climate*

An immediately appreciated bonus resulting from shifting research work to the Northern Colony at Cape Bird from Cape Royds was a
much better summer climate. This was evident in two features especially.
There was much less wind, particularly cold and strong south winds, and
an almost complete absence of blown snow, snow picked up by the wind
from where it had been deposited earlier. On days with strong southerly
winds over much of McMurdo Sound, easily seen through the occurrence
close offshore of a dark boiling fog and violent seas accompanied by a
roaring sound, the Northern Colony might well be experiencing warm,
sunny conditions and light north to northeasterly wind. We speculated
then that we were enjoying the benefits of a wind eddy deflected about the
northern side of Ross Island, a flow strong enough to keep the main
southerly flow offshore. This speculation has been confirmed recently in
the study of wind flows in McMurdo Sound by Sinclair (1982). The
separation of the prevailing southerly stream lines to the east and west of
Ross Island is very clear in his figures. The balance between the two flows
in the Cape Bird region was clearly an unstable one and shifts between
light north winds to strong or gale force southerlies could occur very
rapidly, almost instantaneously, several times a day. It was always an eerie
sensation to be working in sunshine on the colony, in shirt sleeves and sun
hat, with a gale force southerly crashing and roaring past just a kilometre
or so away. The lines of this wind, marked by the iceflows streaming to the
north, were from the tip of McDonald Beach, so that opposite the colony
they were just 1–5 km offshore, depending on the specific wind flow at the
time. The Middle and Southern colonies were not so well protected and
suffered much more from these fierce storms. Even so the Northern
Colony did get blasted with strong gales from time to time. For example, a
southerly storm beginning at midday on 4 November, 1969 ran continously
for six days, at times with winds strong enough to shake the hut vigorously
and to prevent anyone being able to stand in the open. Penguins not tied to
the nests by incubation left the area for shelter. But at the colony there
were quite long breaks of calm or north winds when the storm veered
offshore. Even in this exceptionally fierce storm the colony lay in calm
weather as much as 30% of the time.

As will be detailed later (Chapter 9) the strong southerly winds were,
nevertheless, crucial to the survival of both penguins and skuas at this site.
Without their action in pushing the ice to the north, the Sound would have
been packed tightly with ice flowing in on the water currents from the east,

which would have precluded access to the colony by penguins and prevented foraging by skuas. In one year of this study (1968–69) dense, rough pack-ice occurred well into summer, with a dramatic impact on penguin occupation and incubation success. Ainley and LeResche (1973) have described how important strong south winds were for the breeding birds at Cape Crozier. Much of their account applies generally to Cape Bird as well.

The second major difference in the weather between Capes Royds and Bird colonies was the absence of blowing snow at the latter. Southerly winds at Cape Royds always seemed to give miserable conditions because of the amount of snow they carried onto the the colony area from the surface of the sea-ice. Following even moderate storms skua nests could be invested by packed snow and whole areas of the penguin colony could be covered with a granular pack that gave way to guano mud as it melted. Without blown snow Cape Bird was both a more pleasant and drier place in which to live. Neither place received much direct precipitation and even in the depths of winter substantial areas at both places must have been clear of any snow. Only the wind-sheltered areas, the smaller basins and gullies, and the protected slopes of ridges collected snow drifts.

Standard meteorological records have been made at Cape Bird over the summer since the field station was established. The recording station with its standard screen was on the terrace occupied by the hut, 15 m above and parallel to the 50 m wide beach of H Block and backed by gently rising moraine slopes lifting to the permanent Mt Bird ice cap. The terrace faces out to the west across McMurdo Sound. Records for the first seven years (including the study years) have been summarised in Sinclair (1982). The prevailing winds were found to be from the north or south, with few winds from any other direction. Katabatic (downslope) winds from the Mt Bird ice cap were clearly of little significance. Northerly winds were overall twice as frequent as southerly ones, but almost all the strong winds were from the south. Mean wind speed over these summers was 6 knots ($3.1 \, \mathrm{m\,s^{-1}}$). Sinclair reasons that this low mean wind speed compared with other Ross Island sites reflects the sheltered position of the colony from the prevailing southeast flow. Wind in excess of 20 knots ($10.2 \, \mathrm{m\,s^{-1}}$) occurred only 3% of the time. In contrast, there were calms during 14% of observations.

Temperature

Sinclair's synopsis confirms the general feeling that the Northern Colony was a warm place – in Antarctic terms. The presence of open

Table 2.2. *Summer temperatures (°C) at the Northern Colony, Cape Bird, 1965–1970*

Date	Mean temperature	Daily maximum	Daily minimum
1 Nov.	−9.5	−8.5	−12.3
Mid Nov.	−8.9	−6.1	−11.2
1 Dec.	−3.5	−0.6	−6.8
Mid Dec.	−1.5	0.8	−3.6
1 Jan.	−0.5	2.3	−2.0
Mid Jan.	−1.1	0.8	−3.2
1 Feb.	−0.3	1.1	−2.7
Mid Feb.	−0.5	0.5	−4.2

Mean temperatures and daily ranges are the averages for the five days about the dates shown, pooling the data from all five seasons except for the first and last dates which are for one and two years respectively.

water, the shelter from south winds and the northerly aspect combine to give temperatures on average 3.8 °C warmer than at Scott Base, just 65 km to the south but facing onto the Ross Ice Shelf.

There are no temperature data for the colony in early and late summer so that it is necessary to use those from Scott Base to give a measure of the temperatures encountered by the penguins when they first come ashore in October and by the skuas when they are leaving these breeding areas in March and April. Air temperatures are very cold in October when the penguins arrive – the mean temperature for this month recorded at Scott Base is −22.7 °C – but warming occurs rapidly with above-zero temperatures occurring by the end of November. Cooling is equally rapid at the end of summer. Mean temperature for March is −20 °C, falling to −24 °C in April.

The mid-summer temperatures at the colony (Table 2.2) are characterised by their uniformity, by narrow seasonal and diurnal ranges. The second factor is especially noteworthy with the mean maxima and minima temperatures differing only by c. 4 °C. Although daily maximum temperatures are regularly above freezing, in the different seasons peaking between 3.3 °C and an exceptional 7.8 °C, minimum temperatures did not reach zero in any year. In spite of the narrow diurnal temperature range, nights were nevertheless appreciably colder, marked by frost and the long shadows of a lower sun. At this colony it made great sense to work during the normal hours of temperate daylight – the afternoons were generally the warmest part of the day. This difference between day and night conditions was more obvious during early and late summer.

2.4.2   *Impact of environmental factors on the birds*

Environmental factors impact at all levels on the association between the skuas and penguins and their significance can scarcely be over-emphasised. At the most basic level it is the environmental conditions – the access to open water, the overall wind and snow levels – that determine whether occupation is even possible. Temperature has the most subtle impacts, from direct effects on the metabolism and in responses to extremes – evident in such obvious ways as in the huddling of penguin chicks during cold spells and in the panting seen in both skuas and penguins during hot, still afternoons. But behavioural responses to temperature act in a much more subtle way also to affect the way the two species interact.

Of prime significance for this interaction is the fact that both species require similarly sheltered breeding areas, throwing the two species into close association irrespective of any possible relationship based on prey and predator. The number of suitable breeding places in Antarctica is limited, so that it is not surprising that the two species are often found breeding together. The effect is apparent even within sites, with the highest concentrations of nesting birds being in the most sheltered parts of the general area. At the Northern Colony, most skuas are within the sheltered basin occupied by the penguins and few occur on the wind-swept slopes beyond its margin.

The eggs of both species must be incubated at all times to prevent chilling. This requirement constrains the freedom of skuas to forage at the penguin colony, even in situations where egg predation from other skuas is a minimal risk. The need to stay on the nest restricts the ability of single skuas on a territory to evict others landing there. Although newly hatched chicks establish homeothermy within a few days of hatching (Spellerberg, 1969), they are nevertheless brooded during colder periods – extending the time the parents are held at the nest.

The accessibility of penguin eggs and chicks to skuas is also affected by weather, a major factor in skua predation success. Eggs and young chicks are brooded and guarded by the parents at the nest. During this stage wind force seems of greater significance than temperature, with incubating and brooding penguins facing into the wind. As the chicks become larger they cannot be covered entirely by the adults, so they then tend to stand with their backs to wind, placing the chicks in shelter. The uniform alignment of incubating or guarding penguins gives some advantage to foraging skuas by reducing the overall alertness of the breeding group to attacks. Under low wind conditions the penguins are aligned randomly within the breeding groups. The relationship of wind force and attack success is considered in some detail in the later sections. The relationship between temperature and penguin chick accessibility to skuas does not end here, however. During the post-guard stage the chicks are far more accessible to skuas when they are dispersed on the colony during mild conditions than when huddled tightly together in cold ones. These different dispersions provide very different opportunities for foraging skuas.

However, the most significant environmental factor impinging on the well-being of the penguin and skua colonies was the extent of sea-ice cover. This single factor alone determines whether the area could be colonised by either species in the first instance and once colonisation had been successfully achieved, determined the annual breeding success. Both species require access to foraging areas at sea during the breeding season. Without access breeding fails; as Parmelee *et al.* (1978) have described for the skuas on the Antarctic Peninsula.

Ross Island is an exceptional breeding place in Antarctica providing remarkably secure breeding conditions at an extremely high latitude. This is possible because the sea-ice within the Ross Sea, especially the central and western parts, consistently breaks out and decays each summer forming an immense open sea polynya reaching well into McMurdo Sound and extending far to the east along the Ross Ice Shelf barrier (Fig. 2.7). Its widest extent is not reached until February, by which time the penguin breeding season is over. The newly fledged chicks leave the colony into the open sea. Conditions are more difficult for the adult penguins migrating southwards towards the colony in September and October; they face winter ice conditions with dense pack-ice extending far to the north of Cape Adare.

Stonehouse (1967a) was the first to comment on the implications of polynya for Antarctic birds and mammals, particularly for the colonisation of Ross Island. More recently, seasonal fluctuations in the timing of the decay of the sea-ice cover and of its day to day changes have been examined by Yeates (1975) at Cape Royds and by Ainley and LeResche (1973) at Cape Crozier. In each place the changes were related to the breeding success of the penguins. Fluctuations in the amount of pack-ice near Cape Crozier was ascribed to wind pattern, with south winds clearing

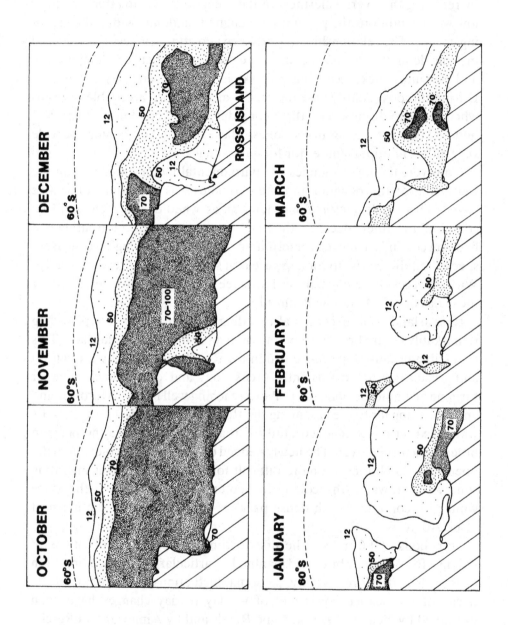

the pack-ice away to the north. This topic is taken up in greater detail in Chapter 9 in an attempt to explain variation in the times skuas spend foraging.

2.5 *General methods*

The approach followed during most of this research was of direct observation of undisturbed birds in order to describe the predatory and defensive behaviours, to place these within a daily and seasonal context and to determine the impact of predation. At the same time limited experimental work was attempted, notably in excluding skuas from selected penguin breeding groups. The relevant techniques used in the work are described in some detail in the appropriate chapters and need not be recounted here. There were, however, a number of methods and systems for observation and analysis that had general application, under-pinning most of the research, and these require explanation and definition.

2.5.1 *For research on penguins*

No attempt was made to develop a banded study population in which the ages of birds were known and their breeding success across the years recorded. Although clearly desirable, the effort involved and the disruption caused were considered to outweigh the potential benefit. Also, at the time the programme was simply limping along from year to year so that a massive investment in banding could not be justified. Anyway, we were convinced that the very fine banding and recovery study being undertaken at the same time on the penguins and skuas at Cape Crozier by the Johns Hopkins University was bound to provide the information needed on age-specific breeding factors. Thus, the penguins of this study were not recognisable as individuals. The detailed maps of breeding groups used in the analysis of breeding success and in the accounts of skua attacks were thus of pairs and nests. It was not possible to distinguish between males and females in pairs. In the first year the individual penguin breeding groups were mapped and the numbers of pairs counted. In each

Fig. 2.7. The sea-ice regime and the development of the Ross Sea polynya over summer. The breakup of the solid pack-ice of winter begins in November with erosion on the northern edge and the formation of the polynya in the south-west corner along the edge of the Ross Ice Shelf (hatched). By December the southern polynya of clear water has extended far to the north into the Ross Sea. From Sturman and Anderson (1986).

year the total numbers of nests (differentiated between nests with and without eggs) were recorded for each breeding group at all the Cape Bird colonies. These counts were made in early December when there were minimum numbers of penguins on the colonies. In the first year the timing of the count was taken from Taylor's (1962a) graphs of the numbers of penguins on the Cape Royds colony in relation to date and breeding cycle stage. Once this date had been found appropriate for Cape Bird, subsequent counts were timed for the same date. Most nests at this date, about 10 December, were attended by a single parent. These counts were used as the basis for estimating breeding success and biomass production for each colony as a whole by scaling up the detailed records made throughout the season for selected breeding groups.

In each year between 10 and 19 breeding groups of penguins were mapped and the fate of the individual nests followed throughout the season. Most were isolated groups but samples of nests on the periphery and in the centres of large groups were also included. Nest checks generally took place at five-day intervals but pressure of other work or bad weather sometimes upset this routine. Using this interval it was possible to interpret differences found in the location of nests and to determine the loss of nests and contents when new pairs come into an established site. Of course, the closer together the nest checks the more precise the records; and the interval used is a compromise between accuracy, disturbance and freedom to carry out other work. Except for study nests in the centre of large breeding groups (in which central and edge nests were compared) all observation was carried out from outside the group margin. This technique goes some way to ensuring that the breeding success measured represents the natural state. It is, however, a very time-consuming method of nest checking as it requires the birds to lift naturally so that the nest contents can be spotted. Once the chicks have grown, nest checking in this way is on the contrary very quick. A few minutes using a good map can check colonies of 50 or so nests. Nest contents of central groups of penguins could not be recorded in this way and these nests were checked by lifting the penguins momentarily by the tail. Birds lifted in this way invariably dug the beak into the ground to prevent being pushed from the nest, as they do in normal pushing contests at the nest, and are much quieter and settle again more quickly than when ejected bodily from the nest. No penguins deserted nests during the regular nest checks.

The growth of penguin chicks was recorded in two ways. In one year growth was recorded in a sample of individually marked chicks to provide information for estimating live weight from carcass remains left by skuas after feeding. Individual chicks were weighed daily from hatching to fledging (for penguins, on leaving the breeding group for the shoreline). In each year a quite different parameter, the individual weights of all chicks on selected breeding groups, was also recorded as they grew and matured over summer. This was done for two reasons. First, to describe the range of age and size of prey available to the skuas on any date. These data allow assessment of possible selection of prey by skuas. Second, these figures allowed the total prey biomass for the colony to be estimated for each weighing date. In conjunction with the counts of numbers of penguin nests within each territory these records also allowed estimation of the prey available for each skua pair through the season.

2.5.2 *For research on skuas*

In contrast to the penguins it was important that skuas could be individually identified from season to season, to measure changing predation skills for example, and that meant that they needed to be banded. Fortunately, many of the skuas nesting about the Northern Colony had been banded in the year before this work began by Bob Wood as part of his study at Cape Crozier of skua breeding and dispersal, so that only selected birds needed to be caught and banded during this work. By the end of the first season all the skuas on H and EF blocks had been banded and many others about the colony area had also been banded. To facilitate recognition of the individuals of pairs during the observational logs one of each pair was also colour banded. All pairs were numbered once the pattern of territories had been established during the first season.

Considerable attention was given to habituating the skuas to people working and moving about the colony area. The reasons for this approach were threefold. First, to ensure as far as possible that the numbers and dispersion of skuas were maintained in this area and that the natural relationship between the two species was maintained. Second, to ensure that the breeding success of the skuas would not be jeopardised by stress or by the enhanced egg and young-chick predation that is likely when skuas abandon nests through disturbance. Third, to allow movement about the area without the noise and distraction of attacking skuas. This programme was most successful. By the third season many skuas were sufficiently tolerant of people to remain tightly on the nest until approached closely. Some needed to be lifted by hand to check the nest contents (P. M. Young, 1971). For the programme as a whole, the best measure of its success, however, was the high stability of skua territories and occupation in the penguin colony area and the very few instances of egg and chick predation induced by disturbance (Young, 1990b).

All skua territories throughout the penguin colony and surrounding areas were mapped in each year and the numbers of penguin nests within the territory estimated. Nest checks were made routinely of all pairs at 4–5 day intervals but this work was not given undue emphasis so that the breeding statistics as a whole are rather weak. (The ones for penguins are much better.) Although the total production of chicks could be recorded each year with some certainty, at this checking interval the statistics of egg and chick loss, and of egglaying and re-laying are uncertain. In two years, however, much more detailed records were made from nest checks at least once a day. It is these latter records that are used for comparing the success of skuas in different sites, especially of birds with and without access to penguin food.

2.5.3   *Observation*

The study relied principally on direct observation of penguins and skuas under natural conditions. Two colony areas were selected for direct comparison. In one, H Block, the skua pairs shared out more or less equally a small number of penguins. In the other, EF Block, a few skuas nesting at the margin of the colony had large territories embracing large numbers of penguins. The comparison was thus between skuas with small and large prey numbers respectively. Although these two areas were chosen principally because of the contrast in prey access an important consideration was that they were both overlooked by steep slopes to provide excellent unimpeded viewing. The penguin breeding groups and skua territories of each area were laid out in front of the observation point as in a map (Figs 2.8 and 2.9). Most observations could be made directly by eye, and field glasses were needed only to confirm identity or to record the detail of an attack, or of the prey being taken or fed to the chicks and mate.

Direct recording of the behaviour of the skuas interacting with penguins, of attack and scavenging activities, of their territorial and breeding behaviour and of their movements away from the territory, were made regularly throughout summer. These were planned for each area at five-day intervals but were disrupted at times through other work or exceptional weather. The two areas were watched on contiguous days. Most routine watches were of several hours between 0800 h and 1800 h but special watches were carried out for longer periods extending throughout the 24-hour day.

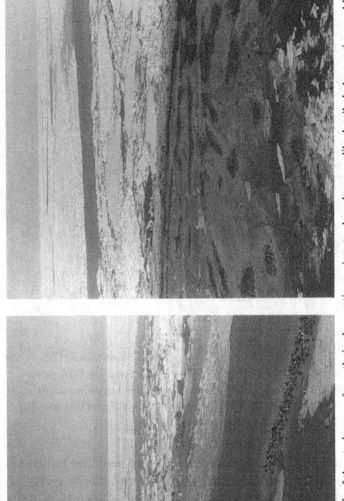

Fig. 2.8. Views over each of the study areas from their observation points to show how readily detailed observation could be made of skua behaviour when they were interacting with the penguins. H Block (on the left) with H4 immediately below the observation point and H1 to far upper left; EF Block (on the right) showing the whole of F Block (from the Second Stream, filled with the snow of early summer, to F1 to the left) and A Block along the shoreline. The clear views out to sea demonstrate also the opportunity to follow skuas foraging there.

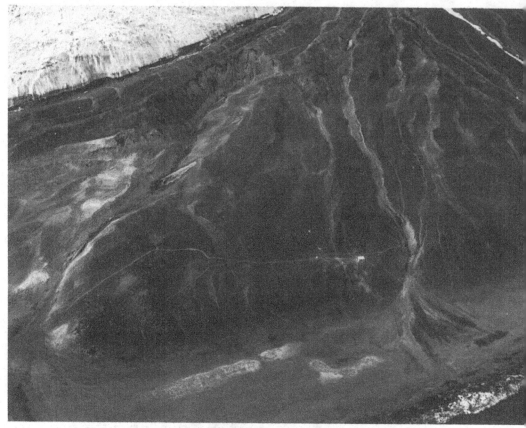

Fig. 2.9. Aerial photograph of H Block along the beach line and C Block on the slopes leading up to the ice cap. The accommodation hut shows clearly on the bench above H Block beside the Lab Stream. The walking tracks from the hut to the main body of the colony also show up clearly.

In the first year long-hand records were kept of all activities but these were coded at the end of summer and coded records were made in subsequent years. The advantages of coding have been given by Young (1970) as facilitating complete recording over long periods, especially when there is little activity and attention lapses, for providing a defined set of terms so that comparisons between birds and between seasons is possible and to ensure that other observers would provide comparable records.

2.5.4 Breeding cycles and key stages

The most important factor in the availability of food to skuas, and their success in exploiting it, is the penguin life-cycle stage. Both the amount of food and its accessibility to skuas changes remarkably through the summer. A second factor is skua demand, the changing nutritional requirement of the parents and chicks throughout their breeding cycle.

Because of these variables it was important to be able to relate the results gained from observation and other sources to clearly specified life-cycle stages. Unless this was done there could be no rational interpretation of the changes in success or behaviour of the skuas through the season – and no chance of comparison between years.

Defined stages in the penguin breeding cycle

Definition of stages used throughout this account are related to the nature of the prey (e.g., eggs, small or large chicks), its volume and accessibility by skuas. It is related to the eggs and chicks, rather than is more usual to the adult activities. These stages are as follows:

Pre-egglaying stage. The life-cycle stage from first dates of occupation of the colony in early October to the date of first eggs.

Egg stage. From the date of the first egg in mid November to the date the first chicks hatch in mid December.

Chick guard stage. From the date the first chicks hatched to the date the first chicks are left unguarded by parents.

Chick post-guard stage. From the date of the first unguarded chicks to when the chicks leave the natal breeding group. This stage was generally split into an early and late stage; the first part extended to the last date for guarded nests (set arbitrarily in most analyses as 9 January).

Dispersal stage. The stage at the end of the season during which the chicks leave the colony.

These stages are of course paralleled by the cycle of adult breeding activities. The breeding cycle stages of occupation and incubation run from pre-egglaying into the chick guard stage and chick rearing begins from the hatching of the first chicks to their dispersal from the colony.

Because the breeding cycle at Cape Bird was both highly synchronised and laying dates were constant from year to year it was possible to relate these cycle stages directly to calendar dates. Through doing this it was then possible to assign data collected on different days in the different years to the same stages for comparison.

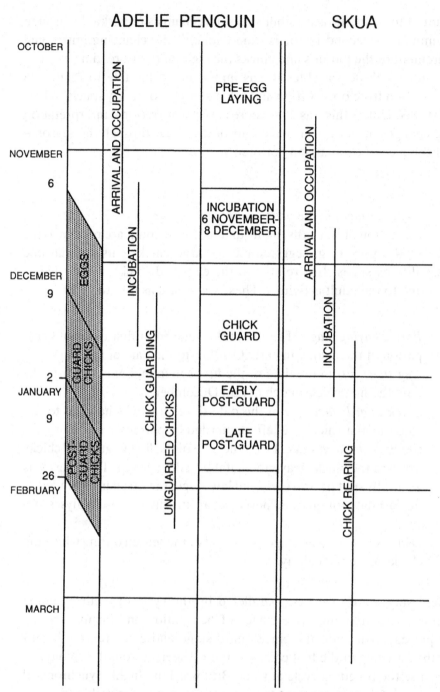

Fig. 2.10. The life-cycle stages of the Adélie penguin and south polar skua at Cape Bird. For the Adélie penguin the egg and chick stages (shaded) give a more precise definition of the cycle than the activities of the adults themselves.

These stages in relation to the adult activities and to the skua cycle are shown in Fig. 2.10.

2.6   *The study areas within the colony*

From the outset it was recognised that the numbers of penguins enclosed within the skua territory could be an important factor in both the form and intensity of skua foraging. In order that this factor could be investigated, two very different groups of skuas and penguins were selected for detailed study. These groups represented the two extremes of the prey–predator ratio continuum as far as it existed at Cape Bird. On H block the skua pairs shared four penguin breeding groups with a total of 296 to 570 nests over the five years and giving an average predator to prey ratio of between 1:32 and 1:40 at the time of the December count in each year. In this area the skuas nested on each side of the penguin groups with territories running from the sea edge on one side and the cliff slope on the other to meet on the penguin breeding areas. This study area provided exceptional viewing of skua and penguin behaviour from an observation point 20 m above the beach on the cliff mid-way along H4 breeding group. The farthest penguin nest was 155 m distant and all of H4, the largest penguin group, was within 55 m. The second study area selected included part of E Block and most of F Block. This area lies on the lower moraine slopes bordering the colony with the individual penguin groups located on the low spurs running from the moraine towards the beach. The skuas here hold territories running across the colony from nests on the higher ground of the moraine above the breeding groups. These are larger territories on average than those of H Block and contain many more penguins. The seven territories that covered the majority of the area contained between 50 and 900 penguins nests each, with five containing more than 250 nests in each year. Further away from the colony on the higher slopes of the moraine a number of pairs held small territories without penguins and these were also observed to provide comparative data on sea foraging flights and territory occupation. The numbers of skuas and penguins are contrasted for the study areas in Table 2.3. For the greater part of the breeding season the skua attacks are confined to the margins of the penguin breeding groups. Because access might be just as significant as penguin number the total length of breeding group margin in each skua territory is also shown in this table. Differences from year to year reflect changes in the shape of the territory and changes in the numbers of penguins within breeding groups and the density and outline of groups.

Table 2.3. *The numbers of penguin nests (=pairs of penguins) and the
length of breeding group margin (in metres) for each skua territory in the
two study areas*

Skua pair	Number of nests (and breeding group margin (m))					
	1965–66	1966–67	1967–68	1968–69[a]	1968–69[b]	1969–70
EF Block						
144	450(230)	450(225)	391(225)	276(225)	276(219)	445(219)
145	73(68)	90(63)	84(80)	95(80)	173(145)	337(158)
146	312(126)	364(138)	267(119)	151(119)	—	—
146 new pair	—	—	—	—	350(208)	337(160)
148	622(277)	827(203)	693(201)	402(221)	—	—
149	32(17)	26(15)	42(20)	15(10)	19(12)	66(40)
150	10(5)	0(0)	5(5)	5(5)	5(5)	6(5)
151	37(29)	54(26)	36(22)	20(22)	20(22)	21(39)
152	524(275)	497(264)	408(264)	285(264)	285(264)	354(262)
153	880(273)	885(243)	642(642)	423(139)	423(139)	585(163)
154	0(0)	11(6)	87(33)	118(93)	118(93)	174(93)
H Block						
8	10(10)	0(0)	4(10)	4(10)		0(0)
9	165(45)	140(43)	113(40)	55(38)		74(74)
11	79(55)	81(50)	67(48)	42(48)		69(50)
12	85(27)	95(34)	54(15)	53(15)		33(15)
13	155(30)	185(27)	180(46)	80(40)		90(33)
14	15(11)	10(8)	5(4)	10(9)		54(17)
19	25(13)	37(22)	19(8)	5(8)		12(21)
20	10(10)	15(9)	10(8)	15(9)		10(6)
21	15(11)	15(10)	10(8)	10(7)		23(10)
22	40(30)	30(25)	40(23)	30(19)		30(22)
24	20(20)	60(19)	35(11)	20(11)		12(12)
25	120(27)	105(27)	100(27)	54(27)		100(37)
26	0(0)	0(0)	0(0)	15(8)		10(4)
27	125(33)	105(30)	54(30)	54(27)		65(24)

The numbers of nests are from counts made between 8 and 12 December each
year. In each column the numbers are shown first, followed by the length of
breeding group margin. The two columns for 1968–69 refer to different stages of
this season. [a] The early season before pairs 146 and 148 were removed from the
area. [b] The end of the season after 146 new pair established a new territory area.

An important requirement for the study areas was that they should be subject to minimal scientist and visitor disturbance. As far as possible no other research was carried out in these areas and they were both protected from casual visitors to this colony.

3

The range of foods available to the skuas at Cape Bird during the breeding season

3.1   Introduction

All food for the skuas, as indeed for all the Antarctic vertebrates, is of marine origin, either plankton or fish caught directly, or bird or mammal material which has originated as sea food. Fig. 3.1 is a schematic outline of the trophic paths for skuas at Cape Bird inferred from the literature and from direct observation. The components are analysed below for interpreting the foraging behaviour of skuas during the breeding season. The partial dependence of skuas on the Adélie penguin for food suggests that it is prudent to consider the two species together. Although skuas are reliant on marine food, especially fish taken directly, the penguins provide the major alternative trophic pathway – and transform a diffuse food resource at sea into a concentrated one at the breeding colony.

3.2   The foraging range of penguins and skuas at Cape Bird

In reviewing the foods that might be exploited by the penguins and skuas at Cape Bird it is necessary to take into account both the stages of their breeding cycles and their foraging range. This has greater signifi-

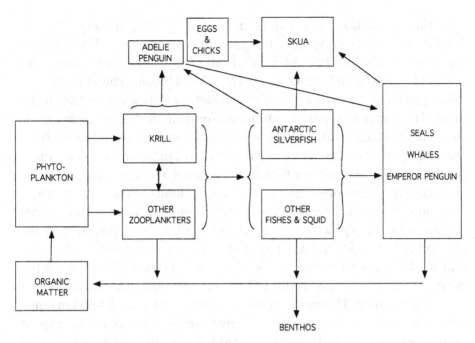

Fig. 3.1. Schematic food web for pelagic species in the southern Ross Sea. This design emphasises the important roles of krill and Antarctic silverfish within this ecosystem.

cance for the skuas, which could exploit foods at the seal and Emperor penguin colonies at the beginning of the season, than for penguins. The following account applies more particularly to the period the birds are resident at Cape Bird.

Observation from the shore of birds foraging and from the distribution of birds at sea suggest that both species have substantial foraging ranges. This supposition is largely untested for skuas but has been validated recently for penguins by a number of studies using radiotelemetry. Fortuitously, major studies of penguin foraging have been located at Cape Bird and this information provides a valuable insight into the ecology of this population. Because Adélie penguins fast during the pre-egglaying and incubation periods, the foraging and nest attention pattern during these life-cycle stages is quite different from that during chick-rearing. The longer times available for foraging during the incubation stage, when fasting parents alternate at the nest, provide an opportunity for penguins to exploit a wider sea area within a much greater foraging range. Thus, the two stages of the breeding cycle must be clearly distinguished in assessments of foraging range.

During incubation each parent in turn, beginning usually with the female, is away from the nest foraging for long intervals. Davis (1988) recorded averages of 19, 12.6 and 3.1 days sequentially for a sample of birds at the Northern Colony. In contrast, birds feeding chicks during the chick guard and post-guard stages are away for only a day or two at the most. The pattern of nest attention and foraging of the incubating birds suggested that the birds could be foraging at considerable distances from the colony at this time, but even so the distances travelled by these penguins were much greater than ever imagined and a high proportion of birds carrying transmitters moved beyond the limits of the radio horizon. The most comprehensive study was carried out for the penguins in this colony in 1986 (of parents feeding chicks) and in the following year (of incubating birds) (Sadleir and Lay, 1990). The records made during chick rearing (from early December to early February) showed foraging in the northwest within 15 km of the colony but exceptionally ranging out to twice this distance. This range is considerably less than the 83–95 km range determined for Adélie penguins at Signy Island on the basis of time spent away from the colony (Lishman, 1985a) but is of the same order as that obtained by employing speed–distance meters on penguins at Anvers Island, Antarctic Peninsula (R. P. Wilson *et al.*, 1989).

From the experience of the first year the radio receiving stations were re-located with one on each side of McMurdo Sound in order to track foraging ranges during the longer foraging spells of incubation. Even with the knowledge that Adélie penguins swim at an almost constant $7.9\,\mathrm{km\,h^{-1}}$ (R. P. Wilson *et al.*, 1989) the distances covered each day were nevertheless surprising. By the third day at least one of the birds tracked during the incubation period was already over 150 km distant. Because of the limitations of the tracking system only the first stages of the foraging of incubating birds could be followed. Birds at this time first travelled out to the northwest from the colony and foraged for several days in mid-Sound before swimming rapidly to the north or northeast and out of receiver range. The maximum range for these birds in this second phase is, therefore, greater than 150 km. In Fig. 3.2 the ranges for incubating and chick rearing have been shown at 200 km and 30 km respectively. Ainley *et al.* (1984) noted high densities of penguins between Franklin and Beaufort islands, well within the range of the Cape Bird colonies, but it should be noted that both of these islands have substantial penguin colonies of their own so that these concentrations could also be of local birds.

Attempting to set foraging ranges for the skuas is more problematical. So far there have been no attempts to track foraging using radiotelemetry so that calculations need to rely on estimated flight speed and duration. Skuas fly at c. 50 km h^{-1}. This estimate is from direct observation of skuas flying across a measured distance and is supported from results for this skua using the Pennycuick (1989) model and from Pennycuick's (1987) records of great skua flight, which gave a mean of 14.9 m s^{-1} (=53.6 km h^{-1}) and maximum range speed of 18.6 m s^{-1} (=66.9 km h^{-1}). Two ranges for skuas are shown in Fig. 3.2. The one closest to the colony is for birds foraging during open pack-ice conditions when mean foraging duration was 43 min. (Table 9.3). This range is set at 20 km, calculated as half the flight distance with no time given to foraging. It is, therefore, the maximum possible range for this duration. Because all observations of skuas leaving and returning to the colony showed them pursuing steady directional flight it is probably not necessary to take deviation into account to the extent of halving the range from the theoretical distance as suggested by Stahl *et al.* (1985) for other sea birds. However, as no skuas seen leaving the colony foraged within the first 5 km the foraging range is really between 5 and 20 km. The second range shown is calculated from the times required by birds foraging in dense pack-ice and is used to show the flight range the skuas can use routinely during unfavourable foraging conditions. The mean flight duration for these conditions was 227 min. (Table 9.3). Using the same method as before the maximum foraging range for these conditions is set at 95 km. In contrast to the penguin ranges the limits for skuas extend for a full circle about the colony. Skuas could of course forage at food sources on the fast-ice and on Ross Island as well as at sea. The different ranges portrayed here for skuas depend on pack-ice conditions, which, as will be shown later, vary widely from day to day through summer. These ranges apply equally, therefore, for most of the summer. Skuas must forage each day. The foraging range given here for fishing in dense pack-ice conditions is simply a measure of the distances skuas may routinely cover. It can be used as well to show the possible ranges for skuas exploiting other food. Thus, this range includes the Emperor penguin colonies on Beaufort Island and at Cape Crozier, the Weddell seal pupping areas along the western side of Ross Island, even the coast and fast-ice on the western side of McMurdo Sound. Both Scott Base and McMurdo Station also lie within this range.

Both skuas and penguins foraged at sea to the north of Cape Bird. Most skuas tracked from the Northern Colony by field glasses flew out to the northwest but during dense pack-ice conditions they flew directly about the glacier cliffs to the north of the colony and were almost immediately lost from view. It was speculated that these birds were flying around Cape

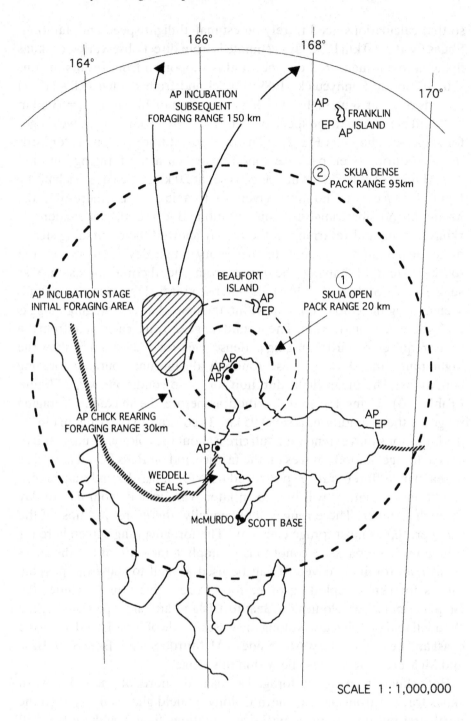

164° 166° 168° 170°

AP INCUBATION
SUBSEQUENT
FORAGING RANGE 150 km

AP
EP ⌇ FRANKLIN
AP ISLAND

② SKUA DENSE
PACK RANGE 95km

BEAUFORT
ISLAND
AP
EP

① SKUA OPEN
PACK RANGE 20 km

AP INCUBATION STAGE
INITIAL FORAGING AREA

AP
AP
AP

AP
EP

AP CHICK REARING
FORAGING RANGE 30km

AP

WEDDELL
SEALS

McMURDO ▮ SCOTT BASE

SCALE 1 : 1,000,000

Bird to the more open water in the Ross Sea to the north and east of Cape Crozier. Cape Crozier is 75 km away by bird flight around Cape Bird.

3.3 *Foods available within the foraging range*

What is clear from this introduction on foraging ranges is that both species utilise an extensive area of McMurdo Sound and the adjacent Ross Sea. Skuas in addition have alternative food sources available to them on the land (Adélie penguins and at the research stations), on the fast-ice (Emperor penguins and Weddell seals) and on the pack-ice (seals). These food resources are listed below and described in order.

Food taken directly from the sea:

1. Zooplankton, caught directly from the sea or washed up on ice or along the shoreline.
2. Fish and squid.

Food taken from other foraging species:

3. Food gained from seals.
4. Eggs and young of sea birds, especially of Adélie and Emperor penguins.
5. Food scavenged from human sources, ships and stations.

(Although many skuas visit the bases at the southern tip of Ross Island at the beginning and end of summer it is doubtful that many go there from Cape Bird once breeding is underway. However, some few birds did travel this far, as was evident from the well-cooked bones that turned up occasionally on the nest areas. This sort of food could also come from garbage thrown overboard from any of the shipping moving through McMurdo Sound.)

Fig. 3.2. Foraging ranges of the Adélie penguin (AP) and south polar skua superimposed on the map of the Southern Ross Sea, McMurdo Sound and Ross Island. Two ranges are shown for each species. For the Adélie penguin the range for birds foraging during the incubation period, which is initially limited to the hatched area, then extends out beyond Franklin Island to more than 150 km. The foraging range during chick rearing is much less, extending out to 30 km. Two foraging ranges are also shown for the skua (bold dashed lines). The smaller range is for skuas foraging under the best conditions of open pack-ice, the larger is for dense pack-ice when the birds are ranging more widely. EP = Emperor penguins. The broad hatched lines show the fast-ice limit within the McMurdo Sound.

3.3.1 *Zooplankton*

In the absence of fish species that feed directly on phytoplankton the primary production of Antarctic oceans is taken up exclusively by zooplankton. A wide array of larval and adult species range up to forms large enough to be taken individually by vertebrates. Zooplankton may be taken directly by birds or form a secondary pathway through fishes or squid. Skuas and penguins feed at both levels.

Krill (Crustacea, Euphausiacea) species dominate the zooplankton biomass over much of the Southern Ocean. Between the Antarctic Convergence and the coastal shelf, an area largely covered by seasonal pack-ice, *Euphausia superba* plays a pivotal role in the transfer of nutrients from phytoplankton to fishes, seals, whales and birds. This role is so dominant that the term 'krill' is often used synonymously for this species, with the ecosystem described functionally as the 'krill system' (Hempel, 1985).

In the Antarctic coastal margins and over the continental shelf the dominant krill species is, however, *E. crystallorophias*, a much smaller species maturing at about 30 mm with about a third of the volume of *E. superba* (Thomas and Green, 1988). This species is widely distributed throughout the Ross Sea. However, it does not have the same overwhelming significance in the local food webs as *E. superba* does in the Southern Oceans and other pelagic species are of consequently greater importance. *E. crystallorophias* is entirely herbivorous.

Adélie penguins take live zooplankton individually, a formidable task when one appreciates the large numbers required to fill the stomach of a foraging bird – 5300 on average were found in the Adélie penguins at Cape Crozier by Emison (1968) and c. 42 000 were counted in one by van Heezik (1988).

Skuas invariably consume the krill in the guts of penguin chicks taken as prey and this is often an important part of the food gained. Weights of stomachs of penguin chicks killed by skuas during the post-guard stage ranged from 60 g when empty to over 800 g, but most weighed between 80 g and 450 g. Estimates of the amount held in the stomach were gained also from weight losses of chicks held overnight. In one experiment on 23 January, two groups of chicks weighing about 3 kg lost 416 ± 140 g (mean and SD) and 324 ± 168 g, respectively. Nine chicks lost more than 500 g overnight, and one lost 870 g. Krill from killed chicks is clearly a substantial food resource.

Krill and other plankters were also foraged on the penguin colony as food spilled by penguins during feeding (Fig. 3.3). Late in the penguin breeding season this food on the colony occupied much of the skuas

Fig. 3.3. Skua picking up spilled krill on the breeding group area late in the season.

foraging interest. Krill could also be scavenged from the pack-ice or washed up on the shore.

3.3.2 *Fish and squid as skua food*

Although squid are eaten by skuas in the Ross Sea area, especially along the continental slope (Ainley *et al.*, 1984), they were not certainly identified in the diet of birds at Cape Bird. Nor were their beaks found in regurgitated pellets of local skuas (Young, 1990a). They would certainly be a food resource for skuas migrating to and from breeding areas. The Antarctic fish fauna is much less diverse than that of the Arctic. It lacks cods, herrings, salmons, smelts, sculpins and flatfishes. It is dominated instead by perciform fishes of the Notothenioidei, evolved from a cod-like ancestral form with benthic habits into a range of pelagic, cryopelagic and benthic species. The structural adaptations associated with neutral buoyancy by the pelagic forms in the absence of a swim-bladder have been of continuing interest to biologists (DeVries and Eastman, 1978; Eastman and DeVries, 1982). Of more direct relevance to this account is their position in the food web between the zooplankton and bird and mammalian predators.

Table 3.1. *Common fishes of McMurdo Sound and neighbouring Ross Sea*

Benthic species	Cryopelagic species	Pelagic species
Nototheniidae	Nototheniidae	Nototheniidae
Trematomas bernacchi	*Pagothenia borchgrevinki*	*Pleuragramma antarcticum*
T. centronotus	*T. newnesi*	*Dissostichus mawsoni*
T. hansoni		*Aethotaxis mitopteryx*
T. nicolai		
T. loennbergii		
T. lepidorhinus		
Bathydraconidae		
Gymnodraco acuticeps		
Artedidraconidae		
Histiodraco velifer		
Zoarcidae		
Lycodichthys dearborni		

There are only a small number of common fish species in McMurdo Sound and in the neighbouring Ross Sea available for higher predators (Eastman, 1985; MacDonald *et al.*, 1987). Only three are pelagic (Table 3.1).

On first sight it appears that although all species might be accessible at times to penguins, only the three pelagic species would be available to skuas surface-plunging or surface-seizing (Ashmole, 1971) at the water surface. Of these species only *Pleuragramma antarcticum* (the Antarctic silverfish) was regularly taken. *Dissostichus mawsoni* is characterised as a mid-water species, and grows to a very large size, and *Aethotaxis mitopteryx* is comparatively rare, so their absence from the skua diet is not unexpected. However, all notothenioids have pelagic juvenile stages (Hubold and Ekau, 1990), which could become available to surface predators or as stranded fish on shores or iceflows.

There is now firm recognition of the key role of *P. antarcticum* in coastal marine food webs throughout Antarctica (Takahasi and Nemoto, 1984). Hubold (1984), for example, considers it to be the key species in the pelagic system of the Antarctic shelf. It is certainly the dominant mid-water fish of the Ross Sea where it represents more than 90% of the total fish fauna in both weight and numbers (DeWitt, 1970). It is, however, characterised by ichthyologists as a mid-water rather than surface shoaling fish, yet it was by far the predominant species in food brought back to the territory by foraging skuas (Young, 1963a and observations at Cape Bird).

Clearly at times it occurs at the surface: otherwise how could the skuas catch them so easily? The alternative explanation, that the skuas are scavenging dead fish from iceflows or along the shore, is not at all supported by numerous observations of birds foraging successfully in sea quite clear of pack-ice. Daniels (1982) records this species shoaling and feeding beneath the ice along the Antarctic Peninsula but in McMurdo Sound it has proved an elusive fish for biologists. It is neither seen by divers nor much taken in nets or on lines, but is nevertheless the main prey of many of the other fish species there (Eastman, 1985). It is also an important component of the diet of penguins. At Cape Crozier Emison (1968) found that fish made up 39% of the diet by volume of Adélie penguins. Although classification of the often well-digested material proved difficult it was considered that most of these fish were *P. antarcticum*. Fish were also important prey in 15 penguins sampled at Cape Bird, making up 44% by volume, but although two species, including *P. antarcticum*, were distinguished the material could not be partitioned between them (Paulin, 1975).

  P. antarcticum is a herring-like fish with extensive lipid sacs providing neutral buoyancy, weighing in seawater about 0.57% of their weight in air compared with over 3% for benthic nototheniid species (Eastman and DeVries, 1982). The skeleton and scales are lightly mineralised, reducing density, with ash constituting only 0.3–0.5% of the total body weight. Hubold and Tomo (1989) found that Antarctic silverfish were slow-growing species in the Weddell Sea with annual growth rates of between 10 mm and 20 mm depending on age and reaching a maximum 245 mm (standard length, SL) by 21 years. The species appear to reach as much as 350 mm in other areas, with the largest fishes occurring in the southern end of the range. Hubold and Tomo (1989) comment that these must be very old fish indeed. Their weights of a sample of fishes from the Weddell Sea ranged to 94.5 g. Hubold and Ekau (1990) note that although *P. antarcticum* is locally abundant mean pelagic fish biomass on the Antarctic shelf seas is unlikely to exceed 0.1 t km^{-2}, much less than that of the demersal (bottom living) fishes. The abundance of this prey in the upper 800 m water has been estimated by Hopkins (1987) from four stations in McMurdo Sound as 0.82 (range 0.18–3.69) g dry wt m^{-2} sea surface compared with 0.21 (range 0.16–0.33) g dry wt m^{-2} for *E. crystallorophias*. Both are shoaling species, however, so that overall estimates made in this way probably bear little relation to their availability as prey.

  Silverfish are an easy shape for skua ingestion, fitting comfortably within the extended crop. They are ingested head first (and are regurgitated with

Fig. 3.4. Skua pair feeding their chicks with disgorged Antarctic silverfish (*Pleuragramma antarcticum*).

difficulty to mates and chicks) so that the heads are invariably partly digested on return to the territory (Fig. 3.4). As Young (1990a) found, the skeleton and scales are so lightly mineralised that they leave little trace in regurgitated pellets.

Four fish analysed by R. Davis (pers. commun.) had an average wet weight energy content of 6.3 kJ g^{-1}.

3.3.3   *Food available to skuas from seals*

The five species occurring in the Ross Sea offer very different prospects for skua foraging. Two, Ross's seal (*Ommatophoca rossi*) and the Southern elephant seal (*Mirounga leonina*) are so rare, the latter really only a vagrant, that they can be ignored as a food resource. Two others, the Weddell seal (*Leptonychotes weddelli*) and the crabeater seal (*Lobodon carcinophagus*), are both very common and could potentially provide considerable volumes of food from scavenged faeces and during pupping. The remaining species in the area is the leopard seal (*Hydrurga leptonyx*). Although occurring in small numbers, probably fewer than a hundred at any time over the summer in McMurdo Sound (Ainley, 1985,

Fig. 3.5. Leopard seal attempting to chase Adélie penguins into the water from their refuge on a small broken iceflow.

estimated 8000 for the Ross Sea as a whole), but its very obvious association with the penguin colonies and its predation on penguins has made this species the best known and most visible of all seals in the region, with the exception of the Weddell seal (Figs 3.5 and 3.6). It makes a quite different contribution to skua diet, not directly as for the others, but indirectly through food lost from its own prey during feeding. Skuas regularly attend any leopard seals feeding on penguins.

But overall, because of the high numbers of skuas that could exploit these foods, seals can make only a small contribution to individual diets. The Weddell seal population in McMurdo Sound has been extensively studied and its biology is well documented (Stirling,1969; Sinif *et al.*, 1977; Thomas and DeMaster, 1983). Most pupping along the western side of Ross Island occurs on fast-ice between Cape Royds and Cape Armitage, from 35 to 70 km distant from Cape Bird colony and well within flight range of skuas. Between 1963 and 1974 on average 510 seal pups were born each year (Sinif *et al.*, 1977) of which c. 10% died shortly after birth (Stirling, 1971; Thomas and DeMaster, 1983). The birth weight of pups averaged 29 kg (Stirling, 1969). In addition skuas might also scavenge on

Fig. 3.6. Adélie penguins running about a lunging leopard seal on ice. The seal has no chance of capturing penguins in this situation.

placentae and membranes. There are no published data on these weights but it is estimated as 2 kg, double the weight of the placentae of the smaller Antarctic fur seal (*Arctocephalus gazella*) (D. W. Doidge, pers. commun.). The total from both sources (dead pups and placentae) for Ross Island is thus c. 2382 kg each year.

Of course this is not the exclusive resource of the Cape Bird skuas. There were c. 320 pairs of skuas in Southern McMurdo Sound in 1965–66 (Spellerberg, 1967) (and over 1100 in 1987–88, McGarry, 1988), and c. 15 000 in total for the Ross Sea region (Ainley *et al.*, 1986). All these birds and any non-breeding skuas could also feed here. Moreover, this food occurs very early in the season. Mean pupping date is about the end of October (Stirling, 1971; Thomas and DeMaster, 1983) and pupping is over by mid November before the first skuas have laid. There is no doubt, however, that skuas scavenge for this food. Stirling (1971) writes that 'skuas quickly tear a dead pup to shreds'.

Ainley (1985) estimated that there were c. 200 000 crabeater seals in the Ross Sea, concentrated in light pack-ice along outer edge of the sea. There are no data for McMurdo Sound. Pupping occurs from late September to early November with a peak in early to mid October (Laws, 1984). Pupping seals could therefore provide food for skuas travelling towards the Ross Island breeding colonies in early summer.

In each season leopard seals appeared at all the Ross Island penguin colonies, lying offshore or patrolling along the beach front. At Cape Crozier four seals were estimated by Penny and Lowry (1967) to have taken in one year 15 000 penguins, 5% of the breeding population. In a second year six seals took an estimated 4800 adult penguins (1.4% of the adults) and 1200 chicks (0.6% of fledglings) (Müller-Schwarze and Müller-Schwarze, 1975). There are no comparably detailed studies for the Cape Bird colonies so that it is necessary to apply the Cape Crozier estimates there. In many hours of watches over the beaches and inshore pack-ice at the Northern Colony kills were seen regularly enough, but not at the rate suggested by the higher Cape Crozier figure. For days at a time there were no leopard seals at the colony at all. Kills are easy to see: the seals thrash the penguin about while stripping away the skin and skuas flock to pick up the small pieces of material thrown away. They are in fact rather difficult to overlook. From this experience the lower figure for Cape Crozier of 1.4% of adults taken is the more reasonable one. Even this one may be too high, taking into account the wider access to the sea available to the penguins at Cape Bird, compared to just four access points at Cape Crozier. On this basis, however, c. 850 adult penguins would have been taken over the 100 days of the breeding season together with c. 250 fledglings.

During feeding the water is strongly discoloured by penguin blood, giving the impression that there is much food there simply to be picked up. In fact, very little seems to be present and the skuas hover and plunge for very small pieces (Fig. 3.7). It is doubtful that as much as half a kilogram is taken by the skuas at any kill, and this may well have to be shared among ten or more birds. Over the whole summer the total amount of penguin flesh gained by the skuas from this source is probably no more than 500–600 kg, some 5–6 kg a day. After the penguin chicks leave in late January and early February there is no food from this source.

Some penguins are attacked by seals and although having gaping wounds manage to reach shore. These may then be attacked by skuas. Few such birds were seen at Cape Bird, no more than a handful in any season, and they made a negligible contribution to skua diet.

Fig. 3.7. Skua overflying a leopard seal feeding from an Adélie penguin.

3.3.4   *Food gained by predation and scavenging on birds*

The skuas are very limited in the range of sea birds that might be preyed on in this area. There are only two breeding species, Adélie and Emperor penguins. In many other places in Antarctica skuas take a range of bird prey and petrels are an important component of the diet (Green, 1986; Norman and Ward, 1990). In some inland breeding sites (e.g. Theron Mts (Brook and Beck, 1972), Rockefeller Mts (Broady *et al.*, 1989), Muhlig-Hofmann Mts (Howard, 1991)) breeding petrels may be the most important food source.

The penguin breeding areas within foraging range are the colonies of both species at Beaufort Island and at Cape Crozier on Ross Island, the three colonies of Adélie penguins at Cape Bird and the single colony at Cape Royds.

Emperor penguin colonies

Emperor penguin colonies may provide food for the skuas when they first enter the Ross Sea area and before egglaying begins at the Adélie penguin colonies. Pryor (1968) found that this happened at Haswell Island when dead chicks seemed to be the skuas' main food source at this time. By the time the skuas begin breeding the Emperor penguin breeding cycle is almost over and the colonies are dispersing. There is no live food for skuas at these colonies over summer, although there may be substantial natural mortality at Emperor penguin breeding colonies. Jouventin (1975) has summarised the breeding statistics of the colony at Pointe Geologie, Terre Adélie to record an average egg mortality of 11.8% (range 4.4–22.0%) and an average chick loss of 19% (range 3.7–29.8%) from an average of 5797 (range 5015–6236) eggs laid. These losses represent a considerable food resource for scavengers in the spring – but only if they remain accessible on the surface of the ice. There are two accounts of visits to the Cape Crozier and Beaufort Island colonies that indicate that considerable numbers of eggs and chicks killed overwinter do indeed last long enough to be scavenged by the returning skuas. Caughley (1960) visited Cape Crozier in early summer 1958 and found 59 dead chicks that he considered had died during September. Todd (1977) estimated a 50% egg and chick mortality for the 1976 season accompanied by a 'disturbingly high' adult mortality. At Beaufort Island in the same year he estimated that chick mortality was between 65 and 70% and counted 1115 chick bodies. These numbers represent a considerable food resource for skuas in these years but there are no other records for these two colonies to judge whether these are typical figures. The colony at Cape Crozier has never been large and since the early 1960s its numbers have dwindled (Stonehouse, 1964; Ainley *et al.*, 1978; Kooyman and Mullins, 1990). Overall it represents only a small resource for skuas, and is anyway much closer to the very large local skua population than to the population at Cape Bird.

Adélie penguin colonies

These colonies represent the second most important food resource for the skuas in the local area, exceeded in volume and seasonal span only by fish. It is not, however, universally available and many skuas are precluded from obtaining food from the colonies by resident pairs. Nor does it occur throughout the duration of the skua breeding cycle. At the time this study was undertaken there were c. 30 000 breeding pairs of penguins at the Cape Bird colonies, c. 1500 pairs at Cape Royds and 175 000 at Cape Crozier (Harper *et al.*, 1984).

Eggs and chicks are the two main components of this resource. The food contents of each will be described at this stage, leading on to the estimations of the overall biomass on the Northern Colony in the following chapter.

Adélie penguin eggs (Fig. 3.8)

Adélie penguin eggs have been weighed and measured by many authors (Taylor, 1962a; Reid, 1965; Astheimer and Grau, 1985; Lishman, 1985b) but there is less detailed information on the egg contents. The eggs measure about 70 mm in length and 55 mm in width with the first egg of the clutch marginally larger. Their volume is accurately described by the relationship $V = 0.5\, ab^2$ (Reid, 1965) in which a and b are the length and width. Egg size is uniform throughout the species range. Lishman (1985b) recorded egg volumes at Signy Island of 81–126 cm^3 with weights of the whole egg ranging from 85–133 gm (means of first and second eggs 120.8 g and 113.2 g, respectively). A sample from Cape Crozier gave mean weights of 122.9 g and 114.8 g respectively for the first and second eggs of clutches of which 15.7 g and 15.1 g respectively comprised egg shell (Astheimer and Grau, 1985). Reid (1965) calculated that the egg shell made up 12.6% of egg weight in a sample measured at Cape Hallett.

As a group penguins have a disproportionately heavy shell (Williams *et al.*, 1982) which is a factor in their undoubted strength. Very few broken eggs are seen on Adélie penguin colonies even though eggs are often displaced from nests and may be pecked strongly by sitting birds when they roll near them. Ironically it is this strength that makes Adélie eggs such good prey for skuas. Because of it they become available to the skuas along the margins of breeding groups or in gaps among nests instead of being broken, with the contents absorbed or trodden into the stony guano of the nest area. They are strong enough also to be knocked or pulled from

Fig. 3.8. Food available on the penguin colonies: eggs.

beneath sitting penguins by attacking skuas and can be grabbed in the beak and carried out of the colony. They are strong enough in fact to be carried in this way for hundreds of metres in flight. At Cape Crozier and Cape Hallett, for instance, skuas nesting at a distance from the colony are able to exploit food there by carrying the eggs back to their nests. The eggs are, however, at the size limit of the skua's gape and many attempts at grabbing them fail. They are also close to the limit for breaking into by skuas. Several to many jabs may be needed before the shell is broken.

Energy content. There does not appear to be any published account of the energy content of Adélie eggs but it can be estimated from information on the yolk and albumen constituents given by Astheimer and Grau (1985) for a sample of fresh eggs from the Cape Crozier colony.

The greater part of Adélie penguin eggs comprises albumen. In the analysis done by Astheimer and Grau (1985) first eggs weighing on average 122.9 g contained 26.0 g (21.2%) yolk and 81.2 g (65.9%) albumen (the remainder comprised shell). Second eggs of clutches were lighter in weight at 114.8 g and on average contained an almost identical weight of yolk (25.8 g, 22.6%) but less albumen (73.1 g, 63.5%). The shells had almost identical weights so that the overall difference between the eggs was attributed almost entirely to the lower albumen weight of second eggs. Yolks comprised 48% solids, 51% by weight lipid and 35% non-lipid (mostly protein). Using the energy contents of lipid (39.5 kJ g^{-1}) and protein (23.6 kJ g^{-1}) the yolk content is estimated to be 358 ± 5.9 kJ. The albumen was found to be 87% water with mean solids 10.5 g and 9.5 g respectively in first and second eggs. The solids were 77% protein with negligible lipid giving on average for the eggs of the clutch an albumen energy content of 181.6 ± 4.3 kJ. In total each egg thus has an energy content (yolk and albumen together) of 540 ± 9.5 kJ (mean $\pm$ SE). This estimate for the Adélie egg is significantly less than that derived from equation 5.5 of Kendeigh *et al.* (1977), which is 780 kJ for a 120 g egg. Perhaps this is because of the disproportionately heavy shell in penguin eggs. The figure of 540 kJ/egg will be used for the estimates of food taken by the skuas. This energy content is assumed to remain constant throughout incubation.

Adélie penguin chicks (Fig. 3.9)

Adélie chicks hatch at c. 80 g and grow rapidly during a six-week period to 3.5–4.0 kg at fledging (Taylor and Roberts, 1962; Volkman and Trivelpiece, 1980)

No analyses of the constituents of the body or the energy content of Adélie chicks of different ages have been published but they have been

Fig. 3.9. Food available on the penguin colonies: penguin chicks.

carried out for the other two pygoscelids, *P. antarctica* and *P. papua*, by Myrcha and Kaminski (1982). The pattern of chick growth of the former is closest to that of the Adélie (Volkman and Trivelpiece, 1980) and is used here as the best available source for Adélie chicks.

These authors considered the water, ash and energy contents of growing chicks. Water content declined from about 84% of body weight at hatching to 64% at fledging. During the same time the ash content rose from 9% to 11.5–12.5%. Energy content determined on a dry weight basis increased steadily during early growth until about halfway through the growth period and thereafter remained constant. Its value at hatching was 22 kJ g^{-1} rising to 27 kJ g^{-1} by mid growth. The energy content measured on a wet weight basis showed a more pronounced increase, linked to the lower water content of maturity. On this basis the energy content rose from 3.3 kJ g^{-1} to 9.5 kJ g^{-1}. For the estimates of energy content of chicks of different ages (and on average for the chicks on the colony at any date) the regressions $Y = 0.762 + 0.000516 \times X$ and $Y = 1.70 + 0.00126 \times X$ (where Y is energy content in Kcal g^{-1} and X is body mass in grams) from Myrcha and Kaminski (1982) have been used, depending on whether the chicks were lighter or heavier than 2500 g.

Skuas can obtain varying amounts of flesh from the chicks depending on chick age. The youngest chicks can be swallowed whole and are almost completely digested; somewhat older chicks are torn apart, by a single bird or between the two birds of the pair, before being entirely consumed. The oldest chicks killed have well developed and ossified skeletons linked together strongly with ligaments and have toughened and thickened skins with dense feathering. These chicks are only partly eaten and much of the skin and skeleton may be left. Thus, although larger chicks contain more material than smaller ones a smaller proportion of the total is consumed.

In the estimates of the amount of food gained by skuas from the chicks both energy density and proportion consumed must be taken into account.

Conclusion

There is clearly a diverse array of foods available to skuas foraging on land in the penguin colonies or at sea within the extreme foraging range of some 95 km estimated for these skuas. This food includes fish and zooplankton, scavenged material at the bases and from seal colonies, and eggs and chicks from Emperor and Adélie penguin colonies. This is not of course the exclusive resource of the skuas at Cape Bird but is shared as well with the more than 1000 pairs on Ross Island and about McMurdo Sound.

In practice, however, the skuas depend on a quite narrow range of foods. By far the predominant items are Antarctic silverfish for those foraging at sea, and Adélie penguin eggs and chicks for those foraging on land. Study of skua feeding has been facilitated by the occurrence here by this low diversity of foods arising from just two, quite distinct, foraging areas.

4

The biomass of penguin eggs and chicks on the Northern Colony

4.1 Introduction

One could be forgiven for thinking that the bustling masses of penguins at the colony represents an almost unending supply of food for the skuas. However, even the most cursory examination of the breeding groups shows that this apparent abundance is largely illusory. Except for a month or so from mid December skuas find that the food available here is limited. Skuas cannot kill adult penguins so that the food resource is made up entirely of eggs and chicks. This chapter documents the numbers of eggs and chicks on the Northern Colony in each of the five years and then uses these figures with the mass of eggs and chicks to estimate the total biomass on the colony during the year. At a later stage the biomass will be partitioned among the pairs of skuas holding colony territories.

4.2 Estimating numbers and biomass for the penguin colony

The calculation of the numbers of eggs and chicks present throughout the season on the colony, of the total biomass and of the amount actually available to skuas was made each year from three sets of data: from total counts of penguin nests in early December; from records of the changing egg and chick numbers through the season in a series of study breeding groups; and from the mass of chicks in representative groups. The colony-wide count of nests can be made with great accuracy, virtually without error, but there is substantial variability in nest contents through the season and in chick mass on the colony on any one date which must be incorporated into the estimates. The methods used to obtain each of these sets of data with error estimation and with comment on decisions that need to be made at each step in the calculations are described in the following sections.

4.2.1 *Numbers of penguin nests on the colony each year*

Colony-wide counts: the season's total

It is not possible to count every penguin nest on a colony of this size repeatedly throughout the season to provide accurate information on the numbers of nesting birds, eggs and chicks on each date. Each count of the colony even in the most favourable period for counting took three to four days with teams of two or three people working together. Counts were made each year between 7 and 11 December, towards the end of the incubation period when nest occupation is at the lowest point for the year, between the abandonment of nesting by unsuccessful breeders at the start of the season and before their return to the colony during the reoccupation period. On this date there are so few auxiliary penguins on the colony that it is possible to separate out the few nests that lack eggs. Distinguishing nests with and without eggs would be almost impossible earlier in the season for colony-wide surveys; there are just too many birds, and the colonies are ringed with rows of naive breeders enlarging them and making the counting immeasurably more difficult. Counts later in the year are also much more difficult. The enlarged groups of the reoccupying stage, made up of both successful and unsuccessful breeders, present an impossible prospect for easy counting – dense masses of birds in continuous move-ment and activity.

Since the early studies by Sladen (1958), Taylor (1962a), and Penny (1968) it seems to have become accepted that the most accurate counts, and the most useful ones for comparing colonies from year to year, should be made during the last week or so of the incubation period during minimum occupation, which is a short interval of stable numbers on the colony (Stonehouse, 1965). An early exception to this practice was by Reid (1964) who counted the penguins at Cape Hallett during peak occupation.

All the colony-wide counts at Cape Bird were made during the minimum period of occupation. It is recognised that this count is taking place nearly a month after the first eggs are laid on the colony and represents the penguin breeding and egg population at the close of incubation and the onset of chick hatching. It is, therefore, in reality a measure of the numbers of eggs hatching, rather than of numbers produced, because egg losses after this date tend to be very low. Overall, it is also a fair measure of the breeding population of the colony except in years where there is catastrophic egg loss early in the season.

In the first year at Cape Bird the individual breeding groups were mapped and these maps were used thereafter for all counts. Counts distinguished occupied and empty nests but did not discriminate between nests with one or two eggs. At the same time counts were made of the Middle and Southern colonies, which were not otherwise disturbed, to provide a check on possible human impact from the work being undertaken in the Northern one. This comparison will be considered later in the book, but it can be noted here that the numbers of nesting birds varied from season to season similarly in the three colonies suggesting that our work was not chasing penguins away from the Northern Colony to the others.

This count each year is the cornerstone for all the estimates of changing numbers and biomass throughout the season. Even so it was inexplicably missed out during 1966–67 so that the numbers in the colony in this year have had to be estimated from the numbers in the study groups in comparison with the years on either side and checked for accuracy in the photographs taken at this time. (This lapse is all the more inexplicable when it is appreciated that the other two colonies were counted – these provide an additional check on overall trends in numbers.) For this year alone, therefore, the numbers of breeding birds on the Northern Colony during early December is an estimate and has an associated error.

All calculations that follow are of numbers of *live* eggs and chicks. A small proportion of eggs failed to hatch, and most of these showed little development. A sample of these eggs was taken at the end of incubation from nests which also contained a large growing chick. By this date, after some 40–50 days incubation, the yolk was generally broken down too far to be certain whether embryogenesis had indeed been started. Certainly, few of these eggs had well developed embryos. Because the yolk–albumen mix of these eggs was generally very liquid skuas could retrieve little of it before it drained onto the ground when the egg was broken. These eggs provided little food for the skuas at this time, especially when compared with that available to them so much more easily in the form of rapidly-growing chicks.

Changes during the season: the role of the study breeding groups

The changes in the numbers of eggs and chicks on the colony during the season were determined by following the fate of nests in a sample of breeding groups. The individual nests in these groups (the study groups) were mapped at the start of each season and monitored at five-day intervals to provide a record of the survival of eggs and chicks for the year. The way this work was done has been described in section 2.5, 'General methods'.

Cape Bird is a hard place to get to at the start of summer and the first nest maps and counts were not made until the season was well under way. In this colony the first eggs are laid at the very beginning of November but first counts were not made until between 12 and 18 November in each of the five years. Eric Spurr has earlier counts in some years which can be used but the best record of the pattern of egglaying comes from the pattern of egg hatching in December, for which there are detailed records for each year from the study groups. The average incubation period for this penguin on Ross Island is 34 days (Taylor, 1962a; Spurr, 1975c). The curve of egg hatching can be transposed to the start of the season for egglaying using this incubation period and assuming that the great majority of all eggs laid survive to hatching without much bias between early and late eggs.

The counts of colony chick numbers can be carried through to the middle of January with good precision but from this date the structure and integrity of the individual groups breaks down. It is not possible to keep tallies of the chicks of study nests in large groups from this date without marking them (causing disturbance and encouraging even greater dispersal than before) while chicks in small groups begin moving to larger ones. From mid January there is a general movement towards larger aggregations and towards the shore. In 1965–66 the first chicks were found on the beach on 24 January and at sea on 27 January. Even though C and H blocks are sited almost at the extremes of distance to the beach chicks left them at closely similar rates in 1965–66. This was even more surprising when it is appreciated that chicks from C block must travel over 200 m through 15 separate skua territories before reaching the next substantial group of penguins in B Block and then still have over 150 m to go to reach the shore. For comparison H Block chicks on their colonies are within 65 m of the shore.

For the estimates of biomass the proportion of chicks remaining on the colony at different dates at the end of the season have been set as:

> 27 January, 100%
> 30 January, 85%
> 2 February, 50%
> 5 February, 20%
> 9 February, 3%
> 15 February, 0%.

4.2.2 *Estimating the total numbers of eggs and chicks on the Northern Colony throughout the breeding season*

Estimates of the total numbers of eggs and chicks on the colony throughout the season were obtained by bringing together the information from the study groups, which recorded seasonal changes, and the total nest count in early December.

The study groups were selected to represent typical situations and consequently, representative survival rates. It is known that some of the small colonies had high skua predation rates. However, small, isolated groups make up only a very small proportion of the colony as a whole (less than 1% of nest numbers) so that their lower egg and chick survival can be ignored when estimating the numbers and biomass of chicks for the whole colony. In any case attempts to stratify the colony into blocks with different productivities were not successful. Attempts considered both group size and situation, the latter criterion indicating probable skua impact. From this exploration it was concluded that the colony could be treated for the purposes of obtaining overall numbers as a single unit. Tests of heterogeneity of the productivity of the study groups in each year failed to discriminate among them and these also can be considered an homogeneous set. This uniformity makes the conversion of egg and chick numbers from the study groups to colony totals a straightforward process. This was done by first calculating egg and chick numbers per nest on each date from the study groups. Weighted means and standard errors were obtained for the series of individual egg–chick ratios for each group on each date. Using these notional nest content figures for each date it is then a simple matter to scale them for the colony totals using the total colony nest count of early December (the reference count). Standard errors of the colony totals for each date were also calculated from the product of the colony total and weighted mean errors.

These estimates give colony numbers for the majority of the penguin cycle, from about mid November until mid to late January, the most important part of the season for skua predation. Estimates for numbers of eggs at the beginning of the season up to mid November have been estimated by transposing the hatching data from the study groups forward and from Spurr's data (1975c) on egglaying. The fall-off in chick numbers at the end of the season as the colonies become progressively deserted are taken largely from counts made of chick numbers on C and H blocks from mid January to mid February in 1966. At this time of the year few chicks are small enough to be killed by skuas so that errors in interpolation of the total numbers at this stage have little significance for the overall study of the skua and penguin association.

4.2.3   *Estimating egg and chick biomass on the colony*

Whereas the numbers of nesting birds establish the basis of comparison with other colonies and allow comparisons among the skua territories, the important variable in the relationship between the two species is not numbers as such but the biomass of food present on the colony and *available* to the skuas. In practice this biomass is from eggs and younger chicks. Adult penguins and older chicks are virtually immune to skua attack. Moreover, in the five years of this study there was almost no adult or older chick mortality to provide food for scavenging skuas.

Once the numbers of eggs and chicks at each counting date have been estimated they can be converted to biomass estimates by using egg and chick mass for the same date.

Until the chicks begin hatching eggs are the only food on the colony. The most valuable ones for skuas are those with well developed embryos that can be entirely extracted from the broken eggs with almost no loss of egg content. The same amount of nutrient is present at the start of the incubation period as at the end (minus the amount used in respiration by the embryo) but its availability to the skua is very different. Much of the egg content is spilled and lost from eggs taken early in incubation.

Conversion of chick numbers to chick biomass

In each year all the chicks of selected groups were weighed and measured (Table 4.1). These weighings were done at about five-day intervals, depending on other work. Between 30 and 80 chicks were weighed in each of several groups on each occasion. As this work was to record the average mass of chicks on the colony all chicks in the selected area were weighed. These chicks were not banded, however, so that their individual ages were not known although the mean age was known from the hatching pattern in the colony. (A study of the growth of known-age

Table 4.1. *The mean mass of penguin chicks on selected breeding groups on the Northern Colony on different dates during the breeding season. Means and* SE *in kg*

	1965–66	1966–67	1967–68	1968–69	1969–70
December					
15–19	—	0.14 ± 0.01	—	—	0.19 ± 0.04
20–24	—	0.23 ± 0.02	0.19 ± 0.02	0.20 ± 0.02	—
25–29	—	0.55 ± 0.04	0.39 ± 0.03	0.54 ± 0.04	0.68 ± 0.04
January					
30–3	—	0.98 ± 0.06	0.69 ± 0.04	1.12 ± 0.07	1.56 ± 0.07
4–9	1.73 ± 0.07	1.57 ± 0.08	1.36 ± 0.07	1.74 ± 0.08	—
10–14	2.69 ± 0.09	2.66 ± 0.08	1.85 ± 0.07	2.32 ± 1.07	2.37 ± 0.08
15–19	3.02 ± 0.06	2.80 ± 0.08	2.59 ± 0.07	3.03 ± 0.07	2.86 ± 0.08
20–24	3.80 ± 0.08	3.49 ± 0.09	3.29 ± 0.09	3.15 ± 0.07	3.19 ± 0.08
25–31	—	—	3.48 ± 0.06	3.48 ± 0.08	—
February					
2	3.51 ± 0.09	—	—	—	—

Essentially the same breeding groups were weighed in each year. These records are for E6, D29 and E10 breeding groups. Between 30 and 60 chicks were weighed on each occasion. The small differences among years for each interval reflects in part the actual weighing date within the four-day interval. The similarity between years is nevertheless quite striking.

chicks was done separately in order to provide information of weight and development against age. These data were used to indicate in the chick numbers and biomass graphs the proportions of chicks young enough on each date to be at risk from skua attack.)

Once the chicks began hatching the biomass present increases very rapidly, with chick mass doubling each 10 days or so during early growth. Within a few days of the first chick hatching the biomass present on the colony is double that present previously as eggs. The product of numbers of chicks and mean mass of chicks on each date give the biomass at that date. This has to be amended at the start of the chick hatching period by adding in the contents of the unhatched eggs.

4.3   *The estimates of egg and chick numbers and the total biomass on the Northern Colony throughout each season*

The data on egg and chicks numbers and colony biomass for each year are shown in Figs 4.1 and 4.2 respectively and tabulated in Appendix

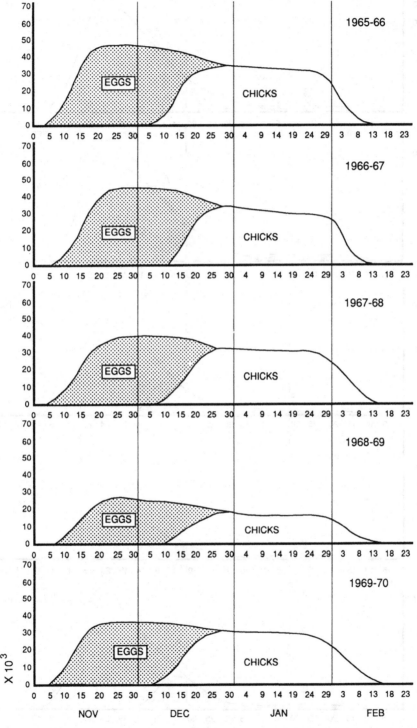

Fig. 4.1. The numbers of eggs and chicks on the Northern Colony for each year from 1965–66 to 1969–70.

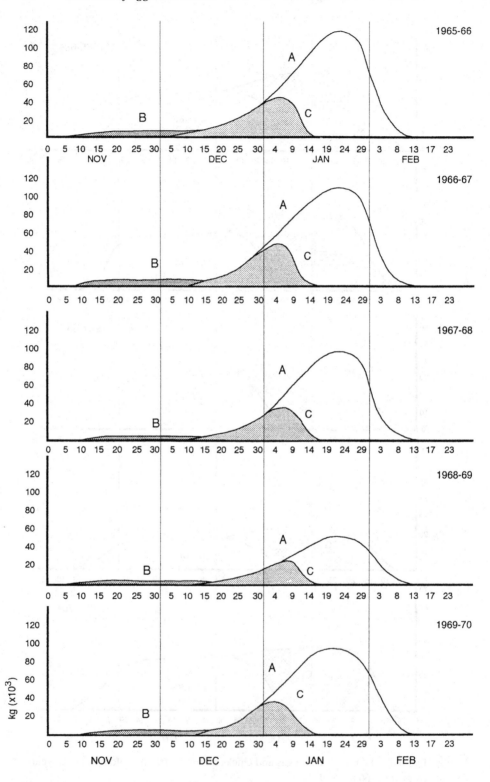

Fig. 4.3. The alignment of penguin nests in rows within the breeding groups. This generally regular pattern of nest location greatly facilitated accurate counting of nest numbers on the colony.

1. The figures for number and biomass in the colony shown in the table have remarkably low estimation error. This estimation reliability derives from the low variability of the estimator components: 1. the nests can be counted with extremely high accuracy, virtually without error (Fig. 4.3); 2. there is high uniformity of egg and chick production – most nests have two eggs and hatch both; 3. there is little variability in egg weight; and 4. the tight synchrony of the breeding cycle together with a uniform chick growth rate gives a small range of chick ages and sizes on the colony at any time. The conjunction of these features, rare in natural populations of any species, leads to the quite astonishing reliability of the estimates portrayed here.

From the skuas' viewpoint the season can be partitioned into five stages depending on kind of food and its availability. 1. From penguin occupation

Fig. 4.2. The biomass of eggs and chicks on the Northern Colony for each year from 1965–66 to 1969–70. A. The total biomass, B. the egg biomass and C. the biomass of chicks available for the skuas by virtue of their small size (less than 2000 g) and immaturity.

in October until the first eggs are laid in the first days of November the colony is barren. 2. Eggs are the only food available on the colony throughout November and the first week of December. 3. From about the end of the first week of December chicks increasingly replace eggs as the major food source and for a few weeks provide readily available food for skuas nesting about the colony. 4. From mid January the maturing chicks grow beyond the skuas' ability to prey on them so that late January and February are again lean days for skua predators. 5. Penguin chicks begin leaving the colony in the last days of January and by mid February the colony is deserted. The skuas' breeding cycle stretches into March and April so that their own large chicks are almost entirely dependent on sea foraging. In theory, dead penguin chicks could become accessible to skuas as the breeding areas are deserted but very few of these occurred at Cape Bird in these years and those that did were flattened into the guano of the colony bed and were ignored by the skuas.

The graphs of colony egg and chick biomass each year (Fig. 4.2) record these changes in food availability very clearly, especially the difference between egg and chick stages. The contrast between the estimates of numbers and biomass for each stage point up how misleading are numbers alone for indicating food abundance. Egg totals are indeed impressive, reaching 50 000 in the first years, but weighing only about 100 g each without the shell they give a colony biomass totalling no more than 5000 kg. It is with egg hatching from mid December that the biomass soars, attaining maximally between 50 000 kg and 120 000 kg by the last week of January. The collapse in mass is equally rapid, within a further two weeks it has disappeared.

Egg and chick survival and chick growth were similar in these five years so that the differences in the volumes of food on the colony reflect numbers of eggs and chicks. There were highest numbers in 1965–66 and greatest volume. Low numbers in 1968–69 produced the lowest volume with chick biomass barely attaining 50 000 kg, less than half the amount of the best years.

Figure 4.2 also shows the biomass of chicks weighing less than 2000 g. These chicks, about 27–28 days old, are at about the upper size limit for skua predation unless circumstances especially favour the skua, or the skua is both desperate for food and an able predator (Fig. 4.4). Chicks are taken that are heavier than this but they come from breeding groups on steep slopes or on sites falling away into gullies, which make it very difficult for chicks to get back to safety when attacked, or are chicks

Fig. 4.4. Chicks at the close of summer. These chicks represent an enormous biomass which is no longer accessible to the skuas.

attacked well away from adult penguins. The special circumstances favouring skua predation will be considered in the following chapter.

The distinction between food present and food available or accessible to the skuas is, therefore, an important one. Not only is the entire adult population denied to them but so too is the greater volume of chicks. In essence, the skuas at the colony have an abundant food supply for less than a month. From mid January little food is available there.

5

Factors of penguin breeding biology that constrain or assist skua predation of eggs and chicks

5.1 *Introduction*

Until the end of the penguin guard stage when the parents desert the maturing chicks almost all attacks on eggs and chicks, the target of predation, must overcome the defences of the parent at the nest and the collective defences of the breeding group. This situation changes with the onset of the post-guard stage and chicks can then be attacked directly, generally still within the breeding group but often without adult penguin interference. For much of the season the eggs and chicks have a passive role in their own survival – the first being immobile and the latter being too small, weak and uncoordinated to offer much resistance to skuas.

The success of skua predation and, conversely, the success of penguin breeding, thus depends importantly on the defensive behaviour of penguins at the nest and on the way individual birds cooperate in colonial defence. This chapter considers aspects of penguin biology that contribute to or thwart skua attack. Some factors are indirect, relating to the breeding biology of the penguins, to their individual competition for breeding success or to adverse weather conditions, but others, such as the forcing of eggs or young chicks on to the breeding group periphery, bear directly on the amount of food available to skuas.

5.2 *Constraints imposed on the skuas by the tight synchrony of the penguin breeding cycle and the relative timing of the cycles of the two species*

Adélie penguins and skuas need to fit in a full breeding cycle within within a short summer period when there is access to the breeding areas, when the nest sites are free of snow and when open sea allows foraging. Penguins can begin breeding as soon as nest sites become free of snow because they can forgo feeding for some weeks by living on fat stores. The end of the season imposes rigid time limits for the breeding cycle and adults and fledglings must be able to move freely away from the breeding area to wintering sites before the seas re-freeze in the autumn. For penguins breeding as far south as Ross Island, this is a very short interval within which to cram the whole process of producing viable chicks. Not surprisingly, the cycle within a breeding area is tightly sychronised. To what extent this synchrony is also determined by skua predation will be explored later in the book. For the skuas there are different restrictions on the breeding season. They may not begin breeding until open sea is close at hand, as they are unable to fast and must forage each day. The end of the season is less rigidly constrained, however. Both adults and newly fledged chicks can fly rapidly away to the north at the end of the season. Because of this they can stay later on the breeding areas.

The incidental effect of the way the two species have adapted to the short polar summer is that there is a relative abundance of food for the skuas on the penguin colonies for a month or so from mid December but very little later in the summer as the chicks mature. The tight synchrony of the cycle means that for much of the season there is only one prey type – eggs, young guarded chicks, young post-guard chicks and mature chicks – present on the colonies at any time. This pattern is important because the best food for the skuas is the younger chicks but these occur for a very limited time at any site. With a less tightly synchronised breeding cycle their availability would extend for a much greater proportion of the skua cycle.

Not only are the cycles tightly synchronised they also seem to be out of phase for a successful prey–predator association. That is, the predator's cycle lags behind that of the prey to such an extent that it runs out of prey at a time when its demands are greatest.

5.3   *Constraints imposed on the movements of skuas in the colonies and on the selection of prey nests because of the colonial nesting habit of penguins*

There are many possible reasons for the colonial nesting habits of Adélie penguins. Volkman and Trivelpiece (1981) list, for example, the shortage of breeding areas, or the advantage of concentrating within the most favourable part of the suitable range; the amelioration of climate because the birds together reduce the strength of wind and so produce a more favourable microclimate; the need to nest colonially to obtain mutual protection from skuas or other avian predators and scavengers; or for social facilitation or communication. Such an array of reasons, or, in the absence of data, range of opinion, indicates that no one of these is paramount. But for whatever reason colonial nesting has developed in this species, once present it undoubtedly greatly constrains the movements of skuas among the nesting penguins, by restricting which nests may be attacked, the breadth of access to peripheral nests and the availability and type of food available within the colonies. It determines as well the type of attack. Peripheral nests may be attacked from the ground and from flight; central nests may only be attacked from flight.

Four aspects of colonial nesting affecting skua access and predation are considered in the following sections.

5.3.1   *Limitations on skua access to nests because of neighbours*

There are now very many accounts of the penguin colonial nest pattern. On uniform terrain the nests are spaced c. 0.70 m apart and generally form into rows (Penny, 1968; Spurr, 1972, and numerous observations at the Northern Colony). This patterning of nests cannot develop in broken terrain and nests there are fitted into the spaces available. On the whole the Cape Bird breeding groups are characterised by regular spacing patterns and in only a small proportion of groups is this interrupted by the topography or through boulders intruding into the generally smooth bed. Most groups with broken nest spacing are in G Block and in the upper parts of B and C Blocks. The dense aggregation of colonies on the lower beach (A, B and H) and in D, E and F are built on low undulations of fine gravels with regular nest patterns.

Until the group structure breaks down following the nest desertion of the post-guard stage (at the beginning of January) skuas are unable to land or move about in the centres of breeding groups and all ground attacks are necessarily restricted to the periphery. Flight attacks and scavenging can occur from flight on central nests, but only on prey small enough to be lifted and carried by flying birds. The ratio of central to peripheral nests in breeding groups is an important statistic when considering predation. This ratio depends both on the size of the breeding group and its outline. Circular outlines give the lowest ratio of peripheral to central nests for any size of breeding group. The importance of group size, however, shows clearly in Fig. 5.1, which models the length of the group margin in relation to nest number, and, separately, the numbers and proportions of peripheral nests in groups of different sizes. The model used estimates the numbers of individual penguin territories (each with a radius of 0.35 m, Spurr, 1972), in relation to the total area and circumference of the breeding group. Estimates of the number of nests on the periphery uses a corrected circumference calculated from group radius reduced by the radius of the individual territories. Not surprisingly, the ratio of peripheral to central nests changes very quickly as breeding group size increases and larger groups have proportionally fewer nests on the margin.

Assessments made of breeding group outline and, from this measure, the degree of accessibility of nests to skuas at the periphery, were done by measuring the *length* of the margin by a tape measure. As seen already the relationship of this length to the number of nests in the group is determined to a great degree by size of the breeding group and this must be controlled for in any comparisons made. This factor has been taken into account in producing Fig. 5.1c, which shows the extent that the breeding groups at the Northern Colony exceeded the theoretical minimum length of periphery for a circular outline. For the majority of groups the increase is between 150–300%, but in some especially complex large groups the increase in the periphery length exceeds 450%. The smallest differences are for B2 and B3, almost circular groups on the upper beach flat in B Block, but E4 and E5 were almost circular also (Fig. 5.2). Consistently, the largest differences are found in the groups spreading over the broken ridges of G Block. In eight of the nine groups these differences exceeded 350%. These differences in the length of periphery translate directly to differences also in numbers of peripheral nests for the groups in this colony, and it is these nests that offer skuas the greatest predation opportunity and reward.

5.3.2 *Protection by being alerted to skuas in the area by neighbours*

The close proximity of nesting penguins provides a collaborative warning system for all penguins of any skua activity in the area. This is a well recognised function of colonial breeding and group foraging species. At all times some of the penguins will be alert and watchful. This

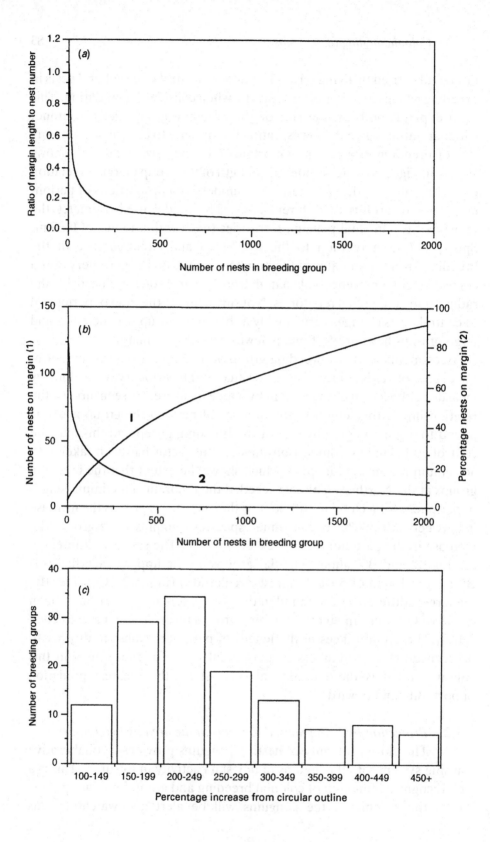

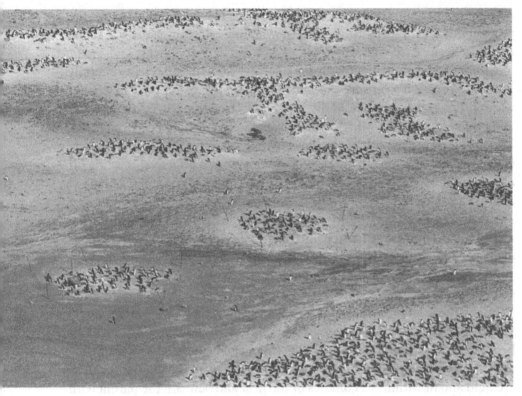

Fig. 5.2. Examples of near-circular outlines of breeding groups on smooth terrain in the foreground (E4 and E5 on the Northern Colony) compared with the complex outlines of the groups on the background ridges.

behaviour makes it very difficult for skuas to approach the nesting birds without being seen.

5.3.3 *Protection of the eggs and chicks in the nest*

The way the eggs and chicks are covered and protected in the nest by the parent strongly influences the likely success of an attack by skuas. The protection given to them by the parent must act both when the parent is alert and when asleep or resting. Two factors influence degree of protection: 1. the position of the eggs and chicks in the nest and 2. the

Fig. 5.1. Modelling the ratios of central to peripheral nests in breeding groups. (*a*) The ratio of length of margin of the breeding group to the nest number. (*b*) The theoretical number of nests (1) and percentage of nests (2) on the margin in breeding groups of different sizes. (*c*) Frequency histogram of the numbers of breeding groups on the Northern Colony in relation to extent of departure from a circular outline.

number of eggs and the number and size of the chicks. Both influence the accessibility of the prey to the skuas, whether attacking alert or resting adults, and the ease with which the parent can deflect attacks.

Although the nest bowl with its layering of small stones is generally judged to function to ensure that the nest contents are kept dry from melt water it functions also to hold the nest contents within the nest perimeter and provides a rim over which skuas must reach to get at the eggs or chicks. The raised rim of the nest also allows the parent to lie forwards across the eggs and chicks, resting the breast on the rim, and so effectively cutting off any attempt by skuas to reach underneath to the nest contents. The bowl and its raised rim of stones is, therefore, an important component of nest protection, minimising the loss of eggs and chicks from the nest during disturbance among the sitting birds (and limiting the ingress of eggs displaced from other nests) and during skua attacks.

Direct observation of the interaction between attacking and scavenging skuas with both resting and alert penguins quickly demonstrates that although there is a general trend for the eggs and chicks to be increasingly vulnerable to skua predation up to the point when the maturing chick can begin to protect itself from attack this trend can be defined by a sequence of stages. Overall, vulnerability depends on the way the eggs and chicks are held and fit within the confines of the nest and beneath the parent with the eggs and smallest chicks being protected best. The following sequence defines different levels of accessibility for skuas, or, conversely, vulnerability of eggs and chicks to being displaced or taken from the nest. In this sequence, measurements of egg and chick position within the nest and sizes of chicks are coupled with accessibility, an appreciation gained from direct observation of the nest and parent and experience of the way nest scavenging and predation occurs. The sequence falls into two series, the first (A) of eggs and small chicks able to be fully covered by a parent and (B) of large chicks which cannot be fully covered.

A. Eggs and small chicks covered by parents (see Fig. 5.3)

Parent able to rest on the nest to cover the nest contents completely so that the eggs and chicks are not visible externally.

This stage applies to the following combinations of eggs and chicks:

A1. Eggs and chicks held within the brood pouch

1. One egg, variably held by the parent. A single egg is usually held on the feet just behind the level of the middle toes – as for the front egg in

(a)

(b)

Fig. 5.3. Illustration of the increasing vulnerability of the growing chicks on the nest to skua attack. Stages as in the text. (a) Stage A1; (b) Stage B1 in the background, B2 in the foreground.

two-egg nests – but may be held as far back as the spur at the back of the feet. The distance from the front of eggs held on the middle toe to the rim of the nest is c. 150 mm, and 180 mm for eggs held towards the back of the foot.

1a. Two eggs. These are invariably held on the crossed feet, one behind the other, with the front egg lying beside the middle toes. At this point the front egg is 150 mm from the nest rim.

2. Egg and small chick less than three days old. Both are held in the brood pouch with the first chick to hatch in front of the egg (for feeding). Eggs that fail to hatch are displaced later by the chick to the front of the nest, and are subsequently lost.

3. Single chick or pair of chicks able to be accommodated within the brood pouch. The limits for this stage are c. 200 g for a single chick and to c. 250 g total weight for a pair of chicks.

4. Single chick or pair of chicks too large to be held within the brood pouch but able to be completely covered in the nest by a resting parent. The size limit for this stage is to c. 650 g for a single chick and to c. 700 g for a pair of chicks but the maximum size depends on the depth of the nest bowl, the way the two chicks pack together and the recency of feeding, i.e. the amount of food in the gut.

B. Chick or chicks too large to be completely covered by the parent reaching forward over them

Parent unable to rest breast on nest rim over chicks. Chicks generally held with heads tucked into the brood pouch with their backs facing out towards the nest rim.

1. Chicks not projecting laterally beyond the sides of the penguin unless sprawling on the nest. At this stage the parent can easily reach over the chicks to touch the ground in front. A single chick standing at the nest never projects beyond parent laterally. Two chicks up to c. 1100 g can just be accommodated within the parent's body width when sitting side by side on the nest.

2. Two chicks that are so large that even with the closest huddling their bodies project laterally beyond the parent. Parent able to reach over the chicks to the ground in front with increasing difficulty as the chicks increase in size.

3. Chicks stand or sprawl beside the parent. This stage begins for single chicks at c. 1500 g, for pairs of chicks when the oldest reaches this weight.

The sequence described here of increasing vulnerability reflects not only the volume of the contents but parental freedom of movement in an attack. For example, one egg can be held further back from the nest rim, and the probing or hooking beak of a skua, and birds holding a single egg can swivel and turn on the nest more easily and quickly than birds holding two eggs. Also, chicks completely covered by parents resting over them within the nest bowl are much safer from opportunistic attacks by skuas than chicks that stick out to the side of a parent. Again, parents can more readily parry a skua attack when they can easily reach forward over the chicks. Parents with large chicks are held fairly upright and have great difficulty in reaching over them to a skua attacking from on the ground.

5.3.4 *Numbers of adults within the breeding group*

There is a seasonal cycle of adult numbers at the colony related to the penguin breeding and reoccupation behaviour, and, late in the season, the demands of the maturing chicks for food. This cycle appears in all accounts of the breeding of this species (e.g. Taylor, 1962a; Penny, 1968; Yeates, 1968) and is a striking feature of colony change through the season. In general the more adults in the breeding group and moving about the colony area the more constrained is skua activity on the ground. To appreciate the seasonal ebb and flow of adult penguin numbers one must return again to the penguin's breeding cycle and to note especially the singular feature of reoccupation by non-breeding and unsuccessfully breeding birds to the colonies in mid to late summer. The reoccupation behaviour is of immense significance when considering the impact of adult penguins on the success of skua predation, as it dramatically boosts the numbers of penguins on the colony.

There are different measures of adult penguin number as it relates to effect on skua predation. One measure is simply of total numbers of adults on the colony, but this does not bring into focus their freedom to chase skuas away from nests being attacked or the protection non-breeders may confer on peripheral nests. Better measures, therefore, are the proportion of adults not tied to nests, either incubating eggs or guarding chicks, and the proportion on the periphery of the breeding groups. Overall, the impact of non-breeding penguins in the breeding groups and moving about the colony as a whole is twofold: their nests on the outside of breeding groups restrict skua access to nests with eggs and chicks, and because they are not tied by the need to protect their own nest contents they are free to chase skuas away if aroused.

The seasonal pattern of numbers of adults relative to minimum occupation in early December (Fig. 5.4) shows the now well established cycle of peaks in November and early January separated by the minimum of

Table 5.1. *Ratio of numbers of adult penguins on the colony to numbers of nests*

	15 Nov.	29 Nov.	7 Dec.	1 Jan.
Adults/nest				
1968–69	1.44	—	1.05	1.44
1969–70	1.26	1.04	1.03	1.24
Adults/nest with contents				
1968–69	1.80	—	1.12	3.43
1969–70	1.60	1.11	1.07	1.74

Numbers of nests in groups counted on 7 December to give the ratios shown: 1968–69, 1989; 1969–70, 2030.

early to mid December. As noted above these records of the total numbers of adults do not greatly assist in interpreting their possible effect on predation rates and a more focussed measure is required. In 1968–69 and 1969–70 large-scale counts were made of the colony at these key times to record the numbers of adult birds in relation to the numbers of nests with eggs or chicks. The ratios of adults to nests with contents shown in Table 5.1 are a good measure of the numbers free to chase away skuas near the breeding groups. During egglaying and early incubation the ratio of birds to nests with contents is in excess of 1.5 and falls steadily to around unity in early December before lifting again with the reoccupation in January. In the first year fewer than usual penguins reached the colony at the start of the season and only 56.6% of eggs laid hatched. In the second year good numbers of birds began breeding and the hatch rate was 72.3%. This much better breeding success is reflected in the smaller reoccupation peak in January.

The changes during the season are clearly evident in the maps of the breeding groups and are well illustrated in the series for breeding group A17 shown in Fig. 5.5. In the first map at about peak occupation there is a mix of nests with eggs in the centre and empty nests with one or two adults on the periphery. By mid December most of the unsuccessful pairs have abandoned the group and there are only two birds on empty nests, so that almost all the periphery comprises nests with contents. Two weeks later the peripheral nests are again occupied, and the breeding birds are again well protected from skuas on the margin (Fig. 5.6).

During the reoccupation period most of the adults flocking on to the breeding groups take up nest sites on the outside of the group. For the

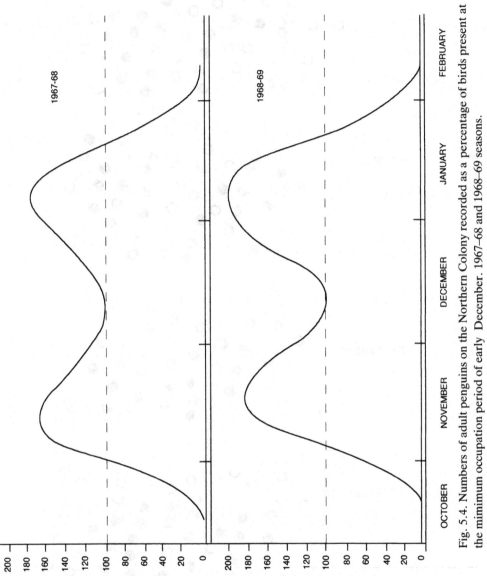

Fig. 5.4. Numbers of adult penguins on the Northern Colony recorded as a percentage of birds present at the mimimum occupation period of early December. 1967–68 and 1968–69 seasons.

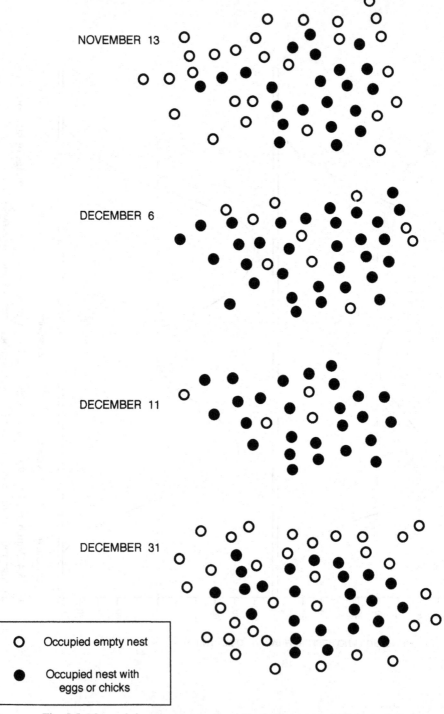

NOVEMBER 13

DECEMBER 6

DECEMBER 11

DECEMBER 31

○   Occupied empty nest

●   Occupied nest with
      eggs or chicks

Fig. 5.5. Maps of changing occupation of breeding group A17 through the
season to show the extent of protection of the breeding birds by other penguins
taking up nests on the periphery of the breeding group.

Fig. 5.6. The effectiveness of penguins on the margin in deterring skuas from attacking pairs with eggs or chicks.

group as a whole, and quite incidentally, this has the dual advantage of confining the disturbance of re-nesting to the periphery and of ringing the group with free birds – effectively reducing access to the nests containing eggs and chicks. This latter effect is clearly demonstrated in Table 5.2, which is compiled from counts made during the period the birds were returning to the colony in late December. Both the numbers of nests with eggs and chicks on the periphery and their access by skuas has been greatly reduced by the accretion of returning birds to the group. In these groups the numbers of peripheral nests with eggs or chicks fell over this period from 317 (57% of all nests) to 114 (23% of all nests) and the numbers of nests allowing skuas an attack access greater than 180° fell from 268 to 43, from 48% of all nests in the breeding group to 8.5%. As will be seen later the greater the access about the nest the greater the advantage to the skua during an attack. These figures of the changing structure of the breeding group are of considerable significance. During the reoccupation period the penguin chicks are quickly growing to a size that precludes aerial attacks on them into the group centre, so that increasingly attacks become confined to the periphery – which is protected by newly arrived penguins.

Although the table of changes to the periphery given here is for the reoccupying period, a similar effect occurs at the start of the season and only during late incubation with minimum adult numbers is the periphery

Table 5.2. *The impact of reoccupying penguin adults on the accessibility of nests on the periphery of breeding groups to attacking skuas (1968–69)*

	Defended nests on the margin			Number of nests with contents allowing skua access from the margin through		
	With contents					
Date	Number	% of total	Number empty	180°	270°	360°
8 Dec.	317	57	0	268	178	59
19 Dec.	223	42	70	137	70	33
2 Jan.	114	23	237	43	23	7

Nests in groups F1–F4 and E12–E18.

of breeding groups made up almost exclusively of nests with eggs or chicks on which skuas have almost free rein.

Later in the season the breeding group format breaks up as the nests of successfully breeding birds are deserted and the chicks are left on their own. During January there is a major fall-off in the ratios of adults to chicks on the group periphery resulting from the fall-off in total numbers of adults in the group and a tendency for chicks to congregate about the group periphery. The impact of these effects is well demonstrated by counts made of groups in E and F blocks through the second half of January (Table 5.3). In mid January there were on average 3.45 adults to each chick on the periphery but only 0.46 adults at the end of the month. This change is not as significant for skua predation success as might be suggested from the bald figures given here as the chicks are maturing so rapidly across this period that few at this date are in fact vulnerable to anything but a sustained and determined attack by a skilled and aggressive skua. The rapid fall-off in adult penguin numbers recorded here is the first phase of their desertion of the colony and its continuation through January and into early February means that chicks might be almost alone on some breeding areas at the end of summer.

In this section so far the increase in numbers of adult penguins on the colony has been recorded as conferring a straightforward advantage to the breeding birds. However, it has long been recognised that their influx is associated with an upsurge of fighting (Fig. 5.7), of shifts in territories and nest locations and the displacement and loss of eggs and chicks from nests. Spurr (1977) has shown that the incidence of attacks and fights involving flipper beating among the birds in breeding groups is closely related to numbers of penguins in the group without eggs or chicks, so that their

Table 5.3. *Relative numbers of penguin adults and chicks on the margins of breeding groups during January, 1969*

Date	Total adults in groups	Total chicks in groups	Adults on margin		Chicks on margin		Ratio of adults to chicks on margin
			Number	%	Number	%	
13 Jan.	2196	1024	1296	59	376	37	3.45
21 Jan.	1330	1020	844	63	540	53	1.56
26 Jan.	402	1020	305	76	657	64	0.46

For breeding groups F2–F17 (minus F15) and E12–E18.

frequency through the year reflects penguin numbers. It is high at the start of the year and in the reoccupation period and negligible during the adult minimum occupancy of mid-incubation. At the start of the season fights are equally common among birds that eventually laid eggs as in non-breeders, but in the reoccupation period most fights are among non-breeders and unsuccessful breeders. At its highest level Spurr recorded an attack per bird each 100 h. This at first sight seems an insignificant figure but applied to an average breeding group of say 200 adults this translates to two fights per hour on average for the group, a very significant level of disturbance. Because most fights are among the reoccupying birds, and most of these will be on the periphery, they will, however, be less disruptive to breeding birds than if they occurred generally throughout the group. Some fights will certainly spill over several territories and may involve ten or more adults, fighting and chasing across nesting penguins and over chicks and eggs. But it is important to keep these fights in perspective. They undoubtedly appear enormously disruptive and destructive when in progress and chicks and eggs may be lost. Certainly they engage the close interest of the skuas and loafing birds will generally come down to them in the expectation that food may become available to be scavenged. But in the final analysis from the figures and maps of breeding success and egg and chick mortality of the study groups overall they had negligible impact. Table 5.4 analyses the causes of mortality for chicks found dead on the study groups from the evidence of nest displacement, the appearance of reoccupying birds and the condition of the chick and the nest. In the three years, between 2.9% and 10.7% of chicks lost from the groups could be attributed to the direct effect of fighting among adult penguins. In each year this was a less significant mortality factor than starvation on the nest.

Fig. 5.7. Penguins fighting on the breeding groups. The upper photograph shows a low level confrontation between neighbours, the lower photograph shows much more serious fighting involving several of the neighbours as well.

Table 5.4. *Impact of disturbance caused by reoccupying birds on chick mortality in late December and early January compared with other mortality factors*

		Chicks lost 20 Dec.–5 Jan.		Dead chicks left on the breeding areas[a]		Killed by reoccupying birds		
Year	Number of chicks at 20 Dec.	Number	%	Total	Number starved	Number	%	Unknown factor
1966–67	1352	305	22.6	71	30	16	5.2	25
1967–68	1392	240	17.2	40	11	7	2.9	22
1968–69	806	186	23.1	52	25	20	10.7	7

[a] Dead chicks left on the breeding groups were assumed to have died through causes other than skua predation. Data are from the study groups in each year.

5.4 *Constraints imposed on skuas as predators through the size and strength of the prey*

5.4.1 *Difference in the size of adult penguins and skuas*

The relative sizes of penguins and skuas impose constraints on the way the skua acts as a predator, on the prey that can be taken and on its availability through the year. The very large differences in size and strength between skuas and adult penguins protects penguins from direct attacks on them as prey. Skuas simply are not powerful enough to overcome penguin adults. Even those coming ashore wounded from leopard seal attacks or exhausted from territorial fighting are generally able to protect themselves from skua molestation. Because of this limitation skuas can only utilise penguin food for a short period during the breeding season. For most of the year there is no food to be gained from penguins, whether on the colonies or at sea.

Both species show cyclical variation in weight through the year, and especially during the breeding season, but the differences between them are always large. Skuas weigh between 900 g and 1500 g; Spellerberg (1970b) weighed 285 breeding and non-breeding skuas at Cape Royds over one summer, recording an average mass of 1263 g ± 7.6 g (mean and standard error). A sample of breeding birds weighed at Cape Bird had the slightly greater average mass of 1315 g ± 11.7 g. These breeding skuas did not vary significantly in mass through the breeding season [1].

Sladen (1958) determined the average mass of penguins in mid season as 4.4 kg and at the end of the season as 4.1 kg. Whitehead *et al.* (1990) report a mean mass between 6.09 kg and 6.39 kg in different years for birds arriving in Prydz Bay, East Antarctica. Volkman *et al.* (1980) give 4.5 kg and 4.2 kg for males and females respectively. There are surprisingly few other records and none at all for Ross Island. Collection of adult weight statistics has not been done routinely in any programme here, in contrast to egg dimensions, which have been taken by almost everybody.

5.4.2   *Constraints imposed on prey availability through the growth and increasing maturity of the penguin chicks*

Mature chicks as they leave for sea are too strong to be easily overcome by skuas and except for special situations where they can be exhausted and killed well away from other penguins they are immune from skua attack. The process of growing up and being able to fight off skua attacks is related to three changes: increasing mobility, increasing mass and increasing strength. Each of these is linked as well to the changes in the skeleton as it hardens and becomes fully articulated. A key indicator of the growing stature and independence of the chicks is seen in the development of skeletal rigidity of the flippers. There is a subtle and significant change in self-confidence of chicks once the flippers have hardened into rigid paddles.

The following section relates the ability of skuas to lift prey in flight or drag prey along the ground to the growth of the penguin chicks. The first limit determines the size range of chicks safe from skua attack within the centres of groups, the second puts an upper limit on skuas being able to drag chicks quickly away from the margin or from pursuing adults.

Skua strength

The limits to lifting prey or carrion from the ground in flight

These limits depend firstly on wind speed and steadiness and on whether the chick is merely bounced along the ground by a flying skua or lifted cleanly into the air. In general skuas can lift greater mass in a wind than in calm and greater mass can be dragged and bounced than cleanly lifted. To retrieve prey from among penguins the prey has to be cleanly and quickly lifted from flight. The records below were obtained by watching skuas attempting to retrieve penguin carcasses from neighbouring territories and in recording the mass after the attempts had been made. Under exceptional provocation skuas may be able to lift heavier loads.

In calm conditions skuas can lift about 450 g from the ground, and drag and bounce up to 600 g. Tests in steady winds of 10–15 knots showed that under these conditions skuas could easily lift 550 g, could drag and lift with difficulty 700 g but were unable to shift 800 g. Higher or more turbulent winds would increase these weights marginally, but at the risk of the skuas being buffeted, and a real risk of being caught by the penguins if attempting to lift heavy prey from breeding groups.

Skua pulling strength on the ground

Tests were made using tame skuas by getting them to pull against food linked by a cord to a spring balance. The skuas pulled back hard in these trials, aiding purchase from the legs and feet with wing beating. Under these conditions skuas could tug up to 1600 g and could maintain a sustained pull of 1100 g to 1350 g. Four birds were tested.

Overall these seem rather modest lifts for such powerful-looking birds. Especially modest is the limit on load lifting by flight of about 30% of body weight. The figures confirm nevertheless the general observations of skuas having difficulty in holding and shifting prey on the ground and on lifting prey or scavenged material out of the middle of penguin breeding groups.

Strength and mobility of maturing penguin chicks

The growth of the penguin chicks has been catalogued for individual chicks and for the colony as a whole, as the food resource, by the repeated weighing of a sample of chicks. The increasing weight imposes an important constraint on skuas by limiting their ability to pull them over or tug them away from parents. From the records given above of the mass skuas can tug on the ground it is very clear that skuas would have great difficulty in pulling a struggling mature chick very far. Other important considerations of the ability of skuas to prey on chicks, and on the likely success of attempts to capture and hold chicks once an attack has begun, are the mobility of chicks and their strength to resist the skuas attacking them. Strong mobile chicks can drag the skuas with them back towards the penguin group if attacked outside and can resist being pulled away in attacks on the margin.

Young penguin chicks with their swollen bellies and stubby, flaccid appendages are unable to run upright and flop along the ground in series of skitters pushing with the legs and flippers. They are soon exhausted. As they mature they develop an upright gait and much greater speed. If these older chicks escape from attacking skuas for only a moment they may be able to run to shelter before being caught again. Thus, mobility is a key

Table 5.5. *Strength of penguin chicks in relation to age and mass. Mean and standard error of strength measured using spring balance (in grams)*

Age (days)	Mass (g)	Chicks upright and running away			Chicks prone on the ground	
		Force generated		Force needed to pull over backwards	Force needed to pull forwards	Force needed to pull backwards
		Sustained pull	Jerk			
15	880–900	170 ± 6.2	360 ± 12.5	240 ± 24	525 ± 12	462 ± 31
20	1340–1675	440 ± 37	725 ± 60	380 ± 18	960 ± 51	812 ± 64
25	2100–2800	990 ± 144	1775 ± 190	490 ± 76	1635 ± 110	1865 ± 20

'Force generated' by the chicks is that measured when they were tugging against a spring balance in trying to escape from capture. The 'force needed' measurements are those needed to pull the chick over onto its back from an upright position or to pull it along when lying on the ground.

factor in skua success and penguin survival. Running speeds of chicks released outside a breeding group and measured as they raced back to its shelter varied widely in chicks of different ages and for the same chick in different trials. Young chicks were too disoriented to escape in this way and generally were reluctant to move at all. Chicks between 20 and 25 days old ran at between 0.9 and 1.2 m s^{-1}, those at 35 days reached 1.4 m s^{-1} and those of 40–45 days reached as much as 2.2 m s^{-1} (c. 8 km h^{-1}) in some trials.

The increasing strength of chicks with age was measured by allowing them to pull against a spring balance attached to a body harness. Four measurements were made representing the different ways they are attacked and dragged about by attacking skuas. The first two measures were of upright chicks attempting to run back into the breeding group with the balance attached from behind. The first records the pulling strength of the upright chicks, the second their ability to resist being pulled over on their backs. The second two measures were of chicks prone on the ground. The first records the force needed to pull the resisting chick forward, the second to drag it backwards. Skuas drag chicks in both these ways during natural attacks.

It is clear from the data in Table 5.5 and the information on skua strength described above that although the youngest chicks tested (15-day chicks weighing on average up to 900 g) could be easily subdued and dragged away from other penguins if necessary, those same chicks just 10 days older at 25 days of age are reaching the limit for the skuas to handle readily. The strongest of these are able to resist dragging along the ground with a force closely similar to the maximum achieved by the skua in the strength trials. They are not very fit or resilient, however, and their strength soon evaporates under the battering the skuas impose. In many attacks on older chicks the skuas simply hang on to prevent the chicks regaining the breeding groups and do not attempt to drag them away.

5.5 *Aspects of penguin behaviour and biology apparently facilitating skua predation*

This final section considers how the behaviour of the adult penguins seemingly contributes to the success of skua predation. In all instances no doubt there are conflicting requirements on the penguins, for example, the need to shelter the chicks from cold winds even though this exposes them to skuas at the margin.

5.5.1 *Penguin breeding behaviour*

Within this category fall such factors as distraction of sitting birds by neighbours, the distraction caused to sitting birds by fights and attacks among neighbouring pairs, and the distraction from watchfulness occurring in nest changeover behaviour (Fig. 5.8). Skuas are quick to exploit any lapse in attention and protection and penguins leaning back away from eggs or chicks towards other penguins or lifting the body up off the nest are an easy and safe target for skuas.

5.5.2 *Rejection of eggs and chicks outside the nest*

Eggs or young chicks becoming displaced from the natal nest are strongly rejected by neighbours and pecked away. In this way eggs are rolled and buffeted to the group periphery and chicks pushed into gaps among the nests, or onto abandoned nests or onto the margin. (Tenaza, 1971, has shown that eggs removed from beneath a sitting bird and placed on the edge of the nest will be retrieved but that if they rolled away while attempts were being made to drag them back into the nest then they will be pecked. The invariable observation on the breeding groups is that any eggs rolling towards an incubating or brooding penguin will be pecked away.)

5.5.3 *Orientation of incubating and guarding penguins in relation to wind direction*

Stonehouse (1967b) suggested that the reason that nesting penguins were seen to orientate into strong winds was to preserve the insulative properties of the feathering. Comment on orientation appears in many accounts and its general occurrence can be easily established in

Fig. 5.8. An example of the distraction from nest protection exploited by skuas. In this photograph the pair are engaged in mutual display, while at the same time being distracted by another penguin standing closely by their territory.

visits to breeding groups during high winds. This generality has been confirmed in detail by Yeates (1971) and Spurr (1975a) who have shown that penguins orientate head to wind when prone on eggs or young chicks but that the orientation to wind is reversed during the late guard stage when the chicks are too big to lie across and the adult crouches over them. At this stage the adult faces away from the wind, exposing the back, to provide shelter for the chicks. In undisturbed breeding groups, therefore, the birds are orientated in a set direction so that birds on different margins present different aspects to skuas. During incubation and early guard stage those on the upwind side face the skuas, those on the downwind side face away. During the late guard stage with larger chicks the directions are reversed and upwind birds face away from the margin while those on the opposite side face towards it. This precise orientation gives advantage to skuas in two ways. Chicks and eggs are better exposed to attack and the uniform orientation of penguins reduces the likelihood of the group as a whole being alerted to an approaching skua. During calm weather or in light winds the penguins are randomly orientated within the group (Spurr, 1975a) and by chance some will be in a position to see skua movements. Not only does the regular orientation in winds provide advantage for skuas skulking along the margin on the ground it also provides a great advantage for birds attacking from flight. Birds flying downwind can drop onto chicks from behind without warning and those hovering and beating into the wind face an unbroken array of chicks crouched in the nest, each protected solely by the parent, as the nearest neighbours are facing away.

5.5.4 *Breakdown of group structure at the beginning of the post-guard stage*

The changeover from the guard stage to when the chicks are left unguarded seems to be driven by two factors. The first is the hunger of the chicks provoking the guarding adult to leave the nest to obtain food. The intensity of begging is no doubt greater in pairs of chicks than in single chicks of the same age and one would expect pairs to be abandoned earlier, but this prediction surprisingly does not appear to have been tested in any published research. The chicks will no doubt also be abandoned earlier in the event of the other parent being late to return or being lost altogether. The second factor influencing the timing of the onset of this stage is the disturbance of the breeding group caused by the reoccupying birds usurping nest sites and displacing parents with chicks. If the chicks survive this fighting they may simply have to be abandoned if there is no space in the immediate vicinity for the parent to rebuild a nest.

The very first days of the post-guard stage provide great opportunity for the skuas as the first chicks left unguarded have few places within the group area where they are safe from being harassed by nesting adults. A high proportion of these first unguarded chicks are forced out onto the margin. Chicks abandoned by parents evicted from the nest site by reoccupying birds will not have even the doubtful occupancy of their nesting area. Some of these chicks manage to force themselves under a neighbouring parent and nests with three and four chicks are at first quite common (Fig. 5.9). None of the unguarded chicks leave the group area entirely and the

(a)

(b)

Fig. 5.9. Common observations at the beginning of the post-guard stage. In (a) a chick has managed to force its way under an adult already protecting its own chicks; in (b) a chick stands on its own in a space among territories.

Table 5.6. *Total numbers of unguarded chicks and numbers on the breeding group margin accessible to skuas on each day at the beginning of the post-guard stage in 1969–70*

Date	Total number of unguarded chicks	Unguarded chicks on the margin	
		Number	%
31 Dec.	18	9	50
1 Jan.	30	13	43
2 Jan.	31	23	74
3 Jan. (1400 h)	160	31	19
3 Jan. (2000 h)	233	35	15
4 Jan.	396	62	15
5 Jan.	517	91	17

furthest movement is to the margin. The changeover in a breeding group from guarding parents with chicks to the post-guard stage with unguarded chicks is very rapid. In each year it began about the end of December and was largely completed by end of the first week in January. Table 5.6 records this changeover in 1969–70 for 24 breeding groups with 1875 chicks. By 5 January, six days after the first unguarded chicks were seen, 517 chicks were unguarded (27.6% of the total number) and 91 were located on the margins of groups within easy reach of skuas.

The mass of these first unguarded chicks ranged from 520 g to 2350 g (mean weight 1487 g ± 66.6 g). There are some very light unguarded chicks at this stage. For example, there were nine chicks weighing less than 1000 g, of which three weighed between 500 and 550 g, among those on the margin on 2 January. All of these smaller chicks had been displaced from nests by reoccupying birds.

Later in January nearly all the chicks are unguarded and their protection comes from their own increasing strength and maturity and the presence on the breeding groups of reoccupying nesting birds. Both these factors have been considered earlier in terms of their impact on the success of skua attack. The behaviour of the chicks themselves may also contribute to skua attack success. The most important contributing factor is failure to be alert to a sudden skua attack. The first moments of an attack on a large chick are critical to the survival of the one and the success of the other because it determines how far the skua can drag or knock the penguin chick away from the group and away from any adult penguins there. It is interesting to observe, therefore, the numbers of chicks that stand and sleep about the margins of the breeding areas. Indifference to the prospect of skua attack

Table 5.7. *Numbers of unguarded chicks standing or lying on the margins of 12 breeding groups during calm warm conditions on 12 and 13 January, 1970*

	Total chicks on margin	Chicks sprawled on margin	
		Number	%
12 Jan.			
1000 h	134	70	52
1230 h	155	61	39
1700 h	284	178	63
13 Jan.			
1630 h	289	205	71

For chicks on E12 to E18 and F4, F8, F9, F10 and F12 breeding groups; in total 737 chicks at this date.

seems greatest during calm, warm days when a high proportion of chicks are sprawled about the breeding area. Counts made on 12 and 13 January in 1969–70 illustrate this point. In four counts between 9.5% and 28% of all chicks (but between 52% and 71% of chicks on the margin) were sprawled sleeping about the margin (see Table 5.7).

5.5.5   *Chicks outside the breeding group area*

The prime reason for chicks being outside the shelter of the breeding group is the feeding chase, in which the feeding adult runs off with the chicks following in order to separate them so that they can be fed one at a time or can be fed without disturbance from other penguins (Fig. 5.10). Chicks dropped off from these chases outside the breeding group or left in another group at the end of feeding must make their own way back to the natal group before the next feeding visit. Such travel may require chicks to move 20–50 m through open space on the colony, under the direct surveillance of skuas. In addition to the risk to unprotected chicks from the chases, parents and chicks are often attacked together when outside the breeding group. Once the chicks are separated in the excitement and trauma of an attack by aggressive skuas the single parent may not be able to defend them both. As often as not the parent may escape with one chick to the shelter of breeding penguins leaving the second to be held and subdued by the skuas. On their own, well away from other penguins, even adult penguins can be frightened off by a pair of skuas attacking with intimidatory flight jumps at the head.

Fig. 5.10. A common reason for chicks being out of the breeding group. A feeding chase with a pair of chicks and a parent.

The first feeding chases were recorded at this colony each year just before mid January. In 1970 the first were seen on 12 January. Counts of numbers of chicks being at any time more than 10 m outside the breeding groups were made on 15 January for 2250 chicks on EF study block and 699 chicks on H block. These counts gave means of 6.1 ± 0.61 for EF block and 1.4 ± 0.21 for H block. Penguin chicks were much less inclined to follow parents too far away from the H block groups, which are far more isolated among skua territories, than are those in EF. These records are for unusually long feeding chases away from the group but most run through and about the margins of groups. Counts made at the same time of the numbers of chicks more than 1 m outside the breeding group give higher figures. A chick even this little distance away from other penguins gives the skuas an immediate advantage in an attack. From these counts the six skua pairs on EF block have on average between one and ten chicks this far from other penguins at any time. The differences in the availability of these chicks to the individual skua pairs depended mostly on the size of the territory and the numbers of chicks falling within it.

Chicks deserted by parents in a feeding chase well away from their breeding area or abandoned in an attack by skuas will run to any other adult penguins in the immediate area. Even passers-by will chase off skuas if there is enough noise and disturbance to attract their attention. Few of these offer more than a temporary sanctuary but by keeping with them the chick may regain the safety of a breeding group. Solicitous attention to chicks rescued in the open ground by groups of wandering adults is more likely to lead to attempts by them to tread and mount rather than brood. It seems a hard fate to be rescued from death by skuas to be attacked by one's own species!

5.6   *Constraints on skua activity because of the need to protect their own eggs and chicks*

Although pairs of skuas are more successful in foraging at the penguin colony when acting together, for most of the season successfully breeding birds can only forage individually. Skua nests are not left unattended for more than a few minutes at a time – as the eggs need to be brooded continuously and protected from other skuas. Although skua chicks do not need to be brooded after the first few days of their life (Spellerberg, 1969) they must be protected from other skuas until fledging and one parent is almost invariably found at the nest area. The nest may be deserted by skuas seeing an apparently easy prey target or likely food from penguin fighting. Mates may leave the nest to join an attack on a chick out of the breeding group but they seldom leave to join an attack on a defended nest. The nest will certainly be deserted, however, by a hungry parent when the other gains an egg or chick.

As will be shown in the assessment of advantage of colony territories, the attraction of penguin food to hungry skuas often overrides the prudent need to protect their own eggs and chicks, and these skuas have high brood mortality as a result.

STATISTICAL TESTS

[1] Variation in mass of breeding skuas during the breeding season at Cape Bird. (Mean and SE)
1. 7 December, $n = 37$, 1292 ± 17.4
2. 28 December, $n = 41$, 1313 ± 18.8
3. 15 January–1 February, $n = 19$, 1365 ± 25.7
Single factor analysis of variance. $F_{2,94} = 2.61$, $P = 0.076$, ns.

6

Description of scavenging and predatory behaviour of skuas and the defensive behaviour of penguins

6.1   Introduction

The study at Cape Bird was established to describe the interaction between the skuas and penguins breeding together on this Antarctic site and determine the consequences of this interaction for both species. Very little could be achieved until the behaviour of the two species had been described and the range of elements of scavenging and predation tabulated for use by different observers and in different seasons. This behavioural listing was developed and tested during the second year and underpinned the collection of all observational records of the prey–predator interaction (Young, 1970). By using this listing comprehensive records could be made in the field of the activities of groups of skuas (each activity needed only a few seconds to record) and analysis of the records was facilitated by decisions made at this point on the categories of activities observed. Except for a detailed study of the responses of penguins to skua overflight (Müller-Schwarze and Müller-Schwarze, 1977) the behaviour of skuas attacking penguins has been described to date in only very general terms (e.g., Young, 1963b; Trillmich, 1978).

The repertoire of skuas foraging among penguins is rather more varied than first suspected; but only because of the very different requirements of scavenging and preying on the margins of breeding groups and on the centres and because of the changing nature and availability of the prey through the breeding season as the chicks mature and the regular nest pattern of the breeding group collapses.

At any one time, however, there is only a small range of possible attack methods. During incubation and early chick rearing when the the eggs and chicks are guarded there are four basic attack forms: from flight into the colony centre or the margin; or from the ground into the centre or margin. Later in the year when the chicks are unguarded they may be tackled from flight or from the ground when in the crèche or out in the open. On each date, therefore, there is a matrix of possible attack forms related to the type of prey and its location. It is not possible for a skua to walk or stand among the nests of a breeding group, they are too tightly packed together for this, so that ground attacks and ground scavenging are essentially restricted to the margins of breeding groups, whereas flight attacks reach both central and marginal nests.

Most attacks at the nest are in reality mock attacks at the parent. Only at the highest attack intensity are adult penguins directly assaulted.

The following account attempts a detailed description of this behaviour and of the responses elicited in the penguins. The account considers in turn the sequence from searching through scavenging to the various attack forms within the different situations of nest defence and location at the different stages of the penguin breeding cycle.

6.2 *Searching behaviour: hunting flights and observation of the penguins*

Skuas spend much of their time at the colony watching the penguins and flying about the breeding groups within their territory searching for accessible carrion and for exposed eggs and chicks (Fig. 6.1). They take an especially close interest in disturbances among the penguins, flying to penguin fights and to any sudden activity that may indicate that eggs and chicks have been displaced from nests. This skua behaviour is clearly an integral part of scavenging and predation. It is described here as an introduction to the later sections on the more specific skua activities which follow from the discovery of carrion or prey.

Most of skua observation of penguins is from roosts overlooking breeding groups, on boulders or ridges or from resting places among the colonies. The birds settle in these places for much of the day and are alert

Fig. 6.1. Skuas searching for food at the colony from flight and from the ground.

and vigilant for most of the time. Roosts above the breeding groups allow easy flight over the penguins.

More active searching is from wheeling flights over the penguin groups within the territory and from walking along their margins. Three categories of flights have been recognised. The first is the flights directly over the penguins by skuas on their way to and from the territory and bathing pools or sea feeding areas. These are at a steady flight speed and direction and are too high above the penguins to provoke a response or interest. These are termed overflights. The second category is for territorial defence flights which are characterised by a much faster flight speed with high amplitude wing beating and which are directed at a target skua on the ground in the territory or flying over or near it. The apparent single-mindedness of these flights and their great speed, far higher than the normal flight speed, marks these flights distinctly. The third flight category is the wheeling, gliding flights of skuas searching for food on the colony. They are termed hunting flights. Hunting flights are between two and ten metres above the penguins (Müller-Schwarze and Müller-Schwarze, 1977 determined 2–4 m to be the optimal height) and generally repeat the same flight path around the territory margin and over the separate breeding groups. They may be carried out for many minutes at a time, flying over the same penguins repeatedly, or, at the other extreme may be a single sweep over the penguins or the slow return flight from territorial defence. During these flights the birds look closely at the penguins with a characteristic head-down attitude, scanning below them by alternating head movements to each side. The discovery of food or a favourable attack situation may result in the skua landing beside the penguins, changing the flight to a more aggressive hovering, or switching to an attack-flight into nests.

Flights by skuas over the penguins of neighbouring territories share features of both territorial and hunting flights. They are more direct than the usual hunting flights, are of shorter duration and are less likely to wheel for more than a minute or so above the same group of penguins. Territorial skuas may tolerate outside birds over their penguins but will evict them quickly when they begin to show close interest, signifying food, or appear to be about to land. Territorial skuas are very alert to the flights of outside birds in areas where they know eggs or chicks have been deserted, even though they are unable for the time being to retrieve this food themselves. No outside birds can land more than momentarily within another territory and can only carry out opportunistic scavenging or predation. Once food is gained from another territory it is carried back quickly to the home territory for feeding, often with the defending skuas in hot pursuit.

6.3   Scavenging

Scavenging is any foraging activity aimed at carrion. Mostly this is dead material. But eggs or young chicks displaced so far from nests that they are unlikely to be recovered by the parent, eggs and young chicks abandoned on the nest by deserting parents, and penguin food spilled on to the ground during chick feeding are also counted as carrion. The young chicks described here as carrion have neither the strength nor the mobility to survive away from parents at the nest. Where carrion is on defended nests the attack form is identical to predation and is considered in that section.

6.3.1   *Scavenging outside the margin of the breeding group*

Small amounts of food become available to skuas in the open areas of the colony outside the breeding groups. Eggs roll free from nests during penguin fights and are then sometimes pecked away by others to come to rest well outside the breeding group margin. Similarly, young chicks that are displaced from the nest are seldom recovered by the parent and may be chased and harassed by other nesting birds until they also end up outside the margin. Later in the year food is spilled during feeding outside the crèche.

All of this food is immediately accessible to skuas and requires little effort and risk in its retrieval.

6.3.2   *Scavenging among nesting penguins*

Eggs and chicks dislodged from nests may end up outside the breeding group but most settle in gaps among the nests whereas those deserted by parents will generally stay on the nest unless this has been taken over by other penguins. The fate of abandoned eggs is fairly clear-cut: they either remain on the nest, are pushed by other parents into gaps among the nests, or end up on the breeding group margin. In the end all will be taken by skuas (Fig. 6.2). These are effectively carrion once abandoned or dislodged, irrespective of whether the embryo was alive. There is less certainty about the fate of abandoned or displaced chicks. Young chicks deserted by starving parents are probably too weak to survive anyway and usually lie in the nest until taken by skuas. Older and stronger chicks may be able to survive the battering of neighbours and even force their way under another parent. If old enough and their own

Fig. 6.2. Scavenging for eggs along the margins of breeding groups.

parents return they may well survive. Others are forced by the nesting penguins into spaces in the group or on to its margin and will probably be taken by skuas.

Scavenging skuas retrieve these eggs and chicks by running in from the margin or by dropping on to them from flight. Chicks weighing more than 400 g cannot be lifted in flight during calm conditions but up to 800 g may exceptionally be lifted in strong winds, at least to allow retrieval from a small breeding group. Carrion heavier than this cannot be retrieved from flight no matter how attractive it is to the skuas and will either be left on the colony until it becomes available, and inedible, when the group disperses, or become the focus of intense activity of the skuas working from the margin. Some skuas will attempt to retrieve food of this sort from within breeding groups by flight-jumping over the outer rows of penguins and then buffeting their way back through the aroused birds towing the prey with them. Penguins can easily catch and hold skuas when they do this and scavenging or preying on heavy chicks in the group centre is enormously risky. Few skuas are prepared to take this risk.

Most scavenging of food within the breeding group is opportunistic in the sense that skuas will generally sit and wait for a favourable opportunity to attempt its retrieval. Once penguins have been aroused by an unsuccessful attempt at scavenging only very hungry skuas, or very impatient ones, will return immediately. It was nevertheless surprising to see how much effort was given by skuas for what seemed little reward. Dead chicks lying in the breeding group may attract their attention and reckless activity for many hours even though weighing only a few hundred grams. If rotten these chicks might not even be eaten after being retrieved.

On the other hand successful first attempts at taking food are often achieved without much disturbance at all. In these attempts the skua runs in quickly from the margin or drops onto the food from a hunting flight and is normally away again before the penguins respond. Once the penguins have been disturbed skuas need to be much more circumspect in their movements among them. This is especially true for hovering flights which provoke a furious response among sitting birds and which attract free penguins from considerable distances, calling and chasing.

To a large extent a skua's success when scavenging is determined by the number of free penguins in the group. The lowest numbers of free birds occurs during the middle of the incubation. At this time, for Cape Bird about the second week in December, virtually all penguin nests have eggs or chicks and are guarded by a single parent and the breeding groups are essentially static with little movement by the parents and the few unattached adults. At this time the skuas have much greater freedom of action.

Scavenging along the margin of the colony exposes skuas to little danger from penguin attack but scavenging flights or jumps into the centre of the group are almost as dangerous for the skua as are opportunistic attacks there on defended nests. Skuas caught by penguins within a breeding group are generally unable to regain flight and must struggle through the penguins across the ground to the outside.

6.3.3   *Mixed scavenging and attack behaviour*

Timid skuas at all times and aggressive ones at the start of a foraging sequence combine scavenging with low intensity attacks on nesting penguins and unguarded chicks along the breeding group margins. Any easily accessible food is snapped up: eggs, dead or living chicks and spilled penguin food. The skuas run at exposed eggs or chicks or carrion but move on immediately they meet alert penguin defence.

This behaviour, like pure scavenging, depends on stealth and speed. The skua is unobtrusively casual and not obviously foraging until the very last moments of an attack. These are opportunistic attacks and are only successful in undisturbed groups, or, on the contrary, where the penguins are critically distracted from nest defence. Success depends on parental neglect of eggs and chicks on the nest exposing them to simple predation.

There is very little risk to skuas during these attacks as the nesting penguins are either distracted by others or are insufficiently alert to be able to respond quickly.

6.4   *Predation on eggs and chicks at defended nests*

These attacks fall into two categories. Those in which skuas attempt to obtain eggs or chicks by a quick sortie are classed as opportunistic attacks. Those in which the skuas attempt to take defended eggs and chicks from alert penguins by deliberate and prolonged activity are termed sustained attacks. Opportunistic attacks last seconds only at each nest, sustained attacks may continue for many minutes. Both attack forms usually occur as part of a longer sequence of behaviour in which cycles of searching, scavenging and active predation occur.

6.4.1   *Opportunistic attacks*

Attacks directed at the eggs or chicks

These attacks are characterised by their lack of warning so that the skua is right on to the nest or thrusting under the parent for eggs or chicks before the parent has begun to respond. They rely on close observation of the prey, a vigilant alertness by the skua to see opportunities for attacks and very rapid movement by running or flying to the nest. Many begin from surveillance points on rocks, ridges or slopes or from a resting or standing place very close to the breeding group margin. Resting skuas habituate penguins to their presence and penguins apparently relax into inattention within a few minutes of a skua settling near them. For skuas using this attack form long periods of surveillance are punctuated by short bouts of activity.

Most attacks are at clearly visible eggs or chicks. Parents may uncover the nest when they lift to stretch and yawn during comfort movements and during ecstatic displays (Penny, 1968; Spurr, 1975b), when the two parents are shuffling about the nest during a nest changeover, during nest building and repair, in the defence of nest stones, and in chick feeding.

The eggs or chicks may be either hooked from the nest back towards the skua or knocked from the nest into the open as the skua crashes from flight or from a jump-flight into the nest bowl. The parent(s) may be strongly buffeted by the skua's attack as it crashes into the nest but these attacks are nevertheless directed at the prey and the buffeting of the parent is secondary. In sustained attacks, by contrast, the parents are attacked to reach the prey.

Under calm or light wind conditions penguins on nests in undisturbed breeding groups are oriented randomly. In stronger winds penguins with eggs or young chicks that can be covered in the brood pouch tend to face into the wind whereas penguins with larger chicks stand with their backs to the wind. This pattern of orientation has now been well described by several authors (e.g. Yeates, 1971; Spurr, 1975a). The regular orientation of penguins during moderate and strong winds facilitates skua attacks on guarded chicks. During these conditions skuas beat into the wind when foraging and can swiftly drop into the nest beneath the standing parent to grab chicks. They have the advantage of an excellent view of the prey and can lift back into the air very quickly merely by spreading the wings again once the chick has been grabbed. Much heavier chicks can be lifted under windy conditions than in calm. A further advantage is that the neighbouring penguins are similarly oriented and are not able to react quickly to a skua landing among them. Because they begin from little more than a metre or so above the penguins they are in any case over in seconds, and generally too quickly for much penguin response. Windy or cold conditions do not favour egg predation in the same way. In these conditions penguins sit closely on nests and the eggs are rarely exposed to skua attack.

Under light wind conditions skuas sweep above the colonies in searching flights that bring them across the penguins from all directions. A common and successful attack form is by skuas stooping into the nest at the front of a penguin after overflying it from behind. These attacks are characterised by a sudden fluttering stoop back into the nest from several metres above the penguin. The pronounced change in flight direction and flight action is very obvious to an observer.

Flight attacks expose the skuas to considerable risk of being caught by the penguins. Unless flight conditions are perfect for the skuas – moderately strong, steady winds – they are in danger of being buffeted into the nesting penguins and thrown off balance with the wings in disarray. The flailing wings of skuas striving to regain flight are easily caught by the penguins.

Attacks directed at the parent in order to prey on the eggs or chicks
    In these attacks the skuas attempt to drag or knock the parent away from the nest so that the eggs and chicks are exposed and may be taken before the penguin recovers its nest position or is otherwise able to give

them protection. These attacks are of two forms depending on the initial stance of the penguin on the nest.

Standing penguins are hit from flight or a flight-jump from the margin so that they are knocked over away from the nest. In these crash attacks the penguins are hit high on the front of the body by the extended feet of the skua from almost horizontal flight and the force of the attack may well knock the penguin a metre away into neighbours, which compounds the disturbance. Although the skua may also end up well beyond the nest its greater agility allows it to recover much more quickly and it can usually grab an egg or chick while at the same time evading the penguins' belated defence. These crash-flights have not been recorded against crouching or incubating birds, which are attacked quite differently. These penguins are dragged backwards off the nest by skuas. In these attacks the skua attempts to jerk the penguin by the tail from the nest in one fierce pull and then scramble across it to the eggs or chicks. This technique is only successful when used against resting or sleeping penguins. The heavier weight, greater strength and formidable weapons of the penguin precludes its use against alert adults.

Skuas may attack both standing and crouching penguins by tail pulling, not to drag the bird from the nest as described above, but to force it to swivel violently in the nest or lift the front of the body off the nest in reaching backwards to chase the skua away. Both actions expose the eggs or chicks. (See Fig. 6.3 for an example of this behaviour during a pair attack on an isolated penguin.) Swivelling may indeed throw them out of the nest altogether. Immediately the penguin has reacted to the attack the skua lunges past it in an attempt to grab an egg or chick before the penguin can resettle over them firmly. In these attacks the skua has to dive forwards underneath the penguin's flipper to reach into the nest and is at great risk from being knocked down.

Attacks by tail pulling may occur on both alert and sleeping penguins. They depend for their success with alert penguins on very quick flights about the breeding groups from target to target with the penguins shuffling about trying to keep the skua in clear view. Because the skuas are moving so rapidly penguins can be caught facing the 'wrong' way, and are then susceptible to attack.

These attacks directed at the adult penguins expose the skuas to very high risks of being caught and badly hurt. Not surprisingly few skuas indulged in them routinely. Crash-flight attacks on standing penguins are rare events – only two birds at Cape Bird were seen to carry them out with much frequency. Tail pulling, causing birds to swivel about on the nest

Fig. 6.3. Tail pulling to distract a penguin during an attack on an isolated nest.

exposing eggs and chicks, was, however, a common attack form of several skuas and long sequences of individual attacks on penguins up and down breeding groups were routinely recorded. It was the most common attack form of the male of pair 25 on H Block, for example. Dragging a penguin backwards from its nest is a desperate measure and was seldom observed.

6.4.2   Sustained attacks

Attacks on defended nests on the margin of the breeding group

This section describes the various attacks and mock attacks carried out by one or both skuas of the pair on alert penguins on nests on the margin of breeding groups. During much of incubation, and at other times depending on the occupation of nests by unsuccessful breeders and reoccupying penguins, nests on the margins of groups can be approached closely by skuas on the ground. In the absence of non-breeding penguins skuas may attack these penguins without risk of interference.

The diversity of the attack behaviour recorded here arises from three features of the skua–penguin interaction on the colony. First, the level of nest exposure to skuas on the margin determines the attack arc – whether confined to a narrow front or from all round the nest. Second, there is great variation in the intensity of the attack behaviour, ranging from a gentle prodding from a safe distance to the frenzied activity of a skua running and flying about the penguin. Third, one or both skuas of a pair may participate in an attack on a nest. The overall pattern of the attack is quite different when two birds collaborate and one can act as a distractor.

Most attacks will be against single parents as the two parents are at the nest together for only very short periods during incubation and chick rearing. The usual pattern in this species is for the female to go to sea for c. 20 days once the eggs are laid leaving the male with the first incubation shift. The second shift of c. 13 days is done by the female and thereafter the birds alternate on the nest with its eggs or young chicks every two or three days (Davis, 1988). Once the second egg has been laid the two parents are probably not together at the nest for more than 2% of the total incubation period or more than 4% of chick rearing (E. B. Spurr, pers. commun.).

Penguin nests have greater or lesser exposure to skua attacks depending on their location in the breeding group. Second only to the interest and skill of the skua itself, degree of exposure is the major factor determining the penguin's ability to survive skua attacks. Those entirely outside the margin allow skua movement near them on all sides, a full 360° arc, and this level of exposure may be approached also at sharp corners and in

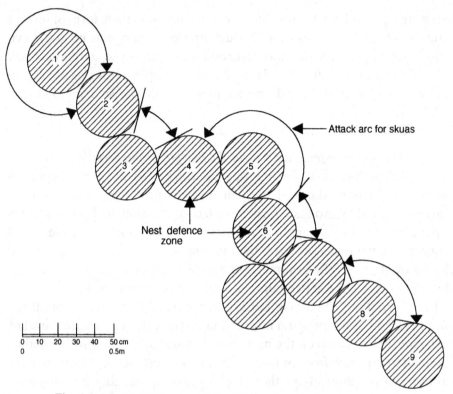

Attack arc for skuas

Nest defence
zone

Fig. 6.4. Schematic diagram of the different levels of access to nests by skuas at
the margin of a breeding group.

broken rows of nests along a narrow ridge for example. Nests in an
unbroken row along the margin may allow up to 180° arc of movement but
it is often not much more than 90° because the nests seldom form a straight
line. Some of these circumstances are illustrated in Fig. 6.4. Nests within
the group are generally not accessible to skuas on the ground.

Nest exposure is, however, the most significant factor determining the
form of an attack and is the basis for the descriptions that follow.

Nests with an attack arc less than 180°

In these attacks a single skua stands just outside pecking distance
and probes with the bill, each probe countered with beak thrusts by the
penguin. Although the skua may be able to run or jump rapidly from side
to side in front of the penguin its restricted arc of movement is easily
followed by the penguin with minimal movements on the nest. In general,
these attacks are low key and have little chance of immediate success.

Attacks by a pair of skuas working together have two common forms. In the first the two skuas stand side by side in front of the nest and thrust alternately at the penguin, causing it to respond first to one and then the other. Unless the penguin shows extreme nervousness (to be described below) or shifts to expose the eggs or chicks the attacks remain at a low intensity and seem to have little prospect of success – at least in the short term. In a second form one bird of the pair carries out a low intensity attack at the front of the penguin while the second bird stands back a half metre or so away watching results. Only if the attack gains in intensity with possible success does the second join in.

Once an attack develops the penguins on neighbouring nests defend their own areas even more aggressively and so further restrict skua movements.

There are four possible immediate outcomes to these attacks.

1. The penguin parries all thrusts by the skuas and stays firmly on the nest in the low brooding or incubating position facing the skuas until the attack peters out.

2. The attack may be terminated by the skuas being chased away by another penguin.

3. The penguin gives ground to the skuas, either lifting up and away from the nest bowl so that it is half standing with the eggs and chicks exposed in front of it, or, in extreme cases, edging back on to the far rim of the nest. In either situation the skuas have a very good chance of taking the eggs or chicks from the nest.

4. The sustained thrust and parry between skua and penguin may draw the penguin forward through the nest towards the skua so that the nest becomes exposed behind the defending bird. In contrast to the previous outcome where the penguin lifted up from the nest bowl these penguins remain in the low brooding position with their breast close to the ground the whole time. Once the penguin has come forward off the nest an excited skua may risk jumping over it into the nest to take eggs or chicks, but such attacks bring them very close to neighbours and are both dangerous and generally unsuccessful.

These outcomes apply to both single and pair attacks. The narrow arc of attack on close-packed nests gives little advantage to pairs over single birds.

On most occasions when single skuas are attacking defended nests the attacks are of such low intensity that they seem unlikely to have much impact on the penguin. The skuas seem merely to be 'testing' the sitting bird. Over-aggressive or nervous penguins found in this way may then be attacked more strongly and with greater frequency later. There is an element of intimidation as well. Certainly a starving penguin at the end of a long incubation shift may well be frightened into deserting the nest by these attacks, which normally would be easily parried. It is because of the possible longer-term effects of such attacks that seem at first sight quite innocuous that it is necessary to distinguish between the immediate outcome of an attack and the cumulative impact of frequent attacks or disturbance. As will be discussed later it is possible that skuas have a direct role in many of the nest desertions that occur.

Nests with an attack arc greater than 180°

In exposed nests single birds or pairs of skuas may attack from almost any direction without interference from other nesting penguins. The skuas clearly hold the advantage at these nests. Their high mobility (they can run and fly about the nest) and the opportunity for two birds to attack from opposite sides puts the penguin in a difficult position as it seeks to counter each attack while at the same time covering the nest.

Attacks by single skuas highlight the mobility of the predator and the restricted freedom of movement of the penguin. Skuas exploit this advantage through rapid attacks from all sides forcing the penguin to shuffle around in the nest bowl to counter each thrust while carefully ensuring that the eggs and chicks remain fully covered, neither trodden on nor thrown out of the bowl altogether. Penguins stand at the nest to follow the skua's gyrations. The degree of nest exposure again determines the range of attack behaviours. Nests along a broken margin of the colony can be attacked only from the front or sides, isolated nests can be attacked from all points by flight and jump attacks, allowing the fullest expression of the predator's mobility.

Attacks on nests with lesser exposure are mostly from the ground with the skua thrusting and probing while running quickly from side to side. Its longer arc of movement compared with that of the penguin negates its greater mobility and penguins can usually counter each thrust readily with only limited movements on the nest. Neighbouring penguins ensure that the skua cannot encroach too far into the group to reach from behind the sitting bird. Excited skuas may 'bounce' with short jump-flights across the front of the nesting bird but these seem more spectacular than effective even though apparently intimidatory. With increasing nest isolation from other penguins the scope for activity enlarges, weighting the advantage towards the skua more and more heavily. Isolated penguins are unable to defend their nests against even moderately proficient skuas. Skuas attack isolated and well exposed nests by flights over and around the penguin punctuated by strong thrusts or dives into the nest underneath a standing penguin or from behind a crouching one. The aim of this often furious activity is obviously to so disorient the defending penguin that the skua is able to reach the nest before the penguin can recover its protective cover over it. This is accomplished by speed of movement.

Three general forms of flight attack on penguins at isolated nests can be recognised. The first is an unstructured furious activity around and over the penguin with rapid thrusts at the nest from every favourable landing point. The flights have no apparent pattern except to take the skua from the front to behind the penguin as quickly as possible. The penguin is usually not touched at any time until the nest itself is attacked. The second form is for skuas to jump-fly over the penguin to grab the tail. Tail attacks cause penguins to screw violently in the nest and are undoubtedly an effective way to unsettle defending birds. The third attack form recognised for these skuas was the the aptly named 'windmill' attack in which the skua flies in a tight circle across the front of the penguin lifting from one side over the head to land on the other and then running across to fly over again. These flights provoke a frenzied attempt by the penguin to catch the skua, to the neglect of nest protection. After a short bout of this hectic activity the penguin is enticed well up from the nest bowl exposing the eggs and chicks to predation.

There are four possible outcomes to these attacks. The most common outcome for an isolated nest is for the skua to beat the defences of the penguin by its quickness and be able to dive into the nest bowl and grab an egg or chick. Less commonly the furious activity so disorients the penguin that it falls over on to its back away from the nest while attempting to counter attacks from above and behind. Once this happens it is easy for skuas to take an egg or chick from the nest. A third outcome is that the eggs or chicks may be thrown out of the nest by the violent twisting of the penguin seeking to follow the skua's movements. The nest itself may be partly flattened. This outcome is only likely where the nests hold two eggs or chicks. A single egg or chick can be held more easily and does not seem to hinder penguin movement very greatly.

Finally, in spite of the skua clearly beating the penguin and gaining access to the nest it is unable to grab an egg or chick before the penguin forces it away. Some skuas immediately resume the attack, most give up for a time – the furious activity that had preceded the dive into the nest being the climax of an attack sequence that takes time to rebuild.

Attacks on exposed nests by both birds of the pair

In some pairs one bird may do virtually all the attacks on penguins and the second bird little more than desultory scavenging. In these pairs it is usually the male that is the foraging bird. In many pairs, however, the birds collaborate in attacks to divide the penguin's attention and its defensive reactions. For much of the season, however, the skuas are also breeding with one of the pair away from the penguins at the nest during incubation or at the nest area guarding their own chicks. Thus, most attacks observed by breeding skuas are by single birds. Nevertheless, skuas will desert the nest and chicks to join mates once an attack is underway or when food is clearly visible.

Two birds work together to divide the attention of the penguin, spacing out on each side of the open arc of the nest to cause the penguin to move quite vigorously to counter their individual probing. In these attacks the birds may take up the separate roles of disorienter/distractor and nest attacker, with one inducing the penguin away from nest covering or protection so that the other is able to lunge at the eggs or chicks. This joint activity reaches its highest expression in attacks at completely isolated nests where one bird (the distractor) concentrates on tail pulling while the other lurks about on the opposite side waiting for a favourable opportunity to take eggs or chicks.

Isolated penguins subject to a collaborative attack soon adopt a half crouched stance with the flippers part extended from the body and the head held half turned away to the side of the body allowing vision both to the back and front. The slightest relaxation in attention in this position will allow one or other of the skuas to lunge at the nest. Skuas may keep up these attacks for many minutes. Either bird of the pair may act as distractor or antagonist; in fact these roles may alternate between the two during an attack. It is usually the male, however, who makes the attack into the nest. This may be from behind or from in front of the penguin and is often extraordinarily straightforward, even anticlimactic, the egg or chick being taken before the penguin even begins to respond, in some cases without it even seeing the egg or chick being taken (Fig. 6.5).

These attacks terminate because the skuas take an egg or chick or because the penguin is so resolute the skuas judge that they are unlikely to be successful. Nesting skuas are constrained also by the need to protect their own eggs and unless the attack is quickly successful one or the other will shortly return to the nest area. These joint attacks will be stopped immediately by flights near their nest area by other skuas. Territory and nest defence have priority over foraging.

Fig. 6.5. Attack by a pair of skuas on an isolated nest. The separation of the two birds, one in front the other behind the penguin is characteristic behaviour in this situation.

Pair attacks on exposed nests are much more static set pieces than attacks by single skuas and lack the hectic activity and the jumping flights which characterise the latter. They seem almost ritualised in their thrust and parry and take place almost as if in slow motion. They are, however, even more successful than are single skua attacks and demand fewer risks for the same result. There is less risk directly from the attacked penguin and because there is less flight they are not at risk either from being blown or buffeted into neighbouring penguins. All the activity of a pair attack is much more precisely controlled.

Sustained attacks on defended nests within the centre of the breeding group

As skuas cannot attack penguins from the ground in the middle of breeding groups, any sustained attacks on nesting penguins there must be from flight. In general, however, flight attacks are opportunistic, at exposed eggs or chicks beneath inattentive or distracted parents. The nearest approximation to a sustained attack from flight is aggressive hovering, but this is usually at carrion and is seldom directed at defended prey. Hovering provokes a frenzied response in the penguins, attracting free birds which crash and tumble their way across the sitting penguins as they strive to catch the skua (Fig. 6.6). As a consequence of this disturbance eggs and chicks may be lost from nests to be picked up later, but skuas do not seem to be acting deliberately towards this outcome. In support of this assertion is the fact that skuas did not carry out these flights unless deserted eggs, chicks or carrion were present.

Skuas may attempt to attack nests within the breeding group by flight-jumps from the margin (Fig. 6.7). Again, these are opportunistic attacks.

6.5   Attacks on unguarded chicks: the post-guard stage

Penguin chicks are brooded by their parents at the nest for approximately 20 days (Spurr, 1975d). From this time on they are left on their own while the parents forage at sea, returning at short intervals to feed them. It is not possible to feed both chicks at once so that feeding chases occur in which the parent runs through the breeding group and about the colony until one chick is left behind. Single chicks can be fed at the nest after being called there from the crèche by the parent.

Penguin chicks are vulnerable to skua attack in the breeding groups and

Fig. 6.6. Sustained attacks on a central nest within a breeding group. Hovering flights.

during and after feeding chases. Attacks may occur on these chicks when they are with parents and when they are on their own.

In contrast to the earlier series of attacks on defended nests which were at adult penguins as the means of reaching the eggs or young chicks, these attacks are aimed directly at the chicks – the real prey. There are characteristically two phases in these attacks. In the first the skua attempts to knock or drag the chick away from the breeding area or from adult penguins. In the second it must be subdued before its release is forced by adults attracted to the disturbance or it escapes into a breeding group. At first the skuas are both heavier and stronger than their prey but later in the season this advantage is overturned with the increasing maturity of the chicks. Skuas subdue chicks by battering and cutting the head. Large penguin chicks are difficult to kill and several to many minutes may be needed before they are completely immobilised. Attacks begun on chicks in small gaps among the breeding groups are rarely successful: it is too easy for the chick to scramble back to safety and too close to penguin adults to avoid interference.

Fig. 6.7. Jump-flight into the centre of a breeding group to retrieve a dead chick lying on an empty nest.

6.5.1 *Phase one. Separating the target from other chicks and adults*

The following sections describe five general situations in which unguarded chicks may be taken by skuas.

1. *Attacks on chicks lying on the breeding area on their own or among adult penguins.*

During warm weather chicks lie about the breeding area with feet and flippers extended looking as though dead. Skuas 'test' them as prey or carrion by running in from the margin and jabbing them strongly on the body. Those that do not respond immediately by scrambling to their feet, with a great show of indignation, and running or backing away towards other chicks may well be further attacked.

2. *Attacks from flight at chicks standing on the margin of the breeding area or crèche*

In these attacks the skua flies at the chick and grabs it by the head in an attempt to knock it away from other penguins. The skuas appear to be attempting to lift the chick, but older ones are far too heavy and chick and skua crash to the ground if the skua retains its hold. Although some skuas earlier in the season carried out crash-flights at standing adults, hitting them squarely with the feet, similar flights were not recorded on chicks and all attacks were directed at the head.

The success of flight attacks on crèche chicks depends on flight speed, the strength of the grip on the chick, the age and agility of the chick, the proximity of adult penguins and the character of the near terrain. Chicks knocked down a steep slope from the breeding area so that they tumble several metres away are much more likely to be held than if knocked on to level ground.

3. *Attacks from the ground on alert chicks in the crèche (Fig. 6.8)*

Skuas can move among unguarded chicks with apparent impunity until nearly the end of the season when increasing maturity and confidence allows groups of chicks together to force skuas away. Attacks on crèche chicks do occur throughout the season but are usually nullified by the presence of adult birds and the ability of older chicks to resist being dragged from the crèche. They are common only in the smaller breeding groups, which may lack adults at times. The usual outcome of an attack on such a group is for the chicks to flee to a neighbouring group.

Fig. 6.8. Pairs of skuas 'working' among a small group of chicks only weakly protected by adults. Within a very short time most of the adults had been frightened away from the group and the chicks were tightly crèched.

4. *Attacks on small, starved or malformed chicks in the crèches*

Even at the end of summer there is a small proportion of smaller or weaker chicks that can be attacked by skuas. These are usually well protected within the crèches but with skillful manoeuvring they may be isolated enough to be attacked. It is not uncommon in late summer to see a pair of skuas 'working' a crèche to get at the smaller chicks. Once these chicks survive one of these attacks they become especially wary and keep near to larger chicks for protection. Nevertheless, few survive to fledging.

5. *Attacks on large chicks outside the crèche or breeding group*

Chicks are caught in the open by skuas during feeding chases with parents or when moving between breeding groups. (Even when chicks move from their natal area to another they must return to their own one for feeding as the parents call chicks for feeding from the nest position.)

Attacks may be directly from flight taking the chick by the head to knock it to the ground away from adults or by flight-jumps from the ground at the head. There is a wide range of intensity in these attacks. At the one extreme skuas may merely fly past the chick, flicking it with the beak or wing before returning to settle back on a roost, while the chick scuttles into a crèche or to adult birds. At the other extreme the attack is strongly forced with the chick caught and held time and time again in spite of the protection given by adults. Parents will keep skuas away until the whole group, which often includes other adult penguins attracted to the disturbance, have all moved on to a breeding area. Adults that are not parents, and especially wandering penguins (young birds that have not yet begun breeding), will usually chase skuas away for a time but may themselves be frightened off by flights or jump-flights at them.

6.5.2 *Phase two: subduing and killing large chicks*

Once the chick has been separated from adults it must be immobilised as quickly as possible to prevent it escaping back into the breeding group or being rescued. By the end of the season the chicks are much heavier than the skuas and are strong enough to drag them across the ground. Although skuas are probably not at much physical risk when tackling these chicks they are nevertheless often cautious, failing to press home an attack that seems straightforwardly one-sided. Chicks caught out in the open will stand up to skuas, even run at them and the skuas seem quite disconcerted by this unexpected aggression (Fig. 6.9).

Fig. 6.9. Attack behaviour by a skua attacking a penguin chick outside the breeding group and the defensive behaviour of the chick.

Attacks on standing chicks begin with a flight-jump to grab the head to knock or drag the chick to the ground. Killing is done through battering and cutting the head. This must be carried out while at the same time preventing the chick escaping. During the first few minutes of an attack on large chicks skuas can do little more than stop the chick escaping or being dragged with it to safety. In these first exchanges skuas hold on to the chick by the beak and use the feet and strong wing-beating to slow the rate they are being dragged, but even so many are dragged long distances before the chick is finally immobilised. Chicks caught well away from adults are not held as tightly and the skua may at first merely peck at it while walking beside. The chicks are subdued by skuas from a combination of damage to the head and exhaustion. As the two birds at this stage are rather evenly matched in strength these attacks can be a drawn-out and gruesome process. Skuas lack an effective killing weapon. They win in the end because they are tougher.

The two skuas of a pair may cooperate in tugging a penguin chick away from adults or preventing it regaining the breeding area. They may collaborate also in flights and flight-jumps about adults with a chick in the open, disorientating the penguin defence and allowing one or the other to grab at the chick. The final phase of the attack in which the chick is killed is, however, usually carried out by only one of the pair with the other watching closely but taking little active part, in some pairs even when the chick is in danger of escaping. The second bird may even hamper the first by beginning to rip at the chick's cloaca before it is fully immobilised. The chick's frantic reaction to this wounding may well break it free and allow it to escape momentarily.

The skuas begin feeding from the abdomen immediately the chick is subdued, often before it is dead, with the two birds ripping at the soft skin of the belly to expose the viscera and then tearing the whole body apart between them.

The outcome of a skua attack on a chick away from the crèche depends on the size and strength of the chick and how quickly it can be subdued, how far away from an occupied breeding area it was first caught, and how much interference there is from adults. Or the chicks may find some other temporary shelter (Fig. 6.10). In general, large, strong chicks take so long to subdue that they can only be killed in the most favourable situations. Few chicks heavier than 3 kg are killed by skuas.

Fig. 6.10. Penguin chick obtaining temporary shelter after an encounter with a skua outside the breeding group.

6.6   *Comparison between skua agonistic and predatory behaviour*

To what extent are the attacks by skuas on penguins like those on conspecifics? It would not be surprising if the actively aggressive parts of scavenging and predation were similar, there are after all only a limited number of ways skuas can fight or overcome rivals. The more interesting point is whether skuas display towards the penguins in attempts to intimidate or distract them. These aspects of skua predation can be investigated by first describing the ways skuas interact with each other and then comparing this behaviour with that of skuas interacting with penguins.

The agonistic behaviour of skuas is overall very similar and is now well described (e.g. Perdeck, 1960; Burton, 1968; Spellerberg, 1971a; Andersson, 1976; Pietz, 1985; Furness, 1987). Its various forms are associated with attacks in the air, attacks from flight at birds on the ground and attacks on the ground. Each has characteristic displays. The principal weapon is the bill with the wings and feet used mainly for purchase and balance during fighting. Displays focus on the attitude of the body, especially on the upright and oblique body postures common also to gulls (Moynihan, 1962); on the orientation of the bill and head; and on the wings.

Birds fighting in the air buffet each other with wings and feet and attempt to catch the wings or tail with the beak. Aerial fights are remarkable for the power and speed of the flying and the intensity and persistence of the pursuit of fleeing birds.

Skuas attack birds on the ground from flight by stoop attacks, diving at them from 10 to 15 m height with strong wing beats, knocking the antagonist with the trailing legs in passage before lifting steeply up again in flight in preparation for another stoop. Birds receiving such an attack crouch closely on the ground as the stoop begins and lift into a wing raising display once it has passed over. Where possible birds caught on the ground like this run quickly towards shelter, crouching behind stones, or behind vegetation in populations on the Southern Ocean islands.

Continuous attack invariably provokes an aerial chase as the attacked bird attempts to chase away the other. These attacks occur under two circumstances. They are used across territory boundaries by neighbours and used by territory holders to chase away birds landing in their territory. Attacked birds are often hit very severely and few are prepared to accept more than one or two stoops without attempting to retaliate by jumping up at the attacking bird or springing into flight themselves. It is important to note here that the attacking bird uses the legs, not the wings or bill to hit the attacked bird.

Fights on the ground, whether beginning from flight or not, are by jump-flights in which the antagonists attempt to peck down on each other. The wings are spread to increase the height and range of the jump. Both the bill and the feet are used to hit and claw at the other bird. As both birds are using the same attack method, and each is attempting to obtain advantage from height, these are vigorous and buoyant encounters. Sustained fights break down sooner or later to 'wrestling' with the two birds locked together tightly, their wings and legs twisted and bent across each other or forced against the ground giving purchase and the bills locked deeply into the feathers of the head, neck or breast. These fights are frightening in their intensity, are carried out in silence and may leave the antagonists gore-splattered. Sustained fights of this sort with the birds locked together on the ground are invariably concerned with territory and mate holding. They have only been observed on territories. Hostile behaviour does occur in the club (the aggregation of non-breeding birds) but seldom with the intensity seen among territorial birds and is usually confined to threat displays and short-lived pecking encounters.

There are few agonistic displays, although each has subtle variation. Of most importance are gaping, in which the opened bill is directed at other skuas; the bend, in which the skua bends the beak strongly down on to the breast; the upright displays in which a stiff-legged gait is associated with upright posture and stretched neck; and the oblique display with or without wing raising. The last of these is generally accompanied by loud calling; the others are silent except there may be some calling with the gape.

The first two displays are seen in birds attempting to deter others from approaching. The first is in common use in the club where it is given by birds resting on the ground at others coming towards them. The bend is commonly used at birds in flight over the territory. Bends generally precede oblique displays in the sequence of displays occurring when birds continue to approach. Both the gape and bend are thus defensive, given in response to the behaviour of another skua. The oblique and wing raising displays have a similar role, but are used also as advertising displays. They are used routinely by birds when returning to the territory and seem identical then to the display used by them at skuas overflying the territory. This then is also a distance display, used by skuas in response to the activities of others, particularly to skuas in flight, although pairs of skuas on a territory boundary may also use this display. The sequence of displays culminating in the oblique with wing raising is termed the long call complex. The aggressive upright display, in contrast to the others, is very much an aggressive threatening display used to intimidate other skuas; to force them to move, to evict them from the territory. It is the usual supplanting display of skuas interacting on the ground. It is seen commonly also in the interactions between the pair, during the initial stages of courtship feeding and mating, for example.

Of all the displays used by skuas in their intra-specific encounters the upright would seem the most likely to occur in attempts to intimidate penguins. Intimidation could occur as well from oblique displays with wing raising because these inflate body size dramatically.

Flight attacks on penguins are significantly different from those on other skuas. Firstly, the flight phase of the attack is not ritualised as it is in the stoop attacks on skuas or other territorial intruders, that is, the skua flies at the penguin directly from the place where the attack begins without first lifting high into the air to allow stooping. Nor does it fly up steeply from the bottom of the attack after hitting the penguin as it does after hitting skuas. There are differences also in the weapons used. Skuas hit each other, and humans, with the trailing legs. They crash into adult penguins with the feet thrust forward and attempt to take older chicks in the bill.

Ground attacks on penguins also differ from those between skuas in two important ways. The first is that skuas do not appear to carry out threat displays at penguins nor attempt to intimidate them by increasing their apparent size with feather fluffing, an exaggerated upright stance or wing raising. The absence of display is most clear-cut in attacks on nesting penguins. When these are approached skuas tend to adopt a low stance. The gape and bend, essentially deterrent displays, are also missing. Nor is there good evidence that skuas ever display obliques with wing raising at nesting penguins; although these are seen often enough during an attack, they are usually found to be directed instead at other skuas overhead or nearby. The second major difference is that much of the attack behaviour directed at adult penguins at the nest (e.g. attacks at the tail, probing with the beak) has no counterpart in skua–skua conflict.

Attacks on grown chicks are, however, very like fights between skuas. There is essential similarity in the way the skuas jump at the head of penguins and peck down at them. Moreover, once the chicks have been forced to the ground the grappling and struggling to subdue them looks just like that occurring between two fighting skuas. Even here there is a difference: in fights for dominance skuas are trying to cause another bird to flee, in penguin chick attacks they must subdue it while ensuring that it does not escape. The intensity of the attack is lessened also by the need to be alert to the approach and attack by adult penguins. Fighting skuas can be picked up by hand. Skuas killing chicks remain alert to their immediate surroundings at all times. The similarity of the behaviour in chick attacks to conspecific hostile encounters suggests that it would be more likely to see displays here than in the more formalised attacks on the static nesting penguins. Nevertheless it is doubted if they occur.

6.7   *The behaviour of penguins in defence against skuas*

There are two very different defence situations for penguins being attacked by skuas. The first is the situation faced by a single parent on the nest protecting eggs or young chicks. The second is where the penguins can move freely to chase the skuas away, as occurs when there are two parents at the nest and throughout the post-guard stage. Their behaviour is consequently very different in these two situations.

The following sections describe the behaviour of penguins during skua attacks. It must be emphasised again that the adult birds are themselves not at direct risk, or even the true target of the attack, but must be overcome in order to get at the eggs and chicks. The parent birds are defending their offspring, not themselves.

Initial observations showed that the penguins were behaving and taking up postures in response to skuas like those at other penguins, that is, a small repertoire of displays was being used in any threat situation. Because of this similarity, description of the responses to skua attack can be based to a large degree on existing knowledge of penguin behaviour.

There is now a fairly clear understanding of communication in this penguin. Detailed accounts have been produced by Sapin-Jaloustre and Bourliére (1952), Sladen (1958) and Penny (1968), refined by Spurr (1975b) and Ainley (1975). Jouventin (1982) has compared the behaviour of the different species within an evolutionary framework, providing for the first time a common terminology. The accounts of Spurr and Ainley are an appropriate starting point for a description of the Adélie penguin's behaviour in relation to skuas.

There is no a priori reason that inter-specific communication between penguins and skuas should occur at all, or that it should occur in the form of displays and calls based on the same range used by the penguins in their own social encounters and activities. However, the options available to a penguin restricted in its movements to the nest area are clearly limited and the response to territorial intrusions by penguins or skuas must be met similarly by a determination to stay put, while at the same time protecting the eggs and chicks from injury or dislodgement. The few parts of the body that can be used in signals are also very limited so that it should not be at all surprising that if threat displays occur they will be identical or derived from those in common use in intra-specific communication.

Penguins have only three options when faced with skua attack. They may flee, abandoning the nest or chicks; leave the nest to chase the skua away by force of superior size, strength and weapons; or stay on the nest and counter the attack through active and sustained defence. In practice the third option is the only one available to penguins on their own at the nest. The question then arises: why should penguins display at attacking skuas, unless they can communicate intention to defend the nest resolutely, causing the skuas to leave them alone? Conversely, any displays that communicate hesitation, inexperience, nervousness or exhaustion would have the opposite effect, and, if read correctly, attract skua attacks. The interpretation of the postures and activities of penguins as threat displays, therefore, depends on whether it is concluded that the penguin is attempting to communicate with the skua. (Whether or not the skua can 'read' what is being signalled is another matter.) The alternative explanation is that the postures used by defending penguins have no meaning within the context of the phases of a skua attack and are, therefore, not signals. They might, for example, merely reflect physiological or psychological state, arousal or boredom. In fact, the different postures and movements are given so consistently in relation to different intensities of skua activity and are so similar to elements of intra-specific behaviour it would be unreasonable not to term them displays. What has not yet been discovered is how the skuas interpret them.

6.7.1 *Protection afforded to individual nests by the colony structure*

The constraints imposed on attacking skuas by colonial breeding were described in Chapter 5 and need only be summarised here as an introduction to penguin defensive behaviour.

Penguins nest together in groups of a few to several thousand. The shape of the breeding group is determined largely by the local topography as nesting begins in spring on the drier, snow-free parts of the colony, on ridges and elevations. The accumulation of debris from nest building and breeding accentuates topographical features so that breeding areas are increasingly raised above the surrounding ground.

The nests are closely packed in central areas with territories approaching the theoretical closest packing of adjoining hexagonals with each penguin surrounded and interacting with six others. This close packing can only be developed on smooth ground and is disrupted by irregularity.

Skuas cannot land or stand within breeding groups as the nests are too closely packed and any spaces present can be reached by penguins on surrounding nests. Spaces large enough to allow skuas to forage within the groups do not begin to appear until late in the season when the group structure begins to collapse as nests are abandoned and chicks begin to crèche. Until this stage large gaps can only arise from extraordinary events such as nest desertion through starvation or fighting. Overall, the colonial breeding habit prevents skua attack from the ground at central nests and restricts skua movement on the margin. The attraction to the margins of existing breeding groups of young breeders and reoccupying birds further restricts access to the nests of breeders. Furthermore, the clumping of breeding groups together has the effect of restricting skua activity within the colony as a whole. In many places there are so many adult penguins moving through these gaps that skuas are unable to use them for predation. They cannot loiter there and attempts at catching and killing chicks are generally hampered by the presence of so many penguin adults. Chicks taken in attacks can also escape easily into any of the surrounding groups.

The importance of non-breeding penguins in the colonies later in the season should not be under-estimated. Not only do they shield the more central nests of breeding groups from the skuas but also once aroused by

skua activity, especially by flight attacks, they become extremely pugnacious towards skuas landing and moving about near them and immediately
chase them away. In this pugnacity they protect much more of the
breeding group than their immediate nesting territory. This protection was
amply demonstrated, for example, in a sequence of flight and ground
attacks along H4 breeding group by the male of pair 25. The attack
sequence lasted in total 204 min. Over this interval the attacking skua was
chased away on landing 40 times and was prevented entirely from reaching
long sections of the margin.

6.7.2   *Nest defence zones*

Penguins respond variably to skuas approaching the nest depending on distance, posture, and speed and direction of movement, earlier
experience of skua attacks and the immediate history of skua activity in the
area. There may also be an element of learning from earlier encounters but
as Spurr (1974) has demonstrated there are wide differences in aggressiveness, which varied also depending on the stage of the breeding cycle. In
spite of this variability it is possible to recognise concentric zones about the
nest relating skua activity to penguin response (Fig. 6.11). The ones above
the nest to skuas in flight have been investigated by Müller-Schwarze and
Müller-Schwarze (1977). These authors recognised three zones on the
basis of penguin behaviour. These are a defence zone up to 0.5 m above
the penguins where active defence occurs; an alert zone between 0.5 m to
12–15 m in which penguins watch skuas, the proportion falling off with
increasing height; merging into a neutral zone (which they set above 14 m)
in which there was no apparent response to skua overflights.

However, the position of the defence–alert boundary depends very
much on the recent experience of the penguins interacting with skuas. The
0.5 m limit set for the defence zone is really a minimum figure. Following
even moderate skua attack the boundary will be considerably raised and
penguins will react strongly to skuas hovering as much as 2–3 m above
them.

Four zones of response radiating out from the nest are recognised for
ground attacks. The active defence zone corresponds approximately to the
limits of the territory, it is the furthest that a nesting penguin can reach out
to a skua without leaving the nest. Immediately outside the territory is a
display zone in which skuas will usually, but not always, provoke penguin
displays. Outside this again is the alert zone in which skuas are watched
closely. Penguins will turn in the nest to face skuas within this distance.
This zone fades out into the neutral zone in which skuas are ignored.

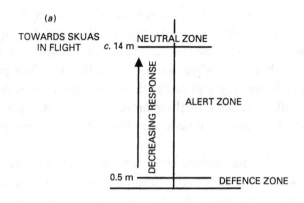

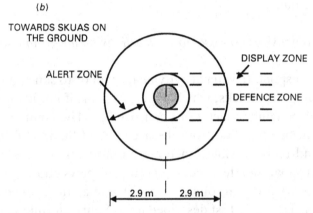

Fig. 6.11. The defence zones about nesting penguins. (*a*) The defence zones for skuas in flight above the penguins (adapted from Müller-Schwarze and Müller-Schwarze, 1977). (*b*) The defence zones about the nest for skuas attacking on the ground.

The dimensions of these zones on the ground cannot be given with the same certainty as for skuas in flight as they are much more dependent on the arousal of the penguins from earlier skua activity and the behaviour of the skua at the time, which is more variable than for skuas in flight. In breeding groups subject to light or moderate skua activity the defence zone is enclosed by the territory boundary, approximately 400 mm from the nest centre, the display zone runs out to a further half metre and the alert zone to another two metres. In total, penguins are alerted by skua activity to about 2.90 m from the nest centre – the distance to about the fourth nest away in a closely packed group. The alert–neutral zone boundary will be closer for penguins little troubled by skuas but much further away for those under regular attack.

Penguin response to skuas near them depends also on the way the skuas are behaving, and obviously differs towards skuas walking past compared with those walking or running at the nest. Beating wings and rapid movement also cause sharpened vigilance locally in the breeding group. Penguins are, however, quickly habituated to skuas standing or resting close to them and will relax back to their more usual somnolence even though the skua may be scarcely outside the defence zone – provided it remains quite still.

The alert zone is easily recognised in groups of penguins subject to attack or scavenging at one point in the breeding group. In time neighbouring penguins shuffle round in the nest so that all the heads point towards the disturbance.

6.7.3   *Penguin defensive responses to skua activity on the ground near the nest*

Penguins respond to skua activity near them or to skua attacks by threat displays or overt attacks, which appear similar, if not identical, to the behaviour of penguins reacting to conspecifics. The components of these displays can be identified from the accounts of Spurr (1975b) and Ainley (1975) and their possible function and motivation inferred both from context and apparent effect. Penguin threat displays utilise a series of specific signalling structures in conjunction with a small range of head and body movements. These are first described using Spurr's terminology and then placed in the context of skua attacks.

The threat displays of penguins

Displays are formed from the combination of a range of quite specific display components.

Feather raising and sleeking on the head, face and body

The feathers may be erected or sleeked. Threat displays are characterised by the erection of the feathers over the back of the head to form a crest, whose effect may be enhanced by sleeking the feathers on the face and forehead. Subtle variation occurs in respect of the feathers on the face. These may at different times be only partly sleeked, or even raised between the eyes (to form a small orbital crest) or may be raised at the base of the beak as a nasal crest. The most obvious effect of feather posture is to form a raised crest across the back of the head. In the relaxed state the penguin lacks a crest. The body feathering may be tightly sleeked during escape or when a penguin is threading its way among tightly packed territories in the 'slender walk'.

Eye rolling

When penguins are relaxed the white sclerae of the eyes are barely visible and the eyes appear black. In threat displays the eyes are rolled down to expose the sclerae which contrast strongly with the black feathering of the face. This display component is termed 'eyes down' or 'eyes rolled'.

Head orientation

In relaxed penguins on the nest the head is held directly in front of the body and the neck is withdrawn. In the various threat displays the head is held turned to the side of the body, waved alternately from side to side or stretched forward or upwards. These postures present the eyes, crest and beak respectively to skuas. The beak may be closed, when the penguin is said to be pointing, or open in gaping or calling.

Flipper movements

The flippers have important roles in balance and as weapons in both fighting and defence. Slow flapping movements, one or two beats a second, commonly occur in penguins standing at the nest but less commonly in sitting birds.

All these components occur in different combinations in specific threat or defence displays. These displays have been listed by Spurr in a graded series of increasing probability of attack and effectiveness as follows:

  bill-to-axilla;
  sideways stare;
  alternate stare;
  point; and
  gape, with the gape the most threatening.

(The alternate stare is termed the fixed one-sided stare by Ainley.)

The major variable in all displays is head orientation and movement. All displays by nesting penguins incorporate the erect crest and rolled eyes. In the bill-to-axilla display the head is turned to the side of the body so that the bill is rubbed to and fro across the side of the flipper or even tucked into the axil. The sideways stare and alternate stare displays have a similar head orientation at the side of the body, which presents the eye towards the intruder or attacker, but in the alternate stare the penguin waves the head slowly from side to side presenting the eyes in turn. The point and gape both direct the bill at the intruder; in the first the bill is closed, in the second it is open. Both displays, or movements, may include head rotation.

Spurr considered that the sideways head displays (bill-to-axilla, sideways stare and alternate sideways stare) were deterrent threat displays, i.e., they deterred intruders from approaching. The forward displays of point and gape (and charge and attack behaviours) were considered to be repellent threat displays, repelling or driving intruders away. The bill is the most frequently used weapon and most ritualisation has occurred in its positioning and presentation. Although the flippers are also used in fighting they are only marginally incorporated into threat displays. They appear in the bill-to-axilla and alternate stare displays, but not in the stronger forward displays of point and gape, except for balance by birds on nests.

Ainley (1975) has followed Smith (1969) in stressing the message function of penguin displays. His listing for the threat displays is therefore a little different from Spurr's and includes bill-to-axilla, fixed stare, sideways and alternate stares and crouch. The gape and point of Spurr are considered to be components of the crouch posture. Both Ainley and Spurr have been able to relate the occurrence of the several threat displays to the proximity of the intruder or disturbance. Their conclusions were similar in indicating that the bill-to-axilla was the most likely display to penguins several metres away and that the two sideway stares were in response to closer activity, giving way within 0.5 m (Ainley) or 'two inter-nest distances' (Spurr) to pointing or gaping at the closest proximity.

> *Displays and other activities used by penguins in response to skua activity on the ground near the nest*

It must be stressed that penguins only take up characteristic threat displays at skuas standing or slowly approaching the nest, or at skuas attacking neighbours. Attacks into the nest or rapid movement about the penguin are met with direct attacks, not with displays.

Penguins respond to skuas approaching the breeding group or standing near the nest by postures and vocalisations that seem identical to the threat displays and calls given to penguins. Within the display zone penguins perform fixed and alternate sideways stare displays and switch into pointing and gaping at closer proximity. The bill-to-axilla display, the least specific and effective of the penguin deterrent displays, is little (or never) used.

The same series of displays is used by penguins lying in the nest over eggs or young chicks and by penguins standing at the nest over large chicks. The sideways stare displays are especially characteristic of skua proximity and appear more ritualised than the others in this attack–defence context. The head is withdrawn to the side, away from the skua, and the bill points downwards rather than directed at the attacking bird. The eyes are moderately to strongly rolled – a posture that would seem to reduce their effectiveness in watching skua activity. The crest is invariably raised, at least during part of the interaction. The wings, together with the bill, the penguin's primary weapons, may be held close to the body or held out to the side. The sideways stare displays take two quite distinct forms in interactions with skuas. One form of this display with the bill pointing downwards corresponds to the one seen in penguin threat. The other has the bill pointed upwards, away from the skua, and appears unique to skua threat.

The 'bill up' display is seen most characteristically in penguins standing at the nest responding to sustained attacks. This variant has the same feather and eye postures as in sideways stares but the bill is invariably raised above the horizontal and may become nearly vertical. The wings are held out from the body and commonly are slowly waved but they may be held by the body or extended and not waved. The eye may be from a quarter to three quarters rolled at different times during this behaviour in the same bird and although the face as a whole is always sleeked the nasal crest may be raised. The bill is always closed. The overall attitude is one of drawing back from the skua and as it exposes the nest contents must encourage further skua attack. Although this seems to indicate a less resolute defence of the nest than the more usual downwards bill postures, the penguin can immediately break into a strong lunge at the skua from this posture and movements into the defence zone are unfailingly countered.

This same attitude is also taken up by penguins attacked simultaneously by skuas at the front and back of the nest. In this situation it is the only possible one that allows the penguin to watch both skuas at the same time while defending the nest. Although the penguin appears similarly lifted away from the nest in these interactions it need not necessarily convey the same message to the skuas of falling resolution. However, the result of the encounter is generally the same in the two situations and skuas are very likely to succeed in reaching the nest contents in both.

An important aspect of all displays in response to skua attack is the stance of the bird on the nest: on whether it is sitting tightly on the eggs or chicks or lifting back from the nest bowl. Although the same display may be given by birds reaching forward or pulling back from the skua its impact appears very different. Similarly, lunging hard at the skua from the front of the nest looks very different from the tentative lunges given by penguins crouched behind the nest. There is the same display or activity in each case – but indicating a very different defence status.

Within the defence zone displays give way to overt attack. In penguins at nests this takes the form of strong lunging at the skua, generally with the open bill and with loud calling. The penguins keep skuas away either by hard pecking or by catching the body or wings. Skuas caught by penguins rear back violently breaking the penguin's grip (the bill is not hooked), before the penguin can use its flippers. It is the flipper, not the bill, which is the most dangerous weapon to skuas.

Analysis of out and out attacks on skuas demonstrates wide variation in the individual components of displays. Although the crest is generally raised and the face partly or fully sleeked the eyes are not usually rolled. Penguins may indeed lunge at skuas without raising the crest at all (Fig. 6.12). Lunging attacks are, therefore, associated with relaxation of the feather and eye postures, as most attacks will have followed situations provoking full threat displays.

Fig. 6.12. Penguin lunging attack on a skua, illustrating that this intense defensive and attack behaviour may occur without obvious eye or crest display.

6.7.4 *Penguin defensive responses to skuas in flight above the breeding group*

Penguins respond to skuas in flight variably depending on height and flight behaviour. Skuas fly closely above penguins when in flight from one part of a colony to another during a foraging sequence or when trying to obtain carrion or eggs in the group centre. The first flights are generally too quick to provoke a sustained response by penguins but the second ones invariably provoke strong penguin reaction.

The key difference between responses to skuas on the ground and in flight is that penguins do not noticeably display at flying birds, except by pointing or gaping, so that alert observation switches directly into attack at close proximity. In these attacks the penguin lifts the front of the body off the nest stretching up at the skua at first with closed bill but with open bill and calling at higher intensities. Birds already standing at the nest with chicks stretch up towards the skua keeping it away from the nest position. In both situations the flippers may be waved or kept against the body. There is no clear pattern discernible in the photographic records of when the crest is raised in these birds and the same penguin may have the crest up or down at different times. Nevertheless, the impression exists that less confident birds are less likely to have a raised crest in these attacks. The face is usually sleeked in penguins reaching up at skuas but some birds, especially those with closed bills, may raise the nasal crest about the upper bill. Few if any birds have eyes rolled and the eyes are normal in all penguins closest to the skua and lunging at it.

6.7.5 *Aggressive defence by adult penguins of chicks outside the breeding group*

This category includes all situations in which adult penguins attempt to protect chicks being attacked by skuas when outside the breeding group. The intentions of the players in these encounters are quite clear: the skuas are attempting to catch, hold and subdue the chick; the adult penguins are attempting to chase the skua away; the chick is trying to escape from the skua either by scrambling back into the crèche or by sticking closely to adult penguins until the skua gives up its attack.

Although both parents and non-parents are extremely persistent in their defence of chicks caught by skuas in the open there are obvious differences in their activity. The defence of the parent is more clearly oriented to shepherding the chick back into the group or crèche. That of non-parents lacks this orientation and seems to be directed simply to chasing the skua – the greater the skua activity the stronger the chasing. Parents continue

Fig. 6.13. The defence of a chick by adults in the open. In this sequence, although the skua can catch and hold the chick for short periods, it is unable to lift it away in flight and is forced to release it by the pursuing adult penguins.

their defence until the chick has reached the safety of the breeding area, non-parents may give up at any time, and their movements during the encounter may have inadvertently carried the chick further away from safety than when the attack began. Chicks that have been immobilised by the skuas, and skuas that have drawn back a little or are resting, do not provoke a chasing response and sooner or later adult penguins that are not parents drift away.

Adult penguins defend chicks in the open by charging, forcing the skua to release the chick as they run or fly out of reach. In a strong skua attack on a chick the skua immediately lands again on or near the chick and the cycle of catch, chase and flight may go on for many minutes. Penguins in these situations are carrying out pure attack behaviour corresponding to the 'charge' described by Spurr (1975b) for intra-specific fighting. During the first part of the chase the crest is generally not raised and the face is normal or sleeked. The eyes may be scarcely rolled or not rolled. During the last metre or so of the charge, however, when the penguin is rushing and jumping at the skua, the crest is invariably raised. Throughout the chase the wings are held out from the body. Certainly non-parental adults at this stage give every indication of strong aggression, but earlier on they may have indicated 'curiosity' rather than aggression, with sleeked feathers and closed bill and wings.

At the end of an attack sequence when the chicks has been rescued, even if only temporarily, differences show in the behaviour of parental and non-parental adults. Parents maintain an alert, vigilant stance protective of the chick. Isolated non-parents especially, or small groups of adults tackling an aggressive skua, may show 'nervousness', stretching and giving a series of rapid wing beats with sleeked body and head feathering. The eyes may be partly rolled. Their demeanour indicates waning interest and confidence and without the reinforcement of immediate skua activity these penguin will most usually begin to run or sidle off towards others or into a breeding group, without taking further notice of the chick.

6.7.6 *Defensive behaviour by chicks attacked outside the breeding group*

Penguin chicks attacked by skuas when they are outside the crèche or breeding group have only two options: they can immediately run or scramble back into the colony or to adult penguins; or they have to defend themselves so aggressively that the skuas are unable to subdue them. The outcome of an attack is determined by the chick's age, its proximity to other penguins and the strength of the skua attack. Defence ultimately is by escape back into the crèche, to be lost among the other chicks. The likelihood of escaping from an attack is, however, largely determined by the age of the chick – by the age-related characteristics of mass, strength and coordination, the effectiveness of the defensive weapons and the running and scrambling speeds. Younger chicks are less likely to survive than older ones. Attacks on chicks of three different ages are described.

1.   *Chicks up to 20 days old*
      These chicks will have been taken from the breeding group by skuas as they are too young to be normally outside the margin. They are obviously immature, lack mobility and coordination, can stand and balance with difficulty and are easily knocked over. Although weighing up to 1500 g most of this mass is concentrated in a grotesquely expanded abdomen, the other parts of the body are puny by comparison. Neither bill nor wings are sufficiently developed or hardened to constitute effective defensive weapons. The skeleton and body covering of these chicks is so soft, and they have so little resilience, that they are soon killed through head wounding by even moderately aggressive skuas. Although they attempt to flail the skuas with the flippers when caught these have no observable impact. Their only prospect of survival is by escape back into the breeding group before being too severely mutilated.

2.   *Chicks 25 to 35 days old*
      At this age the chicks are rapidly developing strength and coordination; the down is being replaced by feathering and the legs, wings and bill are all lengthening and becoming stronger. These chicks weigh between 2000 g and 3500 g and are already heavier than the skuas. In the oldest chicks a white chest and chin are cleanly feathered. These chicks are mobile, exercising their wings and legs regularly, indulge in short spats of fighting with others, and follow parents in long circuits about the colony. They are rapidly developing stamina. In contrast to the younger ones these chicks are not quickly and easily subdued by skuas. Most of the long-drawn-out attacks seen on the colony involve this age group, which are young enough to attack, but too old and strong to be killed easily.

3.   *Chicks more than 40 days old*
      These chicks are so large and strong and their bill and flippers so well developed that they can 'stand off' a skua attack long enough to escape back into the crèche. Most skuas fail to pursue an attack on these chicks except in the most favourable situations.

Conclusions on chick defence and maturity

Aggressive defence develops early and even the smallest chicks peck and beat their rudimentary wings when held. The older the chick the more effective this defence becomes. The principal weapons, as in the adult, are the bill and flipper but at first these are too soft to be effective. The transformation of the flippers from thin and soft, string-like appendages to hardened blades is particularly dramatic. The changing stance with age of cornered chicks reflects their developing maturity. Very young chicks make little attempt to face the skua and simply expend their total effort in an attempt to get back into the breeding area. From about 15 days old chicks stand upright when cornered and use their bill and wings in an attempt to beat off the skua and allow some freedom of movement to escape. Quite young chicks, even those still in full down at 20 days old, will run at skuas, leaping clumsily at them with opened bill. Much of the encounter between a skua and penguin chick in the open is taken up with the skua jumping at the chick in an attempt to knock it over and the chick moving and turning to prevent the skua from obtaining the advantage of an attack from behind. Skuas are surprisingly reluctant to force home an attack on agile chicks and can be deterred by a strongly aggressive defence.

Young chicks evenly covered in down and still with a black-bead eye lack the signal structures of the adult. Older chicks develop the white eye rim and the ability to roll the eye to expose the sclerae at the same time as the feathering becomes exposed as the down is lost. They, therefore, have the potential for threat display. Eye rolling is commonplace during feeding chases by chicks still in full down.

Without doubt the appearance of the chick, especially its overall body shape, the length and strength of the limbs and the extent of down replacement by feathering provide a rapid index of likelihood of skua attack and the probability of escape. The changes occurring in growing chicks are readily visible to human observers and are, presumably, just as apparent to skuas. Differences in body shape, replacement of down by feathering, mobility and general demeanour must be clearer indications of prey status and ease with which they can be killed than any displays involving the eyes or feathering. The best defence by an isolated chick is vigorous and aggressive attack, while moving all the while towards the safety of the breeding group, not static display. This is an attribute of a mature chick.

Finally, it should be recorded that even the most sustained strong defence may not succeed in saving a chick from tenacious skuas. For example, a chick weighing 1065 g at 14 days old was held by one skua of a pair for the entire 16 minutes needed to subdue it in spite of being pursued for the whole time by three adults. At no time did the adults force the skua to let the chick go. They were clearly hampered and distracted by the belligerence of the second bird of the pair who flew at them aggressively all the while. The general rules of encounter between the two species do not apply to exceptional skuas.

7

The diversity and intensity of skua foraging behaviour on the penguin colony

7.1 *Introduction*

The preceding chapter has described the ways in which skuas attack nesting penguins and penguin chicks and the ways penguins defend their brood. This chapter carries the description a step further to give details of the attack forms used by the skuas at different times of the year and of the differences among skuas. Even the most cursory observation demonstrates that there are wide differences in the interest, skill and determination of skuas as penguin predators and scavengers. These differences can be catalogued through direct observation of the foraging behaviour at the colony using standardised recording techniques. Such records must discriminate between individuals, changes during the year, and different prey and predator circumstances. To ensure that this could be done successfully much time was spent in developing and testing a coded description of the behaviour that could be applied rigorously in the field.

7.2 *The study areas within the Northern Colony*

These have been described in section 2.6 and are mapped in Fig. 7.1. H Block study area was on the shore line to the south of the main

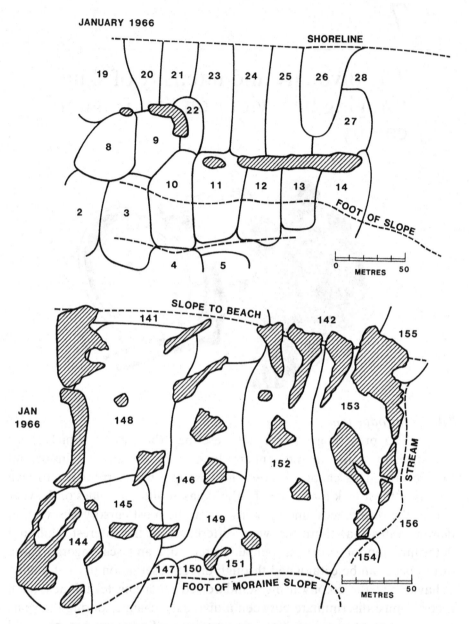

Fig. 7.1. Maps of the H and EF study areas showing the penguin breeding groups and skua territories in 1965–66. In EF area pairs 146 and 148 were replaced by a single pair, 146 new pair, in December, 1968.

penguin colony area and included four penguin breeding groups to which between 14 and 16 skua pairs had access in the different years – giving a predator to prey ratio varying from 1:32 to 1:40. EF study area lay within the main body of the penguin colony. Seven skua pairs had territories covering the majority of the penguin groups in the study area and each had between 50 and 900 nests. These two areas were selected to represent situations where skuas had access to small numbers of penguins (H study area) and moderate to large numbers (EF study area). The numbers of nests in each of the territories in these study areas were recorded in Table 2.3.

7.3   *Observing, recording and analysing skua foraging behaviour*

7.3.1   *The development of the coded ethogram used to record observations of foraging*

During the first season at Cape Bird, and at Cape Royds earlier (Young, 1963b), the behaviour of skuas was described in longhand to give a written (and often garbled and uninterpretable!) record of some of the most significant observations; the '*ad libitum* sampling' method of Martin and Bateson (1986). With experience it proved possible to order the array of behaviours observed into discrete categories and reduce the record taking to a series of numerals coding for each category. This coded ethogram was perfected during 1966–67 and used exclusively for all observations in the subsequent years.

This recording method had a number of significant advantages for this study. It ensured that records could be kept for all skuas very economically with little time lost from observation; it forced immediate decision at the time of activity of the behaviour category to be recorded, removing the bias of memory in attempting later to interpret complex written accounts; it ensured consistency in recording and interpretation from day to day and bird to bird; and it enabled more precise comparisons among the skuas than would have been possible in a free form account. Importantly, under the often cold conditions, it allowed consistent recording to be maintained over long, unbroken watches. The ruled-up but empty spaces on the record sheets 'demanded' that records were kept, irrespective of the working conditions or the inactivity of skuas. For an observer the few short episodes of high drama occurring in each log barely compensated for the tedium of the long periods of nil activity. One's recording was more at risk from going to sleep through boredom than through being exhausted from recording frenzied activity.

The categories of behaviour used in the study are shown in Table 7.1. Thirty-eight descriptors of behaviour are given in this table and at first sight this is a formidable array of categories with which to work in the field. For three reasons this seeming complexity was not a serious shortcoming in practice. First, the list is in two parts, depending on whether the attacks are on eggs and guarded chicks or on crèche chicks, so that for most of the

Table 7.1. *The categories used to describe the behaviour of skuas inter-acting with penguins at the different stages of the penguin breeding season and the numeric codes used in recording them*

 0 Skua absent from the territory and penguin colony.
 1 Skua present in the territory but without evident interest in the penguins.

A Behaviour during the penguin incubation and chick guard stages.
Ground attacks
 2 Standing or sitting near to the penguins, alert and watchful.
 3 Standing or sitting close enough to the penguins to provoke defensive and fear responses (prey items generally observable).
 4 Searching along the margin of the breeding group without penguin contact.
 5 Searching along margin more aggressively than in 4 with penguin contact, stabbing beneath drowsy sitting birds for eggs or chicks. Attacks not sustained.
 6 Sustained attack by single skua from in front of penguins on the margin of the breeding group.
 7 Sustained attack by single skua from the front and sides of penguins on the margin of the breeding group.
 8 Sustained attack by single skua running and flying about penguins on the margin of the breeding group.
 9 Sustained attack by both skuas of the pair at the front of penguins on the margin of the breeding group.
10 Sustained attack by the two birds of the pair from the front and sides of penguins on the margin of the breeding group.
11 Sustained attack by the two birds of the pair from in front and behind penguins, the one behind distracting the penguin by tail pulling and stabbing at the nest.
12 Opportunistic and unanticipated attacks by skuas on the ground, or suddenly arriving from flight, attempting to pull or twist penguins off the nest by tail pulling.

Territorial, search and attack behaviour of flying skuas
13 Territorial defence flights.
14 Territorial defence flights with searching on flight back to roost or nest.
15 Searching flight over the penguins; generally a steady wheeling flight about the territory (but also out of the territory), provoking little obvious response from penguins.
16 Aggressive flight over the penguins, often concentrating on places where eggs or chicks are free from parents or where adult penguins are fighting. These flights at little more than penguin head-height provoke an intense penguin response. Flights both intimidate and disturb nesting penguins.
17 Stooping into a deserted nest with eggs, or dead or moribund chicks or at eggs or chicks lying free out of nests.
18 Return to the territory with eggs or young chicks following flight over the colony away from the territory. (This category is to record those attacks not seen to occur by skuas out of sight elsewhere on the colony. The attack would have been one of catetories 17, 19 or 20.)

Table 7.1. (*cont.*)

19 Stooping from high flight into a defended nest.

20 Stooping from low hard flight into a defended nest on the margin of the breeding group.

21 Crash-flight into a defended nest on the margin of the breeding group in an attempt to knock out eggs or chicks from beneath the sitting penguin. Adult penguin knocked over or severely buffeted.

Foraging behaviour of skuas suddenly running or flying into the breeding group from a resting or alert position near the margin. These attacks are distinguished from ground attacks because the skuas appear to be flying along the ground at the penguins.

22 Running flight at deserted nest or unguarded egg or chick near margin.

23 Running flight into defended nest near margin.

24 Jump-flight attempting to knock adult penguin away from the nest in order to expose eggs or chicks.

25 Running flight into the middle of the breeding group.

Scavenging

26 Feeding on crustaceans or fish spilled during penguin chick feeding.

27 Aggressive flying (16) about feeding penguins causing food to be spilt.

B Behaviour during the unguarded chick stage

26 Feeding on spilt crustaceans or fish.

27 Aggressive flights about feeding penguins.

Attacks on chicks *outside* the breeding group or crèche area

28 Tentative attack from flight deflected by adults before making contact.

29 Chick hit from flight but adult penguins preventing further attack.

30 Sustained attack from ground and flight without adult interference.

31 Sustained attack from ground and flight despite adult penguin defence.

Attacks on chicks *within* the breeding group or crèche area

32 Tentative flight attack at chicks standing on the crèche margin, without contact.

33 Strong attack from flight knocking chick away from margin.

34 Tentative attack by skua running into colony at prone chick, without contact.

35 Strong attack on resting chick.

36 Attack from ground on alert chick on the crèche margin.

37 Attack from ground on alert chick within the crèche.

38 Sustained working of chicks in the crèche to isolate those small enough to be easily subdued and killed.

Table 7.2. *The categories of skua interaction with penguins grouped in relation to stage in the penguin breeding cycle and type of activity*

	Behaviour during the incubation and chick guard stages (Attacks on eggs or young chicks on nests)			Behaviour during the post-guard stage (Attacks on unguarded (crèche) chicks)	
	Ground attacks	Flight attacks	Flight/jump attacks	Chicks outside crèche	Chicks in crèche
Searching behaviour					
	2	14	2	2, 14	2, 14
	3	15	3	3, 15	3,15
	4		4	4	4
Scavenging					
	4	17	22	26	26
Disturbance behaviour					
	3	16	3	3, 27	3, 27
	4		4		38
Opportunistic attacks					
	5	18	22	22	32
	12	19	23		33
		20	24	28	34
		21	25	29	35
Sustained attacks					
(one skua) 6, 7, 8				30, 31	36, 37
(two skuas) 9, 10, 11					

season only one part or the other is in use. During incubation the list reduces to 26 categories; during the post-guard stage to just 16. Second, a fair number of categories are of rare events, for example categories 12, 21 and 24. Third, for long periods the skuas were inactive or carrying out low intensity scavenging or searching among the penguins when their behaviour was monotonously repetitive. There were few occasions over these observation periods when the skuas were so active that it was difficult to keep them all in sight and record what was happening. On the contrary, there were many long hours when virtually nothing happened at all.

The array of skua interactions shown in this table can be grouped into parallel series of equivalent activities based on the form of the interaction (searching, disturbance behaviour, scavenging, predation) and prey life-cycle stage. Most are confined to one or the other stage of the life-cycle (incubation or chick guarding/post-guard stage) but some, particularly the searching categories, apply throughout the breeding season. Within each group of similar categories there is a gradation of attack force or intensity, and with it, likelihood of success in obtaining prey – and risk to the skua.

These comparisons are shown in Table 7.2. Any attempts to compare the aggressiveness of individual skuas or their success in relation to the risk taken in gaining prey from the colony needs to be based on this comparative table.

7.3.2   *The observation logs and the way the recording was carried out*

Each year it was planned that the logs of foraging behaviour would be carried out at about five-day intervals on each of the two sites and that each would be of about six to eight hours. It was hoped that logs for the two sites would be carried out on succeeding days so that they could be compared through sharing the same weather conditions and through taking place at the same stage of the penguin and skua breeding cycles. These logs were, however, only one part of the study at Cape Bird and often other work intervened, or, more seriously, bad weather interrupted the pattern. Because these logs were to be used to compare the patterns of behaviour throughout the season and from year to year it was most important that as far as possible they were carried out during 'average' weather conditions, especially of wind strength. Skua behaviour is affected by extreme conditions. During hot, still days the birds are lethargic. In high winds they have difficulty controlling precise movement so that walking and short flights about the territory are greatly reduced. Because they are buffeted by high winds, especially when the wings are extended, ground attacks on penguins under these conditions are also markedly reduced.

Special logs of the birds during harsh conditions were of course carried out. It is just these conditions that provoke some of the more interesting behaviour, not only in penguin attacks but also in other activity, such as cannibalism, for example, and this is particularly likely when skuas are prevented from foraging at sea by fog or low cloud. The impact of weather on skua and penguin life will be considered in detail in Chapter 9.

These logs used the rather precisely defined categories of behaviour based both on the activity of the skua and on the apparent goal of the attack. For example, different categories are listed for an attack on a penguin nest containing abandoned eggs or a dead chick – or on an egg free on the margin of the penguin group – (scavenging) than for an attack on a defended nest with live eggs or chicks (predation). Such distinctions were possible in this study because overall the amount of skua activity was generally rather low, allowing close following of those few individual skuas working at any time, and because the siting of the observation points overlooking the penguins allowed the selected prey in the breeding groups to be recorded very easily. On most occasions the skuas themselves indicated by their focus of interest, often for very long periods, where these food items were located. For those readers used to more mobile predators and prey living in vegetated areas this precision may be surprising but it must be remembered first, that there was almost perfect, unimpeded, visibility here of a predator working against a static prey set out in a fixed geometric grid and second, that the levels of activity were both low and spasmodic. It was unusual for more than a single pair of skuas in a study group to be active at any time. Most birds spent most time simply sitting around.

These behaviour logs were importantly augmented by a large number of special watches: of the behaviour of skuas and penguins during high winds, or exceptionally warm, still conditions; of the skuas' ability to force penguin chicks to crèche; of the necessity to forage at sea during different stages of the year; of the behaviour of selected individual skuas, for example. Much of the information recorded during these watches was in the form of notes and tape recorded comment and could not be incorporated directly into the coded behaviour logs. It was, however, of great value in setting the coded logs into context and for providing reassurance that the behaviour being recorded in them was in fact characteristic of the season and time of year.

7.3.3   *Sampling and recording behaviour*

The activities of the birds in each group were followed using a modified 'behaviour sampling' rule and recorded by 'time sampling' (Martin and Bateson, 1986) using a 2.5 minute interval. For each bird this gave 24 behaviour records each hour, 48 for each pair per hour. For scavenging and predatory behaviours only the most active (most aggressive) category noted during the interval was recorded. For example, a bird searching along the margin of a penguin group (category 4) and then attempting to pull a penguin from its nest by the tail (category 12) would have its behaviour recorded as 12 – even if this event occupied only a fraction of the interval. Direct attacks on penguins, in which the skuas were both exposed to risk and likely to achieve prey, always overrode searching or low-key contacts. The only difficulty found in following this logical rule was where skuas searched both by walking along the group margin and by flight. Where these behaviours occurred during the same interval both were recorded but only the flight behaviour, the more energetically demanding, was used for the calculations of activity. This recording rule was needed to allow a full record to be kept of the amount of searching carried out by each bird and the extent that the penguins remained under surveillance.

In addition to the records obtained in this way four other categories of behaviour were also monitored and noted. These were the occupation of the territory; the times skuas left and returned to the territory (for bathing and for sea foraging); the times mates and chicks were fed and the food given; and territorial and breeding behaviour, especially of incubation and brooding changes at the nest. These secondary records were valuable aids in interpreting the foraging behaviour recorded in the logs and provided important ancillary information for the study. For example, a single observation of fish being fed to the chicks was an unambiguous statement that the skuas had not been able to obtain food from the penguins.

7.3.4 *The analysis of the coded behaviour data*

The major advantage of recording the behaviour strictly by categories is that it is accessible to precise comparative analysis. The decisions on the form of the behaviour are made in the field at the time the behaviour is taking place and there is no need for re-interpretation before analysis can begin. Nor is there uncertainty about the duration of the various components: each is given for the 2.5 minute interval and the whole record for an individual skua or for the group as a whole shows the proportion of intervals in which each of the scavenging or predatory activities occurred; remembering that it was the most aggressive of the activities in that interval that was in fact recorded.

7.3.5 *Taking the skua behaviour at different stages of the penguin breeding cycle into account in the analyses*

The stage of the penguins' breeding cycle has two aspects for this analysis: the form of the prey (eggs, young chicks, guarded older chicks and free-standing mature chicks, each of which provides a different challenge, and reward, for the predator); and the numbers of adult penguins that are on the breeding areas. In the present account the penguin cycle has been broken into six stages, namely, 1. pre-egg, 2. egg-laying and incubation, 3. chick hatching and chick guard, 4. early post-guard, 5. late post-guard, and 6. leaving the breeding area. In practice food was available to the skuas only in the middle four stages, none was present during the first (pre-egg) and last (leaving the area) at Cape Bird. The lengths of each of these stages and the delimiting dates at Cape Bird are shown in Fig. 2.10. The limits of each of the first three stages are rather precisely marked (the first egg, the first chick hatching, the first unguarded chicks, respectively) but those of the the last three are arbitrary. There is a marked transition in the appearance of the penguin breeding areas between early and late post-guard stages. The early post-guard stage contains both unguarded and guarded chicks but late post-guard groups are composed almost exclusively of unguarded chicks most of which are too large to be taken by skua attacks on the breeding area itself. This maximum size for general skua attack is set at 2000 g, (at about 24 days old) and at Cape Bird this was the mean chick weight at about 10 January (see Spurr, 1975c, Table 5 for hatching dates). The last stage covers the period when the chicks are leaving the natal areas to move down to the shore. Where breeding groups are crowded together in the colony the maturing chicks begin to move among them at a quite early date and it is difficult to decide when chicks are in fact moving away from them towards the shore. In isolated groups, however, this stage is easily recorded and can be quite precisely dated. At Cape Bird it began on about 25 January. From this date fewer and fewer penguin chicks were available to skuas with territories on the inland margin of the colony.

7.4   *The proportion of time spent by the skuas in interactions with the penguins*

If the numerous examples of behaviour recorded on each occasion are summed within the four categories of 0. absent from the territory, 1. present on the territory but showing no interest in the penguins, 2. interest in the penguins manifested by watchfulness, and 3. searching, scavenging and attack behaviours, then a broad picture of the behaviour of the skuas can be obtained for each year. As shown in Fig. 7.2, the average amount of time skuas devote to each of these broad categories of behaviour is remarkably similar both within and between years. On average, irrespective of whether the skuas were on H or EF blocks, about 15% of the time was spent away from the territory, about 75% on the territory with little interest in penguins and in only about 10% of the time were they interacting with penguins. This interaction comprised watching and searching as well as active scavenging and predation so that the time spent by the skuas in attacks on penguins was, therefore, a very small proportion of the day. It is this small proportion nevertheless, that provides much of

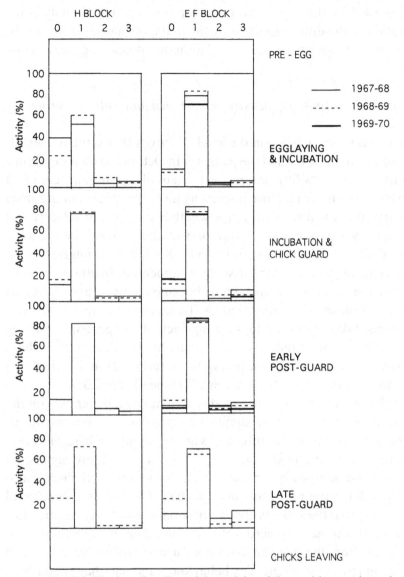

Fig. 7.2. The broad outline of the activity of skuas with penguin territories on H and EF study areas during each stage of the penguin life-cycle. Three years, 1967–68 to 1969–70. 0. absent from the territory; 1. present on the territory without showing interest in the penguins; 2. interest in penguins (watching and searching categories); 3. active scavenging or predatory interaction with penguins.

the food needed by these pairs and which requires detailed analysis for interpretation of the differences among the pairs, of changes from season to season and of success rates in relation to attack and scavenging methods.

7.4.1 *Changes in foraging activity on the penguin colony during the season*

There is wide variation in the level of activity across the full length of the season from the arrival of the penguins in October to their departure in February. Skuas show little interest in them until eggs become common in mid to late November and lose interest again as the chicks mature from late January. What is not so certain is whether the level of interest and activity varies across the shorter duration of the season when food is common. There seemed, for example, to be a heightened interest in the penguins by some skuas once the first chicks appeared. In others, interest seemed to increase once feeding chases became common and the skuas could pick up krill spilled during feeding. To examine this question mean activity levels, taken from the logs during each of the penguin breeding stages, have been compared. Two levels have been calculated for these tests: firstly, the total level, combining the interest (2) and activity (3) figures, and secondly, the activity levels (3) alone (Table 7.3).

The results of these analyses are readily summarised: depending on the year there may or may not be significant differences in interest levels during these stages but where differences are apparent it is in significantly higher rates during the post-guard stage [1]. Nor are there significant differences if the analysis is confined to the incubation and chick guard stages, which have the best sequences of data. Overall, it is concluded from these data that these skuas appeared to maintain a fairly constant low rate of interest and activity across the full breeding cycle. Similar conclusions follow from the same analyses done using activity levels alone, an analysis designed to eliminate the behaviour of skuas that spend long periods watching the penguins but do not actively attack them or scavenge among them. This analysis showed that the skuas were more interested in the penguins during the chick guard stage in 1967–68 but it did not discriminate significantly between the stages in the other years.

7.4.2 *Differences among seasons in skua interest and activity*

There are data for three seasons for the skuas on EF Block and two seasons for those on H Block. In the third year on H Block attention focussed more specifically on the behaviour of some individual birds there

Table 7.3. *Comparison of skua interaction rates with penguins at different penguin life-cycle stages on EF and H Blocks in different years*

Data are mean numbers of 2.5 min. intervals pair^{-1} h^{-1}. For 7 pairs on EF Block and 13 pairs on H Block. Categories used here are the combined interest (2) and activity (3) categories, and the activity category (3) alone; see section 7.4 and Fig. 7.2.

	1967–68		1968–69		1969–70	
	2 + 3	3	2 + 3	3	2 + 3	3
EF Block						
Incubation	2.38	1.80	2.49	1.71	2.81	2.16
Chick guard	4.83	4.24	3.37	1.79	2.28	1.57
Early post-guard	5.55	4.25	3.89	2.39	4.53	1.52
Late post-guard	10.09	6.92	3.00	1.91	—	—
H Block						
Incubation	1.12	0.59	4.75	0.78	—	—
Chick guard	2.68	1.54	1.72	0.79	—	—
Early post-guard	1.95	1.16	—	—	—	—

so that the data for this year are not comparable. Is the behaviour of these skuas similar in the different seasons? The best data sets for this test are again those of the incubation and chick guarding stages, avoiding possible bias from exaggerated differences in logs made variably early and late in the season. (The high levels recorded during the late post-guard stage in the logs on EF during 1967–68 are significantly higher than in other years.) For the bulk of the season, however, whether compared for total interest and activity or activity alone no differences between years could be established [2]. From this analysis it is concluded that the small changes in penguin numbers available to the skuas in each year and the few changes in skua identity within the pairs did not impact significantly on the rather uniform levels of activity that continued from year to year.

7.4.3 *Comparisons between the two study areas*

This examination of the broad categories of skua interaction with the penguins has established so far that activity and interest of skua pairs was rather constant through the season (measured for each of the stages in the penguin cycle when eggs and chicks were present) and that the levels of interest were on average also fairly constant from year to year. An important point of the design of the field work was in the comparison between H Block with its high predator–prey ratio and EF Block with its much lower ratio. Are these differences in ratios reflected in differences in skua activity? Although the differences [3] are not significant for the total interest and activity categories they are significant for activity when tested on its own. That is, the results appear to show that although the skua pairs in the two sites showed the same interest in the penguins those on EF Block carried this interest further into the more active interaction of scavenging and predation. This divergence might be explained by the greater freedom from penguin interference skuas on H Block had when sitting near the penguin breeding groups, so that these pairs could spend more time there. These groups were located on an open beach; on EF the breeding groups were more tightly packed together and there were many more penguins moving among them. In short, skuas could rest more easily close to penguins on H Block than on EF Block. Skuas sitting close to the penguins like this would be recorded in the logs as 'showing interest', and accumulating sequences of 2s on the record sheets. The small difference in foraging rates between the two areas has implications for penguin survival: similar rates of skua interest are focussed on fewer pairs of penguins on H Block.

7.4.4   *Were the pairs equally interested in the penguins?*

The analyses so far have considered only the average levels of interest and activity taken across all of the skuas of each of the study groups. Of major interest is whether the individual pairs had consistently different foraging levels. This study is a prerequisite to assessing how differences might be related, for example, to such factors as numbers of penguins within the territory or the geography of the territory in relation to penguin breeding groups. Alternatively, differences might merely reflect differences among the individual skuas, through variation in aggressiveness and experience, or the different food demands of the family.

Wide differences in aggressiveness and interest were in fact immediately apparent as soon as systematic observations of the skuas began at this colony. Some pairs were never seen foraging among the penguins within their territory whereas neighbouring pairs might be completely preoccupied with gaining food there. Pair 107 for example, had one of the largest territories on the colony, containing several thousand penguin nests, but was never seen to do anything more than timid scavenging. By way of contrast, pairs 150 and 151 with no more than 10 nests apiece worked their penguins assiduously. This being the case the value of the statistical

Table 7.4. *Differences among the pairs of skuas on EF Block in their levels of interaction with the penguins*

Rates are numbers of 2.5 min. intervals pair^{-1} hour^{-1}. Mean and SE.

Pair	1967–68	1968–69	1969–70
144	10.13 ± 1.96	1.85 ± 0.35	2.93 ± 0.94
145	6.13 ± 2.77	3.36 ± 0.94	3.27 ± 1.45
146	5.57 ± 0.87	—	—
148	7.11 ± 1.38	—	—
146 new pair	—	3.76 ± 0.96	4.80 ± 1.46
149	1.55 ± 0.46	1.18 ± 0.81	0.91 ± 0.34
152	5.17 ± 1.33	3.41 ± 0.86	4.39 ± 1.05
153	3.37 ± 0.68	5.39 ± 1.00	3.79 ± 1.27

analysis is to systematise this variability and examine whether differences were maintained consistently throughout the season, and from season to season.

These differences have been examined in two stages. The first considers differences among the pairs for their combined interest and activity rates (categories 2 and 3 above). The second considers differences in the proportions of time spent by the pairs in various interactions with the penguins. This latter analysis is included in section 7.6.

The largest data sets are for the EF Block skuas (Table 7.4) which show that there were significant differences among the pairs in the first two years but not in the third [4]. The table also shows that there were higher involvement rates in the first year, especially by pair 145. Pairs 146 and 148, which were also especially active in this year, were replaced for the second two years by 146 new pair, which was less active. The other pairs maintained a rather uniform activity throughout the three years. This uniformity is especially pertinent in pairs 152 and 153, which had the largest territories and gained much of their food from the penguins.

There is less information on the H Block skuas, with data available only for the two years 1967–68 and 1968–69. Analysis shows nevertheless that there were wide differences among them, with pairs 9, 13, 25 and 27 significantly more active and pairs 19, 20, 21 and 22 showing nil or low interest in the penguins in both years. Overall, there were significant differences among pairs but not between the years [4].

The different interest and skills of the skuas as predators has great implication for the penguins within the territories. Penguins in a territory of a timid pair will be much less harried than those in that of an aggressive one.

7.4.5   *Were the two birds of the pair equally interested in the penguins?*

Up to now the discussion has been concerned with *pairs* of skuas. It is important to determine at this stage whether the male and female showed the same interest in the penguins and whether this interest was translated into similar levels of attack and scavenging. During the logs themselves it *seemed* quite clear that males were both more active on the colony, more aggressive and more effective predators. This view was, however, coloured by the superb ability and high levels of activity of two males, those of pair 153 on EF Block and pair 25 on H Block. These two birds were so superior as predators to all their neighbours that it was easy to translate their individual accomplishment to that of male birds in general. However, when the full records of the behaviour of all birds in each of the areas is compared then a quite different picture emerges. Analysis of these records fails to demonstrate significant differences in the behaviour of the two sexes, whether compared for total interest and activity or for scavenging and active predation alone [5]. In only one year, 1967–68 on EF Block, are significant differences between male and female birds apparent. This difference shows up in activity levels rather than in overall interest.

7.5   *The proportion of time away from the territory*

The converse of interest in the penguins is being away from the territory altogether. Once the breeding season is well underway from late November skuas are absent from the territory for just two reasons: for bathing and loafing at the fresh water ponds on McDonald Beach or along the shore line between the Northern and Middle penguin colonies; or for foraging at sea. Additionally, they may forage elsewhere on the penguin colony and during exceptionally high winds skuas on exposed territories where there is blowing sand and grit will leave the territory and sit out the storm at sea. But these are uncommon events compared with the first two.

As the two birds of the pair are absent bathing for about the same time each day, overall differences between them, and between pairs, should reflect different feeding habits. Interpretation of any such differences is not at all straightforward, however, because of the way skuas feed during the different phases of the breeding cycle. In this skua only the male forages away from the territory during the pre-egglaying and incubation stages so that in pairs unable to gain enough food at the colony the female is provided with food throughout this period. Not until the chicks hatch does the female begin foraging at sea for herself and the chicks. In pairs failing to produce eggs the male of pairs dependent on marine food feeds the female throughout the season. In pairs losing eggs or chicks the male feeds the female for the remainder of the season (Young, 1963a and observations at Cape Bird). This difference between the sexes is not as pronounced in pairs on territories with penguins because the female may also forage there. In the absence of penguin food during pre-egglaying and incubation it is the male who forages at sea. Of course the female is perfectly capable of foraging for herself during the breeding season and single females on territories will do this, although they will also beg and be fed from males attracted to them. The complication in attempting to interpret the records for average times away from the territory is in reconciling these records with both skua feeding behaviour and different stages of the penguin breeding season. Fortuitously, the breeding cycles of the two species come together about 8 January. At this date the penguins begin to switch from guarding to leaving their chicks; for skuas this is the mean date in this area for chick hatching – and the advent of sea foraging by females. Analyses from the beginning of the skua breeding season until this date in January should, therefore, be uniformly of males feeding females. This uniformity should apply to all pairs, those successfully hatching chicks as well as those in which breeding has failed during incubation. The analyses that follow are thus confined to the penguin incubation and chick guarding stages, to test whether males were more often away and whether males on territories with few penguins were away more than males on territories with many. Intuitively, one would expect that both differences should exist.

Overall, these skuas were away from the territory about 15% of the time. At both the start and end of the season there is little or no food on the colony and all birds must feed at sea – at Cape Bird even those with the largest numbers of penguins within the territory. Consistent occupation of skua territories does not occur, however, until the end of November so that prior to this date there can be very long absences that are not concerned with either bathing or foraging. Because of this uncertainty analysis was of data from later in the season.

The first point to establish is whether there were differences in the time spent away by each of the birds in a pair. It is expected from observations on the foraging behaviour that the males will be away more often, but this effect could be offset if they were also more intensely territorial and were away for shorter times for bathing or loafing at the communal areas. In the

Table 7.5. *Mean times away from the territory by males and females*

Numbers of 2.5 minute recording intervals each hour. Data are means and 95% confidence limits.

	Males	Females	Number of pairs
H Block			
1967–68	7.82 ± 2.07	3.96 ± 1.17	13
1968–69	6.98 ± 1.76	3.50 ± 2.33	13
EF Block			
1967–68	3.38 ± 2.62	1.46 ± 1.02	7
1968–69	4.52 ± 4.32	1.98 ± 1.03	6
1969–70	4.17 ± 3.46	2.16 ± 1.98	6

event the results are unambiguous. Despite considerable variability the males of the 13 pairs on H Block and the six or seven pairs tested on EF Block in the three years were on average away for about twice as long as females in all years; for 18 minutes each hour compared with 9 minutes on H Block and 10 minutes compared with 4.5 minutes on EF Block (Table 7.5 [6]).

Of equal significance for the study of the feeding behaviour are the statistics on the proportion of time males from different pairs – pairs with few penguins compared with pairs with many penguins – were away feeding, as these figures might well indicate the degree of food independence derived from the penguins within the territory. As noted above the time away for males is made up of time foraging and time bathing and the two cannot be separated within these records. It is recognised, however, that most of the time away will be for foraging so that the time away should be inversely related to the amount of food gained from the penguins. EF Block has more records of behaviour and a greater range of penguin number within the territories than H Block so that it is not surprising that the results of the analyses of these records more closely support the prediction that males of pairs with least penguins were away longer than males with large numbers of penguins. In each year there were significant differences among the pairs, with males 152 and 153 being away least but with different males of the other pairs in each year being away significantly longer on average than the others – in 1967–68 the male of 145, in 1968–69 the male of 149, in 1969–70 the male of 146 [7]. It was not possible to establish significant differences among the males on H Block in either year tested.

7.6 *The behaviour of skuas interacting with penguins – the behaviour of predation and scavenging*

The earlier sections have given a broad outline of the way these skuas spend their time on and away from the territory and in relation to the penguins. Surprisingly, for most skuas the penguins are not their predominant interest and much of the day is spent at the nest area. The charts in Fig. 7.2 indicated that only about 5–10% of their time on average in the different stages of the breeding season is spent in scavenging and predation. This section describes how this small portion of the day is utilised.

As in the earlier section the logs of behaviour recorded at intervals across the three breeding seasons between 1967 and 1970 provide the majority of the data on the behaviour of these birds. These logs were deliberately split across short intervals during the season between EF Block with its high prey–predator ratio and H Block with a much lower ratio.

As will now be familiar to readers the amount of time skuas spend in direct attacks on the penguins, either on adults in order to obtain eggs or young chicks or on the maturing chicks crèched without parents, is so small that it is difficult to make statistical comparisons among the pairs or during the different parts of the season, for example, using the categories listed in Table 7.1. The categories of the behavioural repertoire of skuas interacting with penguins are, however, very readily grouped into five classes of attack behaviour and these can be used to provide a realistic description of the way skuas react with penguins and still allow differences among individual birds to be assessed.

The classes used in this section are:

1. Searching behaviour (categories 2, 3, 4 (ground) and 14, 15 (flight)).
2. Disturbance and/or intimidating behaviour (categories 3 (ground) and 16, 27, 28 (flight)).
3. Scavenging behaviour (categories 4 (ground) and 17, 22, 26, 27 (flight)).
4. Opportunistic attacks (categories 5, 12 (ground), 18–21 (flight), 22–25 (running and jump-flights) and 28–35 (attacks on unguarded chicks)).
5. Sustained attacks (categories 6–11 (ground), 30, 31 and 36–38 (ground and flight attacks on unguarded chicks)).

Skuas assigned behaviour category 4 (i.e. those working about the margin of breeding groups of penguins) were both searching for prey to attack (class 1) and scavenging (class 3). These records are listed in both these classes. With this single exception categories are assigned exclusively.

As described earlier the records produced were of the behaviour carried out by each skua during each 2.5 minute interval of the log, giving 24 individual records each hour. For comparative purposes this record has been converted to percentage frequency of activities or to rates per hour. These conversions accurately represent low-key activities that tended to carry on for long periods – searching flights, scavenging, etc. – but over emphasise many of the isolated attack events, events that occur very suddenly and are over in a few seconds. This bias is defended on the grounds that it is the violent attacks that achieve most of the food taken by the predator. Their recording in this way ensures that the obvious differences apparent from watching skuas as predators are highlighted.

7.6.1   *The frequency of the different attack forms*

The overwhelming predominance of watching and searching behaviours is clearly evident in the average levels of activity shown in Fig. 7.3 for each group of skuas, from records made during the penguin incubation, chick guarding, and early post-guard stages. Even so, these are quite low levels when measured against the time available – with each pair averaging only two to three intervals of activity in each hour. All other activities are nevertheless of even lower incidence – scavenging being recorded for between one half and one interval per hour; and disturbance flights and attacks each averaging less than half an interval each hour.

There is a similar pattern in the order of frequency of the different behaviours in each year and in both study groups. Although the skuas on EF Block appear in these figures to be more involved in attack behaviours than those on H Block this overall difference was only substantiated in the first year [8]. Much of the difference in this year occurs in the very much higher levels of searching and disturbance behaviour in the post-guard stage of the EF skuas, but even so opportunistic and sustained attack levels were also significantly different [9].

Figure 7.4 records for the skuas on EF Block how the different categories of activity vary during the penguin life-cycle stages. As expected, searching behaviour predominates in all stages each year, varying at different times from 45% to over 80% of all interactions. Disturbance behaviour is consistently the next most important interaction but also varies markedly among the stages. It is, however, the single category showing trends through the different seasons, although these are

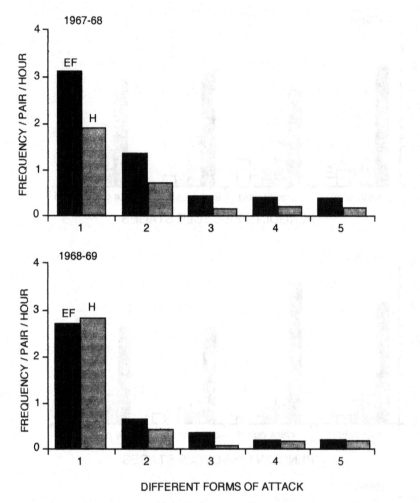

Fig. 7.3. Frequencies of different forms of interaction with the penguins by the skuas on H and EF study areas in two years during the incubation, chick guard and early post-guard stages. 1. searching and watching behaviour; 2. disturbance behaviour; 3. scavenging; 4. opportunistic attacks; 5. sustained attacks. (Categories of behaviour defined in the text.)

not consistent from year to year. This behaviour rises in importance through 1967–68, remains constant in 1968–69 and falls away during the season in 1969–70. The other categories are such minor components of skua activity that the variability indicated is only to be expected. Moderate changes in levels of these activities at any time will impact strongly on their overall significance.

The data on the combined levels of activity by the skuas have been summarised and compared for each of the three years in Fig. 7.5. They

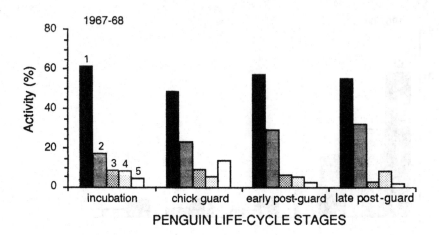

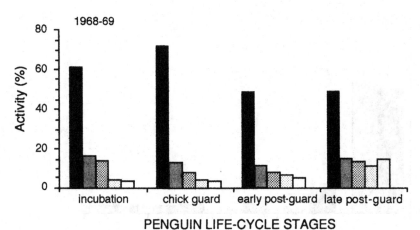

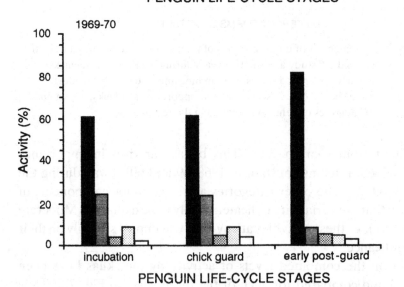

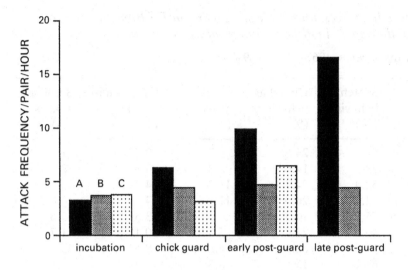

Fig. 7.5. Comparison of the total levels of predatory and scavenging activity of skuas interacting with penguins during each stage of the penguin breeding cycle stage for the three years 1967–68 (A), 1968–69 (B) and 1969–70 (C). Data are the means for each stage of the total interactions recorded during the logs of that stage as rates pair^{-1} hour^{-1}.

have been statistically analysed for the two seasons with complete records. In the first year (1967–68) the overall level of activity (the sum of all interactions with the penguins) increased significantly through the season, in the second (1968–69) it remained rather constant [10].

7.6.2 *Individual differences among the skuas*

The data considered so far have been of the average levels for each of the study groups. Although producing similar outlines of behavioural interaction with the penguins these could well mask differences among the different pairs. This section takes up the analysis again, but this time of individual differences.

The skuas on EF Block again provide the best prospect for defining individual differences. For these skuas most of their time interacting with penguins was taken up with searching (40% to 75%) with disturbance

Fig. 7.4. Relative importance of different activities of skuas foraging on the penguin colonies during the different stages of the penguin breeding cycle. For the three years 1967–68 to 1969–70. 1. search and alert watching; 2. disturbance behaviour; 3. scavenging; 4. opportunistic attacks; 5. sustained attacks on the penguins.

Table 7.6. *Differences among the skua pairs on EF Block in the proportion of time spent in different interactions with penguins*

Totals for three years, 1967–68 to 1969–70.

Pair	Searching behaviour (%)	Disturbance behaviour (%)	Scavenging (%)	Opportunistic attacks (%)	Sustained attacks (%)
144	64.2	24.6	5.7	3.7	1.7
145	72.5	14.6	5.8	4.8	2.2
146	52.4	24.7	11.3	4.2	7.4
148	42.7	26.0	11.5	4.8	14.8
146 new pair	75.7	10.4	4.6	3.5	5.6
149	40.4	21.9	5.8	20.1	11.7
152	60.3	22.6	4.8	9.5	2.8
153	64.2	15.5	8.3	10.7	1.4

behaviour next most important (10% to 26%) and with scavenging (4.6% to 11.5%), opportunistic attacks (3.5% to 20%) and sustained attacks (1.4% to 15%) taking up a similar amount of time. In the first year only two pairs stood out as being different: pair 148 with 15% of their time spent in sustained attacks and pair 153 with 17% of their time spent in opportunistic attacks – each markedly higher levels than for the respective activities in the other pairs. These idiosyncrasies were not evident in either pair in the following years. The single deviant in these years was pair 149 with very high levels of both opportunistic and sustained attack, and consequently reduced levels of the other behaviours. The records for the three years (Table 7.6) portray little variability overall, with only pair 149 appearing much different in the way it foraged, but this difference may be an artefact of its small number of records. The core pairs in this area (144, 145, 152 and 153) have remarkably similar profiles. It is not possible to test these differences statistically because of lack of independence of the records of searching behaviour.

There were too few records of H Block skuas to warrant similar analysis. The behaviour of the most successful and aggressive skua there is described in section 7.7.

7.6.3   *Overall differences in searching, scavenging and predation*

Some of the behaviours are both sufficiently common and sufficiently well defined to allow comparisons within and between seasons, between the two groups of skuas, and of individual skuas. Searching

Table 7.7. *The mean times skuas on EF and H Blocks spent searching for prey from the ground and from flight during the different stages of the penguin life-cycle*

Data are mean numbers of 2.5 min. intervals pair^{-1} hour^{-1} during each life-cycle stage. Ground searching is given first in each record. The ratio of the two forms of searching is shown on the following line.

	Incubation	Chick guard	Early post-guard	Late post-guard
EF Block				
1967–68	1.28/0.67	2.47/0.95	4.49/1.12	8.4/0.85
Ratio	1.9:1	2.6:1	4.0:1	9.9:1
1968–69	1.43/0.56	2.28/0.84	2.09/1.25	1.27/0.47
Ratio	2.5:1	2.7:1	1.7:1	2.7:1
1969–70	1.82/0.52	1.37/0.48	3.73/0.62	—
Ratio	3.5:1	2.9:1	5.9:1	—
H Block				
1967/68	0.49/0.17	2.70/0.16	2.87/0.04	—
Ratio	2.9:1	16.9:1	71.8:1	—
1968–69	4.17/0.09	1.50/0.11	—	0.68/0
Ratio	46.3:1	13.6:1	—	

behaviour and scavenging and predation are described in the following sections.

Searching

Skuas search for prey and food to be scavenged both on the ground and from flight. There are several questions. Are there differences through the season as the prey changes from eggs to mature chicks? It would seem reasonable to expect this because eggs can be taken from flight but older chicks can only be taken from the ground at the edges of the breeding groups, being too heavy to lift in flight. It might be expected, therefore, that skuas would spend less time searching the centres of the penguin groups from flight as the season progresses. Moreover, those skuas with the smallest territories and the most compacted penguin groups could also be predicted to have lower flight rates. The data from the logs confirms the second of these predictions but not the first. EF skuas maintained fairly constant flight rates throughout the season in each year, within the narrow range of 0.47 to 1.25 intervals each hour on average for each pair (Table 7.7). These rates were, however, much higher than those of the H Block skuas on their smaller territories, which range from 0.04 to

0.47 intervals each hour. In neither group was there a consistent pattern of less searching from flight during the post-guard stages. Although the ratio of ground searching to flight searching increased through the season in 1967–68 in both EF and H skuas, this arose from increasing ground searching not from a decline in flight times. In each year and in all stages of the year more time was spent in ground searching, ranging from 1.9 to 9.9 times the periods spent flying in EF but from three to 80 times for skuas on H Block.

Scavenging and predation

Distinguishing scavenging from predation was one of the important criteria in the design of this study. Indeed, it was uncertainty about the significance of the feeding methods that stimulated shifting the research from Cape Royds, with its few penguins, to the much larger Cape Bird colony where it was thought that skuas might obtain a much higher proportion of their food by scavenging alone. Later in this account attempts are made to distinguish between these food resources in the sections on food taken and in the analysis of the experiments of protected and unprotected penguin breeding groups. At this point it is useful to attempt to separate out the time given to the two feeding methods from the behavioural records. This is attempted in three stages. First, what proportion of the time on the colony is devoted to each of these feeding methods? Second, is there relatively more scavenging in territories containing larger numbers of penguins than in those with small numbers? Third, do individual birds vary significantly in their predilection for scavenging or predation? The comparisons made here depend on precise separation of scavenging and predation in the records. The data used are the scavenging categories 4, 17, 22, 26, 27 (searching along the penguin group margin, retrieving abandoned eggs and chicks, and feeding on food spilled in penguin chick feeding) and the predation categories describing attacks on defended nests or alert and defended post-guard chicks. Disturbance and aggressive flights are excluded from this analysis as they occur during both scavenging and predation attacks. For the purposes of this comparison all records have been standardised as rates per pair each hour.

1. What proportion of the time is spent in scavenging and predation? Overall the rates are very low (Table 7.8). For EF skuas these average from 0.14 to 0.63 intervals each hour for scavenging pairs and 0.16 to 1.55 intervals for predation. For H skuas the comparable rates range from 0.01 to 0.19 for scavenging and 0.09 to 0.50 for predation. The ratios of the two

Table 7.8. *Comparison of the relative amounts of scavenging and preda-tion during the different penguin life-cycle stages for skuas on EF and H Blocks*

Rates shown are mean numbers of 2.5 min. intervals $pair^{-1}$ $hour^{-1}$. Scavenging records are shown first. The ratio of scavenging to predation is shown on the following line.

	Incubation	Chick guard	Early post-guard	Late post-guard
EF Block				
1967–68	0.28/0.35	0.63/1.00	0.31/0.63	0.47/1.55
Ratio	0.8:1	0.6:1	0.5:1	0.3:1
1968–69	0.45/0.16	0.33/0.20	0.33/0.31	0.46/0.47
Ratio	2.8:1	1.6:1	1.1:1	0.9:1
1969–70	0.14/0.37	0.14/0.26	0.27/0.24	—
Ratio	0.4:1	0.5:1	1.1:1	—
H Block				
1967–68	0.08/0.23	0.19/0.18	0.07/0.50	—
Ratio	0.3:1	1.1:1	0.1:1	—
1968–69	0.01/0.09	0.11/0.19	—	—
Ratio	0.1:1	0.6:1	—	—

foraging methods do not appear to vary consistently through the season and, with the exception of EF skuas in 1968–69, predatory behaviour is generally more important at all times. Specifically, neither when eggs are being abandoned by penguins (late incubation and early chick rearing) nor at the end of the season when most penguin chicks are too big for the skuas to attack, is scavenging significantly enhanced.

2. Is the expectation that skuas on large penguin colonies would be able to obtain relatively more of their food by scavenging borne out by the records of their behaviour on H Block (with few penguins in each territory) and EF Block (with many penguins)? The prediction is not supported by the data from the behavioural logs. Although the amount of time H skuas spent foraging is significantly less than for EF skuas they also spent equivalently less time with the penguins in all activities and the ratio of scavenging to predation was similar in the two groups [11]. This aspect of foraging behaviour will be considered again in the section on foods taken by these skuas.

3. Do different skuas consistently favour scavenging or predatory behaviour?

This aspect of the interaction between the skuas and penguins has been examined by comparing the levels of each attack form by the EF Block skuas over three seasons [12]. None of these pairs could be characterised from this analysis as a scavenger or predator consistently across these three years. Three pairs, however, favoured one or the other method in two of the three seasons: pairs 144 and 153 favouring scavenging and pair 152 predation. Pair 153 were the most successful of these skuas, and the male of this pair the most skilled predator in the local area, yet in two of the three seasons more time was spent scavenging.

In each case where pairs favoured one or the other foraging method this resulted from the preference of one bird, but this was not related to sex. In the first year the males of 144 and 146 preferred scavenging whereas males in 152 and 153 preferred predation. In the same year the females of pairs gave similar effort to both, except for 148 where scavenging was preferred. Nor could the individual birds be characterised from these data. None showed the same significant preferences across the three years, although the pair 153 birds showed the strongest difference, with the male preferring predation and female scavenging. For this pair the outcome of the analysis of the times spent in these activities bears out the opinion formed of them from watching them tackle prey. The male could invariably get eggs or chicks by direct attacks, and was then able to deal with them effectively. By contrast, the female was the more timid of the two; if it managed to get a chick it seemed then at a loss of how to go about killing it. In summary, from these data the two birds of pairs may show quite different preferences. Although they often collaborated in sustained attacks on nests or at carrion, these joint attacks were not sufficiently common to compensate for their individual preferences.

The form of the predation attack

The detailed records of the skuas' behaviour on the penguin colony allow definition of the general technique employed by these birds in their attacks on defended nests and unguarded chicks. Although each attack is a unique event they are nevertheless separable on the basis of the skua's overall interest in penguin food and their persistence in pressing home an attack on alert and aroused penguin adults. The records distinguish between opportunistic attacks, a sudden rush or stoop into a nest or a run at a sleeping chick in a crèche, and sustained attacks when the skuas persist in attacking aroused and strongly defending adults. This distinction occurs for both scavenging and predation. Skuas may be just as persistent in attempting to retrieve a dead chick on or near a defended nest as they are in taking a live chick. The distinction between these two attack forms is also useful later in this account when the success of predation and scavenging is measured by the food taken.

In practice there is little difficulty in distinguishing between these two forms of attack. Opportunistic attacks occupy no more than a few seconds, 10 seconds at the most, generally one to five. They rely on stealth and speed with the skua breaking off the attack immediately the penguin is alerted or retaliates. All attacks from flight into the middle of a breeding group are of this form; skuas cannot settle for more than a very few seconds within a group without being caught. Persistent, sustained attacks on the same penguin may last many minutes and continue in spite of the penguin's defensive responses. Whereas opportunistic attacks are by a single skua, sustained attacks may be by single birds or the two birds of the pair working together.

For the present analysis the attack forms have been divided as follows:

> Attacks on defended nests:
> Opportunistic attacks, categories 5, 12, 18–21, 23–25; sustained attacks, categories 6–11.
> Attacks on post-guard chicks:
> Opportunistic attacks, categories 28, 29, 32–35; sustained attacks, categories 30, 31, 36–38.

The first point to note is that these are quite uncommon events. As distinct from all the interest taken in the penguins and all the time spent searching among them, walking around the margin of the groups or flying over them, actual *attacks*, in which the skua comes into contact with the penguins and attempts to obtain eggs or chicks, occur infrequently. For the whole group of skuas on each study area one would see on average only one or two opportunistic attacks an hour and an attack on chicks every two or three hours. The actual rates per pair of birds each hour are shown in Table 7.9, together with the ratios of the two forms of attack.

With the exception of the first year on EF, which had significantly higher attack rates, the average rates are much the same for each year on both areas. But it must be reiterated that all the rates are really very low. Sustained attacks on alert crèche chicks, whether defended by adults or not, are of least importance when measured in this way, but of course this need not reflect their significance as a way of successfully obtaining food.

The relative importance of the two forms of attack can be assessed from the ratio of the times given to each (measured as number of intervals in which each category was recorded). In all years opportunistic attacks were

Table 7.9. *Mean rates of opportunistic and sustained attacks on defended nests and post-guard chicks*

Rates measured in numbers of 2.5 min. intervals pair^{-1} hour^{-1}. Ratios shown are for the activity totals.

	Attacks on nests			Attacks on chicks		
	Opportunistic	Sustained	Ratio	Opportunistic	Sustained	Ratio
EF Block						
1967–68	0.93	0.18	4.9:1	0.07	0.09	0.7:1
1968–69	0.25	0.02	10.6:1	0.22	0.03	6.3:1
1969–70	0.29	0.05	5.7:1	0.18	0.01	13.3:1
H Block						
1967–68	0.15	0.11	1.4:1	0.32	0.0	—
1968–69	0.12	0.04	3.0:1	—	—	—

the more common form. For EF skuas opportunistic attacks on nests were from three to ten times more common than sustained attacks. It might be argued that this difference arises because sustained attacks must be confined to nests on the margin while opportunistic attacks can occur both there and within the middle of the breeding group. That is, there is a much greater opportunity for opportunistic attacks and it is this factor, not the preference of the skuas, that determines the imbalance. In theory this is quite true but most records of opportunistic attacks are also of those on nests on the outside of the breeding group. Attacks into the middle of penguin groups can only be successfully executed when these have eggs or chicks small enough to be lifted in flight away from the penguins. Few skuas are ready to risk dropping into the middle of a penguin colony, certainly not one with tightly packed nests, with alerted adult penguins or one with a proportion of adults not tied to nests in incubation or chick guarding.

Although it has been established that overall these skuas carried out more opportunistic attacks than sustained attacks on the penguins – and in fact sustained attacks, the ones most often seen and remarked on by casual observers at the colony, are quite rare – are there individual preferences for the way penguins are attacked? Analysis of the totals for the individual skuas and of the individual pairs seems to show this. The six pairs of skuas certainly had individual preferences. Attacks by both male and female birds of pairs 152 and 153 were predominantly opportunistic, whereas those of 146 and 149, much less active skuas, had higher than expected sustained attack rates.

7.7 *Foraging sequences during the day and season*

The skuas did not go through their behavioural repertoire in one sequence and then stop for the day. Their various activities ran throughout the day, alternating foraging with breeding and maintenance behaviours. The pattern of the timing of the different foraging behaviours is the subject of this section. It covers two levels of organisation. The first is about the way skuas break up the duration of a foraging bout, the time between beginning to forage and giving it up altogether (point 7 in Fig. 1.3), by alternating active and resting periods. This information can be culled from the behaviour logs. It is augmented by the more detailed account of the foraging of one selected skua, the male of pair 25 on H Block. The record of this skua illustrates the high activity levels of a very successful skua predator but of equal importance illustrates also the amount of disturbance impact on the penguins within its territory. The second level of organisation is of the way the skuas' behaviour varies throughout the polar day. The records for this analysis are from a number of logs made at different times during the 24-hour day and from a record made of the behaviour of EF skuas during continuous observation through four days in January 1970.

7.7.1 *The sequences of behaviour during a foraging bout*

It is possible to recognise naturally defined foraging bouts in the records of the behaviour logs by a clear switch in activity to and from foraging.

A foraging bout begins when the skua starts showing interest in the penguins and is generally marked by movement to them from the nest area. It ends when they leave the penguins to return to the nest area or in territory defence or when they leave the territory. In skuas nesting near penguins, where the distinction between foraging and other behaviour is necessarily blurred, it was considered to have ended when there were 12 intervals (30 minutes) in which the skua showed no apparent interest in the penguins. Over three years, 137 bouts defined in this way have been taken from the logs of EF Block. In the first stages of bringing these data together the records from incubation and chick guarding stages were listed separately from those of the post-guard stage, but these were so similar that all the records could be lumped together. The shortest clearly defined foraging bout is just one interval; the time, for example, for a skua to leave the nest and carry out an opportunistic attack on a penguin nest or chick, or to fly a single circuit about the territory – a very common occurrence in skuas with large territories. At the other extreme the longest sustained

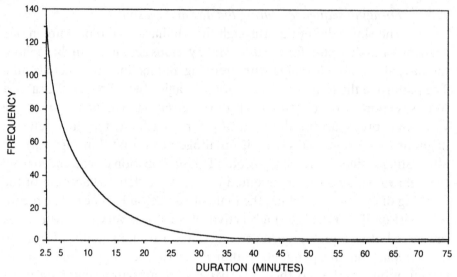

Fig. 7.6. Frequency of distribution of activity periods within skua foraging bouts on the penguin colony.

bout of activity without any of the natural breaks listed above found in this survey was an exceptional 69 intervals (172 minutes). The data on foraging bouts are, however, skewed and over half of all bouts occupied less than 40 minutes.

Once natural foraging bouts have been recognised it is possible to consider the pattern of activity within them. Generally, skuas break up a bout of foraging activity by short rest periods away from the penguins. Overall, there is an enormous diversity of activity within foraging bouts, from a single flight over the penguins to continuous foraging across one to two hours (Fig. 7.6). However, several patterns exist within this diversity related to constraints on activity from the skuas own breeding require- ments and to the food taken. A range of patterns is given in Fig. 7.7. The first (a) is characteristic of skuas scanning their penguins from flight without great expectation of food being available for scavenging or easily taken by attack. This is a common pattern early and late in the penguin breeding cycle and is also seen commonly in the early phases of a foraging bout where the skua is again becoming hungry and beginning to hunt. In the third of these tracks the skua had settled beside the penguins and was attempting to obtain an egg seen on the outside of a nest. The second pattern (b) is that of hungry skuas searching, scavenging and attacking penguins. These birds have bouts of activity interspersed with breaks away in which they seem to be reassessing their tactics and the chances of gaining food. There is a third pattern (not shown in Fig. 7.7.) of skuas involved in sustained attacks on penguins or sitting near carrion. In the latter situation they seem reluctant to come away from the penguins, because of the attraction of the food and because of the need to protect it from outside skuas. This pattern of sustained interest is seen also among skuas foraging for spilled food from penguins feeding their chicks. Some skuas spend hours at a time fossicking among crèche chicks for what often seems to be minute amounts of fish and krill.

These patterns also illustrate differences between pairs of skuas with nests or chicks and pairs which lack them. Nests and young chicks need to be defended, and the eggs incubated, continuously allowing only short breaks away by the skua at the nest area. This constraint on foraging is important and may account for the preponderance of opportunistic attacks compared with sustained attacks seen in some pairs. Skuas needing to protect eggs and chicks are rarely able to cooperate in foraging, although the nest may be deserted when the mate appears to have found a particularly favourable situation to exploit and is becoming active or has managed to pull a chick from the crèche. Before the eggs are laid or if the eggs or chicks are lost the two birds of the pair are free to work together among the penguins.

7.7.2 *The attacks of a strong predator: a sequence of attacks by the male of pair 25 on H Block*

The broad pattern of the skuas' interaction with the penguins has now been established. This section describes the foraging activity of the male of pair 25 one of the most accomplished, aggressive and persistent skuas on the entire colony. Although this skua pair had access to only a moderate number of penguin nests (between 50 and 120 nests in the different years) they nevertheless obtained much of their food from them during December and January and foraged at sea only as a last resort. The female of the pair undertook very little predation. The hide on H Block directly overlooked this territory and its section of H4 penguin group. This was an especially favourable conjunction of an accomplished and aggressive skua, a small, well mapped penguin group and a perfect viewing site so far above the birds that they seemed to be performing on a plan map, and far enough above them that they were not apparently disturbed by observation. The nest of this pair was 25 m distant from the far side of the penguins so that definition of a foraging bout, from when the skua came over to the penguins, was very straightforward. The detailed records of the behaviour of this skua were made for two purposes. First, to follow the

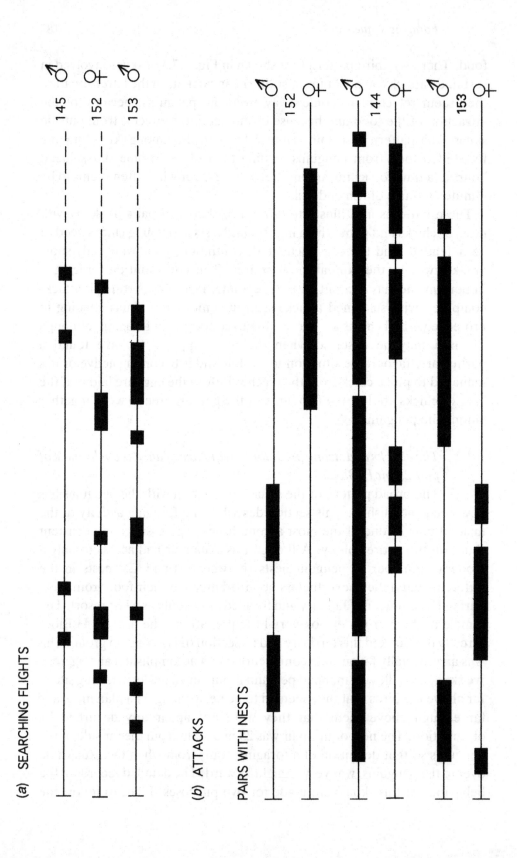

(a) SEARCHING FLIGHTS

(b) ATTACKS

PAIRS WITH NESTS

PAIRS WITHOUT NESTS

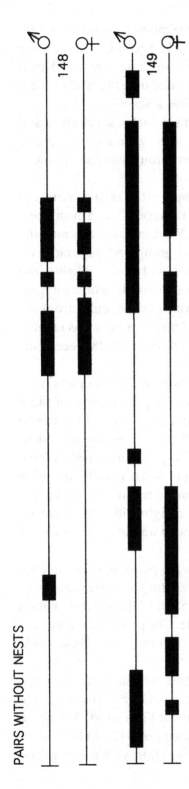

Fig. 7.7. Examples of the timing and duration of foraging bouts on the penguin colony by different pairs of skuas. (a) The attention pattern of skuas beginning to search for food at the start of a foraging spell or maintaining a desultory surveillance over the breeding groups from a roost. (b) Foraging patterns of skua pairs foraging actively on the colony. In pairs with nests the two birds generally alternated at the colony, in pairs with chicks (e.g. pair 144) they may also collaborate. In the pairs without nests the two birds were active together. Duration for each example is 140 minutes.

exact course of a foraging sequence from beginning to successful conclusion, or to when the skua gave up the attack. Second, to record the
impact of the attack on the penguins, to record which nests were selected
and how much disturbance was caused. At the same time the effectiveness
of colonial breeding defence of the penguins was assessed.

The foraging sequences are described at this point; nest selection and
breeding group disturbance will be taken up later when the intensity of
skua overview and searching of the penguins throughout the colony is
described.

In order to achieve the detail needed to pinpoint the skua's interest in
individual nests the nests along the margins of the group, those most likely
to be attacked, were mapped and numbered. Working from this map the
skua's movements up and down the breeding group and its attacks on
individual nests could be accurately recorded and timed. Flights and
attacks were recorded in seconds using a stopwatch, for the whole foraging
bout beginning when the skua first came down to the penguins from the
nest area and ending when the attack was broken off or prey was taken.

Eight foraging bouts were recorded over six days on 23–28 December,
1969.

Although H4 penguin group was divided among eight skua pairs, with
four on each side meeting down the middle of the group, some of these
skuas showed little interest in the penguins and others had lost mates so
that several territories at the penguin margin were only weakly defended.
Because of this unusual situation 25 had access to penguins outside its own
territorial limits, both along the western margin and across the breeding
group on the eastern margin within neighbouring territories. By far the
majority of time was spent along the western margin within the territory
and all periods spent watching the penguins and in walking searches along
the margin occurred there.

Figure 7.8 shows the skua territories on the H4 penguin group and the
area accessible to pair 25, both within its own territory and extending on to
others. This skua therefore, had access to c. 55 m of the western margin
and c. 60 m across the group on the eastern side. The penguin area was too
densely occupied to allow skuas to land near central nests. These could
only be attacked from flight.

At the date these observations were made most penguin nests contained
young chicks. On 27 December penguin chicks on H4 weighed on average
$670 \, g \pm$ SE $38 \, g$, $n = 63$, on average they were about 12 days old. At this age
they are an attractive and easily taken and killed prey. On 23 December
the 63 nests along the western margin comprised four with one egg, one

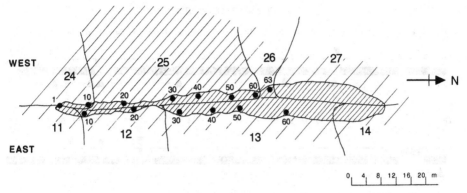

Fig. 7.8. The territory of pair 25 in relation to neighbours on H4 breeding group. The positions of identified nests along the east and west margins of the breeding group are indicated by the numbered closed circles.

with two eggs, 23 with single chicks and 35 with two chicks. On 28 December, the date of the last record, two chicks had been taken from single-chick nests and one chick had been taken from each of three two-chick nests, leaving four with one egg, one with two eggs, 24 with one chick and 32 with two chicks.

The penguins on this breeding group were not banded and could not be recognised as individuals. The attack sequences recorded across these several days are, therefore, to be interpreted as attacks on individual nests and not on individual penguins. At this stage of the penguin cycle the adults feeding chicks could be expected to change over at the nest each day. Those few penguins on nests with eggs might well have been the same individual birds across the full period of the observations.

The eight bouts record the activity of this skua interacting with penguins for 245 minutes out of a total observation period of 920 minutes. For the remainder of the time the skua was at the nest, loafing on the territory or absent from the area.

Fig. 7.9 shows the way this skua divided each foraging bout between ground and aerial searching, and between ground and flight attacks. This pattern is rather different from those described earlier for EF Block skuas for two reasons. First, the bout runs as a continuous activity sequence from the time the skua arrives beside the penguins until it ends. There are no obvious breaks, and the skua is either watching or patrolling among the penguins all the time. Second, because of the small area being searched by the skua, flight times are very short. The ratio of opportunistic attacks to sustained attacks is, however, similar: with 33 opportunistic attacks on the

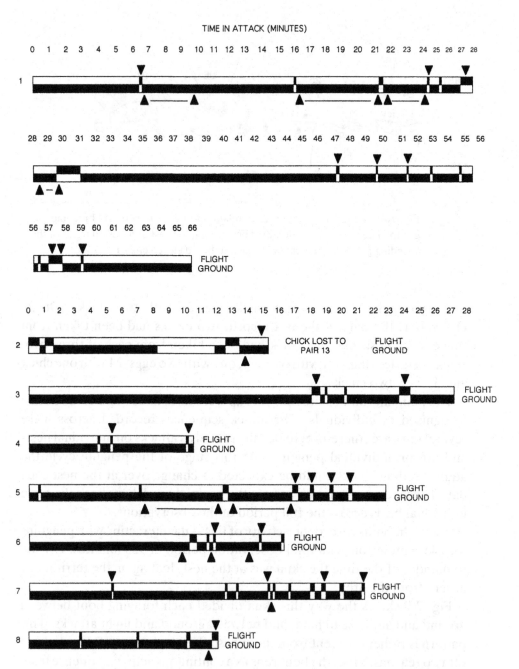

Fig. 7.9. Schematic interpretation of the activity during the separate foraging bouts of H25. Time in flight is shown in the upper half of each bar, time on the ground in the lower, in each case shown shaded. Attack events are indicated by arrows.

Table 7.10. *The record of the activity of the male of pair 25 on H Block during eight foraging bouts, 23–28 December, 1969*

Bout number	Bout duration (s)	Behaviour (% of time)			Number of events		Chased off by free penguins
		Alert and searching	Ground attacks	Flight attacks	Flights	Attacks	
1	3762	69.4	24.5	5.1	22	17	8
2	980	73.9	11.7	14.3	7	3	5
3	1632	96.9	0.0	4.0	6	2	8
4	647	95.2	1.4	3.4	5	2	5
5	1344	64.4	29.6	5.9	98	8	5
6	960	86.4	9.4	4.2	6	5	2
7	2108	95.2	0.1	4.6	14	6	13
8	3004	97.2	1.2	1.6	6	4	6

penguins along the western margin (plus a further 14 on the penguins on the eastern margin giving 47 in total) compared with only six very mild sustained attacks (in categories 5 and 6).

The details of these eight foraging bouts are summarised in Table 7.10, which records durations spent in each major activity as well as number of events.

There were 63 penguin nests along the western margin of the area worked by 25 and a further 45 nests were mapped on the opposite side of the group. This end of H4 was only five to six nests wide at its greatest width at this stage of the season, and was only two nests across towards the south end, from nests 6 to 18. There were few reoccupying penguins on the margins. The difference between the two sides was that the western fell within 25's territory while the other side belonged to pairs 12 and 13. Although these pairs were not much interested in the penguins themselves they did protect their territories, which by chance contained penguins, and prevented 25 from extending his very forceful predation on to this side. All attacks on these penguins had to occur from flight and 25 could not land safely there for more than a few moments. There was also a second constraint inhibiting effective working on this side. Chicks might be knocked out of nests there but unless they could be carried quickly in flight across the penguin group back to 25's territory they must invariably be lost to the defending skuas on this margin. This was precisely the outcome that ended the second foraging bout. A 400 g chick was knocked out of nest 25 on the east side by a fierce stoop but was immediately lost to the two birds of pair 13. On this side then the skua had only limited access. Similarly, 25 had only limited access to neighbouring territories on the west side (territories 24 and 26) and ingress was tolerated only if brief and only if 25 was not seen to be being successful. These constraints accounted for differences between the behaviour within the core of his territory and outside it. There was, however, variation as well in the overall level of attack and interest at nests within the territory. These differences are demonstrated in Fig. 7.10 for the amount of time spent near each nest and

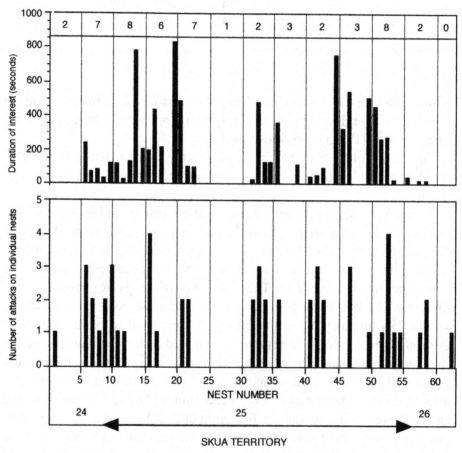

Fig. 7.10. The numbers of individual attacks on each nest and the amount of time near individual nests by male H25 during the foraging bouts. The numbers of times this skua was chased from the breeding group by free penguins is recorded along the upper margin of the figure.

for numbers of attacks on each. Nests 12 to 59 were contained within 25's territory, 1–11 were in 24's and 60–63 were in 26's.

Selection of prey by skuas will be considered in chapter 10 but selection by an individual skua is a different case and may be considered here. Nests may attract the interest of skuas through their contents, of eggs or chicks being exposed for example, through the demeanor or behaviour of the adult penguins, through their position in the breeding group or skua territory and especially because of their exposure to attack. Conversely they may be rejected because they are well protected by others, or through the presence of adult penguins nearby not tied to nests – reoccupying birds or mates. Across the time span of these records the constant variable in prey attraction was nest position, the siting of the individual nests in relation to others along the margin, which determines their degree of protection from attack. A second variable was the nest contents. Attacks on nests with two eggs or two chicks are more likely to be successful than attacks on nests with only one, because the penguin is less able to manoeuvre freely. Relating these two variables with attack and interest levels showed that only the former was a significant factor [13].

It is possible also that the skua was constrained in some sectors by the presence of free adults able to chase it away. This possibility is not supported by the records of the numbers of times the skua was chased off when approaching nests or when attempting to land near them. Fifty-two chases were observed of which only eight fell between nests 23 and 45 and three between nests 54 to 63, the two sectors receiving least interest. In this record of interest and penguin response there is support for the contrary view that the highest levels of interest and penguin response were directly correlated. Those sectors most used by the skua had the highest numbers of penguin chases – the skua provoked penguin defence.

7.7.3 *The pattern of activity across days*

This pattern was investigated by watching the skuas on EF Block over 81 hours, from midday on 6 January to 2100 h on 9 January, 1970.

Location of this log within seasonal and life history stages was as follows:

1. *Light and temperature.* The log took place between two and three weeks after mid summer, so that it occurred during continuous daylight. The study area was in full sunlight throughout the day except from 0100 h to 0630 h when it fell within the shadow of the enclosing moraine. The screen temperatures were quite uniform during this period ranging from a minimum at $-2\,°C$ in the early morning to between 2–$4.4\,°C$ in the afternoon. Ground temperatures would have been much warmer than this

during the sunny afternoons but night temperatures are a fair reflection of the conditions as both the meteorological screen and the study area were in shadow for much of the early morning.

2. *Penguin life-cycle*. At this date the penguins were in the early post-guard stage with chicks weighing on average 1565 ± SE 73 g on 3 January, 2319 ± 75 g on 11 January. Even at this late stage, however, a proportion of nests had guarded chicks.

3. *Skua life-cycle*. The skuas were either in incubation or very early chick stages. Pair 153 hatched their first chick on the last day of the log.

4. *Weather conditions*. The log began at the end of a spell of calm settled weather and uniform conditions continued right through the log with light variable winds and little cloud.

The observation point for this study was the summit of the moraine hills, some 60 m above the colony. Although further away from the colony than the usual observation point for EF Block it was selected for this study to ensure that the birds were not disturbed over the longer period and to provide better viewing during the late afternoon and evening when looking directly across the colony into the low sun.

The observations were shared by two people with the coded activity records for each 2.5 minute interval written directly on to prepared maps of the study area. Detailed records of foraging, both on the penguin colony and for birds feeding at sea, of nest changes and the interactions among individual birds were kept for the eight pairs of skuas that occupied this part of the colony. There were several other pairs of skuas on the slopes above the colony in clear view from the hide and records of their foraging flights to sea were also made.

This log was carried out for two purposes: to obtain a record of the way skuas divided the time between foraging, breeding and maintenance activities and to record the amounts of food obtained, comparing the different skua pairs across several days. The critical problem for this assessment was to be able to distinguish foraging at sea from all other flights away from the territory. Discrimination was facilitated by some very obvious behavioural differences. First, the skuas could be followed by eye and by field glasses in their flight from the area. Those bathing invariably flew down across the lower stretches of the colony towards the south, to the bathing pools on McDonald Beach. Foraging skuas by contrast flew directly out towards the northwest or turned sharply around the ice headlands to the north. These birds flew steadily, gaining height towards the horizon. Second, their behaviour on return was very different. Most birds returning from bathing began preening on return to the territory; those returning from foraging fed mates or chicks. The behaviour of mates on the nest area was also strongly indicative. Birds returning from bathing were largely ignored by mates, those returning from foraging were immediately pursued, begged at and chased. Anticipatory behaviour of the female on the territory invariably warned of the return of the mate and improved the chances of being able to see the transfer of any fish taken. This transfer from the foraging bird to its mate happened very quickly, within seconds of landing, but the shape and the silvery-white colour of fish were generally very obvious through field glasses. Moreover, at the end of feeding the female often pecked at the ground where the fish had been dropped, picking up scales and small digested pieces and sometimes harried the male for a second feeding. Third, there were large differences in the times the birds were away. Bathing flights were generally of no more than a few minutes. During this log foraging took birds more than an hour. In practice, there was little difficulty in distinguishing between bathing and foraging.

Most of the routine logs made during this study were carried out between 0800 h and 2200 h, during the warmest part of the day when the sun was highest in the sky. A number of attempts were made at times also to see how activity varied across the full 24 hours but these required more than the few hours usually available to achieve much more than to demonstrate that both skuas and penguins were active at all times. To investigate more subtle variations in level of activity or the type of activity required more detailed testing. Even now after a number of dedicated attempts to establish diurnal rhythmicity in skua and penguin behaviour at these high latitudes there is still no certainty that it has been demonstrated (Yeates, 1971; Müller-Schwarze 1968; Spurr, 1978). Although this 'long watch' on the skuas was to determine food taken it also recorded activities that can be analysed for a possible underlying circadial pattern.

There are a number of measures that might be made of the activities of these skuas to record pattern. The one shown in Fig. 7.11 marks each 30 minute period in which the skuas were scavenging or preying on penguins or foraging at sea. The intent in this figure is to disclose active periods during the day. To screen out tentative, brief foraging spells a minimum of 25% activity for each half-hour period was set as a lower limit for inclusion in the record. As noted earlier, skuas are clearly active throughout the 24-hour day with bouts of activity interspersed with long periods of inactivity. These bouts of activity do not show a regular cycle in this series of records nor are they apparently related to the daily cycles of higher and lower sun levels and higher and lower temperatures.

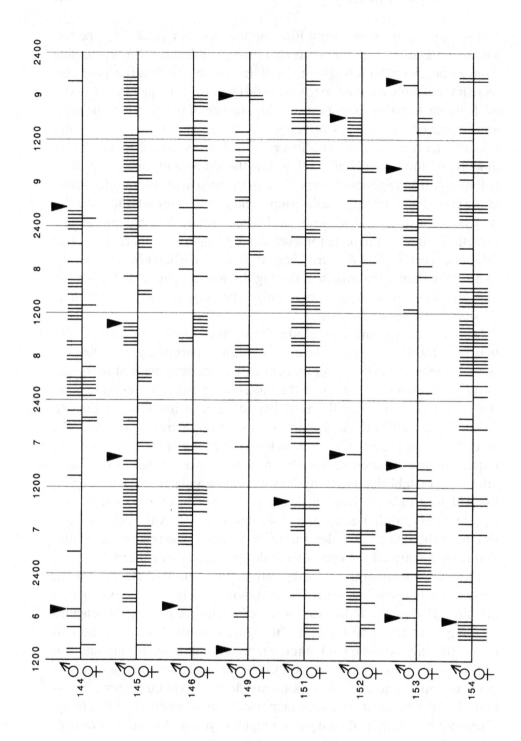

All the skua activity and 'absent from the territory' records have been used to produce the table of activity levels for the skuas in each four hour interval spaced symmetrically about midday [14]. The means shown summarise levels for the three days. There were significant differences among the time intervals in the levels of activity on the territory and in total activity. The birds on these days were more active in the 0200 h– 0600 h period.

Although this result suggests that circadial rhythms are present a similar effect could have arisen simply by chance from the time during the day food was obtained and these skuas might have already been set into individual cycles of activity based on hunger and satiation when the study began because of their earlier foraging. The penguin chicks are so large at this date that they can satisfy a pair for a day. Once one has been taken foraging ceases for a long period. One possible measure of circadial rhythmicity that is at least partly independent of the feeding cycle is the proportion of skuas asleep at any time. Skuas sleep with the head tucked under a wing and there was little difficulty with field glasses seeing this in the pairs under observation. The numbers of sleeping birds in 13 pairs was recorded in scans of the group each five minutes across 23 half-hour periods, allocated to five points of the day. At any time between zero and 23% of birds were asleep but there was no clear relation between numbers asleep with time of day. For example, on average 14.5% were asleep at 0200 h on 8 January, but only 3.2% at the same time on the following day. Similarly, 21% were asleep at 1800 h on 7 January but 0.6% at the same time on 9 January. A second measure is the timing of bathing and here there is some evidence of a preference for afternoon and early evening. Each of these birds bathed about twice a day on average (Fig. 7.12). Even so the timing of bathing flights is not entirely independent of foraging. Birds that have been feeding, or have feed mates by regurgitating fish, will generally leave to bathe and drink.

Although the levels of activity during the day have been shown to be *statistically* significant they are not all that convincing that an endogenous circadial rhythmicity is sufficiently strong at this stage of the season to

Fig. 7.11. The pattern of foraging activity on the colony or at sea in relation to time of day and to obtaining prey by the pairs of skuas on EF, recorded during four days 6–9 January, 1970. The markers show 30 minute periods when skuas were active for at least 25% of the time. Arrows indicated times when penguin chicks were taken. All pairs shown had access to penguin food within the territory.

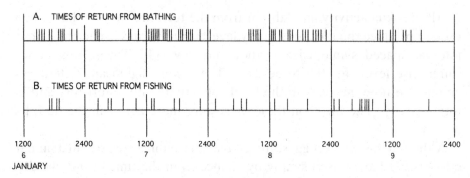

Fig. 7.12. The times of return from bathing and from fishing for the skuas on the EF study area monitored during four days 6–9 January, 1970. This figure includes pairs with and without penguins in their territory.

override other factors influencing the skuas' daily activity. The reason for caution is twofold. First, the records are from a small sample of skuas, 16 birds in all, and ran for a short time. Second, by chance, six of eight pairs were well fed during the first afternoon of the log so that their foraging behaviour was thereafter in part synchronised, as none of these would really need to feed again until late in the next morning, 12 hours away. Testing for any influence of circadial rhythmicity on foraging is better done earlier in the season when the skuas are foraging for eggs or such small chicks that foraging needs to be an almost continuous activity, not so strongly punctuated by gaining large chicks as occurs later. Alternatively, it might also be confined to recording times of foraging at sea in pairs entirely dependent on marine food, provided that flying and sea conditions are not also determining when birds can forage. At the end of severe weather when foraging is curtailed the skuas may take off together, and this synchrony would certainly impose an artificially synchronised cycle of activity across the following days.

An analysis done of skua activity at different times of the day in early to mid December 1969, when the skuas were feeding on eggs and young chicks at the colony or foraging at sea, failed to demonstrate circadial influence at this time. The timing of this record is approximately the same number of days *before* as the other log was *after* mid summer. The skuas should, therefore, have been subject to an identical light regime. Six pairs were examined for a total of 6–10 hours each during four periods, during early morning, midday, late afternoon and about midnight. The levels of activity recorded during these logs were very similar in all intervals, ranging from 9–13 intervals pair^{-1} hour^{-1} and could not be distinguished statistically [15].

The four-day log should also be able to test the proposition that as skuas become more hungry they become both more active and aggressive when foraging. These skuas are not in fact ideal subjects for testing the prediction as they have alternative food available to them at all times by foraging at sea. They are never in a position of having to get prey from the colony or starving. (This situation may occur during extremely bad weather conditions, of very high winds or fog that could stop the birds leaving the territory. But these conditions were rare at Cape Bird.) At the basic level this cycle of foraging and satiation with absence of foraging occurs for all skuas. Once a large chick has been taken or a bird returns from foraging at sea with food little activity does take place for a time. What needs to be tested is the change occurring in the behaviour of skuas foraging at the colony and unable to capture prey after hours, or even days, of effort.

There is a clear example in these records of dramatically increased aggressiveness driven by hunger. Pair 153 skuas were the most dependent on the penguins for food of all the skuas in the local area and foraged less at sea than the other pairs. It was no coincidence then that the example of intensifying aggressiveness with hunger came from this pair.

In mid-afternoon on 7 January the male of the pair took a guarded chick of 1200–1500 g estimated mass and both birds fed from this carcass over the following 10 hours. It was not lost to 152 pair until late the following morning when most of the flesh had been stripped away. By midday on the 8 January both birds had begun tentative attacks again on the penguins with both the male and female separately managing to catch but not hold well grown chicks during the afternoon. As usual the female was especially hesitant, only attempting to take a sleeping chick after a two minute wait when half a metre away and the nearest adults over 2 m away. After this attempt this bird returned to its more characteristic behaviour of searching and watching well outside the margins of the penguin groups. By early next morning both birds were much more aggressive but a series of quite fierce attacks by the male produced nothing more substantial than a single rotten egg, of which most of the contents were lost in feeding. A second bout of attacks began 30 minutes later but this time no food was obtained and the male then returned to the nest area and nest changeover occurred. An astonishing and dramatic change in the behaviour of the female was then evident. Between 0658 and 0750 h a number of strong attacks were made on chicks on guarded nests, crèche chicks on the breeding area and chicks with parents during feeding chases. This frenetic activity culminated in the female getting one of three chicks sheltering with an adult on a well formed nest.

This build-up of aggression over the 40 hours without food may be summarised as follows.

7 January, 1605 h 1200–1500 g chick taken.

8 January
– to 1200 h no interest.
– from 1200 h to 1300 h casual searching (category 2) and flights over (category 15) by male.
– to 1650 h no interest.
Female
– from 1650 h to 1710 h searching and one category 35 attack, chick caught and lost.
– 1710 h to 2135 h no interest.
Male
– 2135 h to 2425 h about third of time in search, one category 32 attack without contact, abandoned egg taken from outside margin.

9 January
Female
– from 2425 h to 0245 h female among penguins almost continuously but only one tentative (35) attack.
Male
– from 0245 h to 0658 h about half the time spent among penguins with one attack (29) on a chick during a feeding chase, three tentative attacks on chicks on a crèche margin and flights (27) about feeding chicks for krill.
Female
– from 0658 h searching and attacking penguins continuously until chick taken from guarded nest at 0750 h. There were 14 separate attacks in this sequence, three chicks were caught for a time and another was knocked over from flight outside a breeding group before the fifth chick attacked was taken and killed. As soon as the chick was taken the male deserted the nest and flew immediately down to assist its mate. The pair collaborated in killing the chick and in tearing it apart.

7.8   *The relative success of the different forms of attack on penguins*

In each sequence of attack types shown in Table 7.1 the force of the attack, that is, the amount of movement, vigour, aggressiveness and determination, increases towards the higher categories. As the commitment increases so also does risk to the skua – and the likelihood of success. Both reflect the same measure of how close the skua gets to defending adult penguins. To hook out an egg or chick from a defended nest demands that the skua enters well inside the penguin's defensive zone. Inside this zone the skua may be caught by the penguin and beaten with the flippers, or beaten without being held. Both are potentially damaging and life threatening.

It is already well established that the predominant activity of skuas interacting with the penguins within their territory is watching and searching, from the ground, categories 2–4, and from flight (categories 14 and 15). The amount of time spent in attacks or in scavenging defended prey is very small. Skuas can spend all day watching the penguins or poking about the margin of the breeding group or crèche, but they will pick up very little food – an egg or two that rolls or is pushed out from the group, a young chick pushed out beyond the margin, some krill or small pieces of fish lost during chick feeding. These are very poor reward for the time spent and grossly inadequate for maintenance of breeding skuas. To feed off the penguins requires far more commitment than this. The prey must be found or produced by disturbance and once found it must be obtained by skill and determination, and what to an observer often seems reckless determination.

In an attempt to describe how the skuas get eggs and chicks all attacks have been extracted from the logs of behaviour for the three seasons 1967–68 to 1969–70. Because of the different attack patterns on eggs and young chicks compared with those on post-guard chicks these were recorded separately. Each attack is considered here as a single event for the interval even though there are clear differences in the proportion of the recording interval each occupied. Sustained attacks run through the entire interval, an opportunistic attack may last for only a few seconds. Each opportunistic attack was in fact separately recorded in the logs, in addition to the background category of behaviour so that their numbers are precisely known. A summary of the number of attack events recorded and their success is given in Table 7.11.

(Category 18 is not included within the table as it was used to record those instances of a skua returning to its territory with an egg or young

Table 7.11. *Summary of scavenging and predation attacks and their success observed during logs of behaviour by skuas during the 1967–68 to 1969–70 breeding seasons*

Attack group	Attack categories	Number of events (or 2.5 min. intervals)	Number successful	Percentage successful
Searching	4	344	0	0
Scavenging flight attacks	17	73	34	46.6
Disturbance flights	16	246	0	0
	27	21	0	0
Opportunistic attacks				
Flight-attacks	19, 20, 21	150	40	26.7
Jump-attacks	22, 23, 24, 25	56	18	32.1
Ground/flight-attack	12	90	15	16.7
Sustained attacks				
Attacks on defended nests				
Ground attacks	5, 6, 7, 8	253	11	4.3
Ground attacks	9, 10, 11	67	4	6.0
Attacks on post-guard chicks				
Attacks on chicks outside crèche				
	28, 29, 30, 31	112	5	4.5
Attacks on chicks within crèche				
	32, 33, 34, 35, 36, 37	92	7	7.6

chick taken elsewhere on the colony. It was generally not known in these instances how the prey was taken but because of the rigid territorial defence it was no doubt taken from a flight attack, probably categories 17, 19, or 20.)

Two general categories of attack are shown here to be the most successful, the attacks from flight into undefended or defended nests (17 and 19, 20, 21) and the attacks made by skuas running with wings raised at nests near the breeding group margin (22–25). Nearly a third of these attacks were successful. The next most successful form was the 12 attack of a skua attempting to catch standing or prone penguins by the tail and pulling or twisting them from the nest in order to expose the eggs or chicks. This attack form was perfected by only a very small number of skuas – the two birds of pair 141 and the male of 25 were the most proficient. In general it was carried out by these birds after becoming frustrated through inability to obtain food in safer ways. These attacks caused so much disturbance that unless they were quickly successful the penguins became so angry and alerted that there was little chance of prey being taken.

All the other attack forms led to much lower success, between 4 and 7.6%. Although this was to be expected for some of these, for example attacks 6 and 7 from the margin against defended nests where the two species soon appear to reach an accommodation and the skua can do little more than make tentative thrusts towards the penguin's beak. The rate was not appreciably better for the much more vigorous attacks in this series (categories 10 and 11) where two skuas have the advantage of being able to divide the penguin's interest. The relatively low levels of success registered for direct attacks on chicks is in part an artefact of observers being able to see (and record) all attack attempts at this stage. Categories 28 and 32 of flights at chicks that were deflected because of the close proximity of adults made up the bulk of these records. Once a chick was knocked from the crèche or caught by skuas in the open the likelihood of completing the kill is quite high. Skuas at this stage are very selective about which attacks to pursue.

STATISTICAL TESTS

[1] Comparison of the activity levels of skua pairs during different stages of the penguin cycle. Single factor analysis of variance within each year.

A All stages in EF Block skuas, total interest and activity levels
1967–68 $F = 9.24, p = 0.004,$ df 3,12
1968–69 $F = 0.384, p = 0.76,$ df 3,12
1969–70 $F = 6.86, p = 0.02,$ df 3,9

B Analysis of levels of interest and activity for the incubation (INC) and chick guarding (CG) stages.

	Year	Total interest	Activity alone	
H Block	1967–68	INC = CG, $p = 0.125$	INC = CG, $p = 0.077$	df 1,7
	1968–69	INC > CG, $p < 0.001$	INC = CG, $p = 0.981$	df 1,4
EF Block	1967–68	INC < CG, $p = 0.004$	INC < CG, $p = 0.029$	df 1,7
	1968–69	INC = CG, $p = 0.352$	INC = CG, $p = 0.876$	df 1,8
	1969–70	INC = CG, $p = 0.530$	INC = CG, $p = 0.379$	df 1,5

[2] Comparison of activity levels by skuas among years in the two sites during the incubation and chick guarding stages.
Analysis of variance of total activity and interest and activity alone.
H Block. Comparing 1967–68 and 1968–69 for levels during incubation and chick guarding stages.
Total interest and activity $F_{1,12} = 3.475, p = 0.083$.
Activity $F_{1,12} = 1.25, p = 0.287$.
EF Block. Comparing the three years, 1967 to 1970 during the incubation and guard stages.
Total interest and activity $F_{2,20} = 0.979, p = 0.385$.
Activity $F_{2,20} = 2.785, p = 0.082$.

[3] Comparison of total activity and interest and of activity alone of skuas on H and EF Blocks.
Mean levels compared for each penguin life-cycle stage across the years 1967–68 and 1968–69. Units are mean intervals ± standard deviation pair^{-1} h^{-1}.
Total activity at penguins
H Block 2.41 ± 1.48; EF Block 3.27 ± 1.55.

$F_{1,28} = 2.364, p = 0.135.$
Activity alone
H Block 1.02 ± 0.64; EF Block 2.35 ± 1.42.
$F_{1,28} = 9.81, p = 0.004.$

[4] Comparison of levels of interaction with penguins among the skua pairs on
EF Block.
1967–68 $F_{1,6} = 3.24, p = 0.007.$
1968–69 $F_{1,5} = 3.59, p = 0.006.$
1969–70 $F_{1,5} = 1.42, p = 0.232.$
H Block
Comparing 11 pairs across the two years 1967–68, 1968–69.
Two factor analysis of variance.
Among pairs $F_{1,10} = 2.49, p = 0.011.$
Between years $F_{1,10} = 0.21, p = 0.657.$
Interaction $F_{1,10} = 1.07, p = 0.392.$

[5] Comparison of the interest and activity of the two birds of a pair. Because of
the changes in pairs and numbers of pairs on EF Block it was not feasible to
carry out a single analysis comparing the sexes across the three seasons. They
have been compared, therefore, within each season separately. Gross
differences in interest in the penguins by several skuas on H Block in the two
years suggested that comparison within each season was also more appropriate for these skuas as well. Means are calculated from total interest and
activity intervals for each bird throughout each season.

		Mean levels			
	No. of pairs	Male	Female	t	p
H Block					
A Interest and activity					
1967–68	13	54	38	0.70	0.49
1968–69	11	45	31	0.89	0.38
B Activity					
1967–68	13	38	19	1.56	0.13
1968–69	11	14	7	1.29	0.21
EF Block					
A Interest and activity					
1967–68	7	261	151	1.80	0.09
1968–69	6	166	99	1.45	0.18
1969–70	6	208	85	2.15	0.06

continued

cont.

| | No. of pairs | Mean levels | | *t* | *p* |
		Male	Female		
B Activity					
1967–68	7	216	111	1.91	0.08
1968–69	6	105	51	1.91	0.08
1969–70	6	110	29	2.56	0.03

[6] Test of the different amounts of time males and females of pairs were away from the territory for records between late November and 9 January in each year. In pairs foraging at sea males feed females throughout this period.
H Block
Two factor analysis of variance.
Difference between sexes $F_{1,48} = 17.73$, $p < 0.001$.
Difference between years (two years, 1967–68, 1968–69) $F_{1,48} = 0.543$, $p = 0.464$.
EF Block
Two factor analysis of variance.
Difference between sexes $F_{2,30} = 6.44$, $p = 0.017$.
Difference among years (three years, 1967–68, 1968–69, 1969–70) $F_{2,30} = 0.24$, $p = 0.783$.

[7] Differences in the time males of EF pairs were away from the territory. Mean number of intervals per hour individual males were away from the territory during logs between the end of November and 9 January in each year. Mean intervals away for each pair.

Pairs	1967–68	1968–69	1969–70
144	3.9	4.9	3.2
145	13.8	1.9	4.3
146	2.3	5.2	10.3
148	4.2	—	—
149	4.2	11.1	4.2
152	0.9	0.4	1.0
153	0.9	0.8	0.5

Single factor analysis of variance among pairs.
1967–68 $F_{5,54} = 3.90$, $p < 0.000$.
1968–69 $F_{5,54} = 9.05$, $p < 0.000$.
1969–70 $F_{5,54} = 5.31$, $p < 0.000$.

[8] Test of differences between H and EF skuas of average levels of activity in each year. Mean and standard error of mean.
1967–68 EF mean level = 5.70 ± 1.05, H mean level 3.02 ± 0.66, t = 2.19, df = 17, p = 0.04.
1968–69 EF mean level = 4.11 ± 0.54, H mean level 3.55 ± 0.69, t = 0.631, df = 12, p = 0.539.

[9] Tests of differences between EF and H skuas for combined levels of opportunistic and sustained attacks. Intervals pair^{-1} h^{-1}. (Mean ± SE).
1967–68 EF mean level = 0.47 ± 0.16, H mean level 0.17 ± 0.05, t = 2.84, df 17, p = 0.011.
1968–69 EF mean level = 0.34 ± 0.05, H mean level 0.03 ± 0.08, t = 0.48, df 12, p = 0.633.

[10] Comparing the overall levels of activity of skuas foraging on the penguin colony during each life-cycle stage, for the skuas on EF block. Mean and SE.

	Incubation stage	Chick guard stage	Early post-guard stage	Late post-guard stage
1967–68	3.21 ± 1.5	6.26 ± 0.2	9.86 ± 0.1	16.58 ± 1.9
1968–69	3.71 ± 0.8	4.43 ± 0.7	4.73 ± 1.4	4.40 ± 2.8
1969–70	3.79 ± 0.04	3.14 ± 0.5	5.32 ± 0.5	—

Single factor analysis of variance of differences among stages.
1967–68 $F_{3,8}$ = 20.65, p = 0.001.
1968–69 $F_{3,9}$ = 0.14, p = 0.93.

[11] Average amount of time spent scavenging in each penguin life-cycle stage by EF skuas (0.38 ± 0.05 intervals pair^{-1} h^{-1}) and H skuas (0.09 ± 0.03 intervals) were compared for the two years 1967–68 and 1968–69 in which comparable statistics were present. t = 4.48, df = 9, p = 0.002.
The ratios of the times spent scavenging and foraging (defined in the text) were similarly compared for the same period. t = 1.826, df = 9, p = 0.10.

[12] Differences among individuals and pairs of skuas on EF Block for preference for scavenging or predation. Tests were by Chi-square for pairs and individual birds of pairs separately in each year. With the single exception of females during 1969–70, for which there were few data, all tests disclosed significant differences (p < 0.001) among the pairs for the numbers of intervals recorded in each attack form.

[13] Relation between the level of interest by male 25 in the individual penguin nests along the west margin of H4 with nest position on the margin and number of chicks in the nest. Nests were scored as +, 0 or − depending on whether they projected from the margin, were in line with neighbours or were inside neighbours. By two-way analysis of variance, nest position was found to be a significant factor ($p = 0.03$) but neither chick number nor the position/chick number interaction were significant.

[14] Analysis of activity levels of skuas at different times of the day during three days continuous observation on EF Block, 6–9 January 1970. The records shows mean values of activity recorded as number of 2.5 minute intervals for eight pairs of breeding skuas.

Period	Activity on territory[1]	Absence from territory[2]	Combined level of activity and absence
1400–1800 h	154.7	152.3	307.0
1800–2200 h	150.7	85.7	236.3
2200–0200 h	118.7	159.0	277.7
0200–0600 h	243.3	240.7[3]	484.0[3]
0600–1000 h	221.3	118.3	339.7
1000–1400 h	232.3	67.7	300.0

[1] Activity on territory = all interactions with penguins. [2] Absence from territory = all absences for foraging and bathing. Activity on territory analysis of variance, $F_{5,12} = 6.37$, $p = 0.004$. Absence from territory analysis of variance, $F_{5,12} = 2.13$, $p = 0.131$. Combined level analysis of variance $F_{5,12} = 2.99$, $p = 0.055$. [3] Significantly different from other periods by the Tukey test.

[15] Levels of foraging activity by skuas of EF Block, 5–18 December, 1969. Records are of mean levels of activity (pair^{-1} hour^{-1}) summed for activity on the territory and absence from the territory for six pairs.

About midnight (2200–0200 h) 10.0; early morning (0400–1000 h) 8.96; midday (1000–1400 h) 11.0; late afternoon, early evening (1600–2000 h) 13.2.

Single factor analysis of variance, $F_{3,6} = 0.688$, $p = 0.589$.

8

The amount of food taken by the skuas from the penguin colony

8.1 *Introduction*

The previous chapter has described a diverse interaction between skuas and penguins at this colony, ranging from seeming indifference to the penguins to intense scavenging and predatory behaviour. In each of the two study areas several of the skuas were shown to be aggressive and skilled predators on penguin eggs and chicks, overcoming the defensive abilities of the adult penguins by stratagem and reckless aggression. But how much food did they get? The purpose of this chapter is to record the amounts of food obtained by these skuas throughout the breeding season and attempt to relate this source to their requirements. For this latter aim the information on foraging at sea is a critical component as it provides certain evidence for at least some skuas that they could not obtain sufficient penguin food from the territory on these dates.

As outlined earlier (Young, 1970) there seem to be just three ways to obtain records of the foods taken from the colony by skuas: by direct observation of skua activity on the colony, from changes in the numbers of penguin eggs and chicks at set intervals through the season, and from evaluation of the prey remains found on the skua territories. Each method has both strengths and weaknesses and an early decision in the plan of this study was to utilise all three approaches as far as possible. The first (direct observation) can provide good records of the food taken throughout the season while the second (penguin egg and chick counts in breeding groups) gives good estimates until the groups begin to disband and merge as the chicks mature. Equal attention was given to these methods. The third (records of prey remains found) is practicable only at the beginning and towards the end of the penguin breeding cycle, i.e., when skuas are taking eggs or large chicks. During early chick rearing small chicks are taken and consumed without leaving any remains. Its most serious shortcoming is that its accuracy and completeness depends on the frequency the study area is visited and on the diligence of searching for prey remains. It also causes much more disturbance and care needs to be taken to ensure that this disturbance does not lead to a more than natural movement of prey remains among skua neighbouring pairs. In practice all three methods share the common deficiency of requiring extrapolation, scaling, from limited observations or records to embrace the populations as a whole: for direct observation, extrapolation of captures by a few pairs during a few hours of observation to the skuas in general for the season; for the records of losses of eggs and chicks, from a few penguin groups to losses for the colony as a whole; for the records of prey remains on a few territories, to extrapolation for all territories. There are, inevitably, wide errors in these transformations. The differences among these methods are summarised in Table 8.1.

Comparing these advantages and shortcomings suggests an overwhelming advantage in direct observation, and in general this advantage is real. Provided sufficient long periods of direct observation are made of a range of skua–penguin situations then certainly a very detailed understanding of this interaction must develop. The method does, however, have one serious weakness. On its own it cannot give information on the prey available, and therefore allow comment on impact on the prey or on selection of prey, nor can it provide accurate data on the mass of prey, as distinct from number of prey items. The disturbance caused by attempting to retrieve prey as they are taken for measurement and weighing would immediately destroy the advantage of the method – of observing the relationship between the two species under natural conditions. Rushing down the hill from the observation hide waving and shouting to chase off the skuas every time eggs or chicks were taken would not foster good observation recording!

Table 8.1. *Comparison of different methods of gathering information on predation*

	Direct observation of skua activity	Records of changes in penguin egg and chick numbers	Records of prey remains
Records throughout entire season?	Yes	Yes, in isolated groups	No
Precision	Potentially high	Very high	High in parts of season
Determination of whether prey scavenged or predated	Generally	Generally	Generally
Foraging behaviour and effort expended	Yes	No	No
Evidence of foraging at sea	Yes	No	No
Discrimination among different skua pairs and individuals	Yes	Yes	Partly
Determination of biomass of prey obtained	Visual estimation only	Yes	Part record

These deficiencies in this method were to a large measure made good by the records of the penguin egg and chick numbers on the study breeding groups. These records were of mapped individual nests so that information was obtained not only on changes in numbers of prey but on which nests were attacked and which probably scavenged. At the same time measurement of the growing chicks provided records of the amount of food available generally on the colony and of the age and mass of prey lost. Judicious interpretation of the records, of the ages and sizes of the chicks, allows identification also of selection of prey by the skuas.

With all this information to hand why was so much time and effort expended also in retrieving and measuring prey remains; a grisly task for a natural historian more interested in observation than experimentation.

The answer is that towards the end of the season when only small numbers of very big chicks were taken these events were so widely spaced in time that most would inevitably be missed during the duration of the normal observational logs. By checking the territories of skuas regularly for prey remains a complete record of feeding could be obtained for this stage to complement the behavioural observations.

8.2   *The numbers of eggs and chicks taken by the skuas*
        This section provides information on the numbers of eggs and chicks taken by the study groups of skuas on EF and H Blocks. It brings together the estimates of food taken recorded in direct observation and from the records of prey remains. At this stage the records are of numbers of prey items, later these will be translated to biomass and compared both with the amounts needed by the skuas and the amounts recorded as being lost from the penguin breeding groups throughout the seasons.

The records from direct observation of skua foraging activity and from prey remains are complementary. Until mid season (early post-guard stage, beginning 3 January) the prey volumes are best recorded by direct observation (little attempt was made in this study to count penguin egg shells on the territories and young chicks are eaten virtually without trace) but from this date records of prey remains become an increasingly complete and accurate measure of prey taken, when direct observation is conversely becoming a less and less efficient method, as prey captures become widely spaced in time. The basic unit for recording prey capture and for comparing different skua pairs is prey taken pair^{-1} day^{-1}. This unit can be used as a basis for estimating the food needed to support the parents and chicks and can be used as well to estimate the eggs and chicks taken by all the skuas from part of the colony (or the entire breeding group). It is also amenable to extrapolation across different time periods, and allows the inclusion of intervals of different lengths where days have been lost from the recording period.

8.2.1   *Numbers of prey items taken at different stages of the penguin breeding cycle*
        The numbers of eggs and chicks taken by the individual pairs of skuas on EF and H blocks are shown in Appendix 2.

These records have variable credibility. They are especially weak for the incubation stage where there were too few observation hours in some years for valid comparisons among the individual pairs. Fewer hours were spent with H Block than at EF Block but pair × time products were in fact similar, because there were twice the number of pairs on H, so that it is likely the differences between the two blocks reflect different activity levels and success rather than different observation effort. The figures shown for both incubation and chick guard stages are estimated numbers for each stage extrapolating from observed captures. The figures for the post-guard stages, however, represent totals for each day determined from the prey remains found on each territory. The success of this assessment method depends on the search effort.

Skuas at Cape Bird did not kill penguin chicks or take eggs beyond the numbers needed to satisfy their immediate hunger and did not waste the resource by eating only part of the prey and leaving the remainder to spoil. Prey taken is hard fought for and fully utilised. This satiation control of foraging activity no doubt provides an explanation for the similarity of numbers of prey taken by the different pairs. The maximum number is a modest one even during the incubation stage when there are very large numbers of potential prey available for skilled predators.

During the incubation and chick guard stages up to 10 eggs or small chicks might be taken by a skua pair each day but the average take is much less than this. As shown in Appendix 2, there are pairs in both areas for which there are no records of predation at all. Fewer chicks are taken during the post-guard stage and the skua pairs average less than a chick a day over this time.

These records are interesting in two regards. First, they demonstrate a wide variation in the numbers taken by the different pairs of skuas, and secondly, they record that only modest numbers were being taken by any pair. In the analyses that follow it is important to record clearly the prey being compared. There are several forms: live eggs and chicks, rotten eggs, dead chicks that are eaten and others that are retrieved and then left. Most analyses are of live eggs and chicks, the food obtained by skuas preying on the penguins. Few of the eggs remaining on the colony in January, and becoming accessible to skuas as the chicks crèche, are eaten. Most are rotten or frozen and provide little food for skuas. Similarly, as the season progresses, the difference in food value between dead chicks, most of which were small, and live chicks becomes wider and wider – 100 g or so for a dead chick compared with 2000 g or more for live ones.

8.2.2 *Comparison of H and EF skuas*

The design of this study allows comparison of the performance of skuas with many and with few penguins nesting in the territory. How do these two groups compare for numbers of eggs and chicks taken?

Table 8.2. *Comparison of estimated numbers of eggs and live chicks taken each year by the skuas on H and EF Blocks*

The records are means and range per day.

	1965–66	1966–67	1967–68	1968–69	1969–70
Incubation stage (6 Nov.–8 Dec.)					
Direct observation records					
H	—	—	0.84 (0–5.2)	1.14 (0–4)	—
EF	—	—	1.78 (0–3.6)	3.57 (0–7.6)	3.13 (0–10.4)
Chick guard stage (9 Dec.–2 Jan.)					
Direct observation records					
H	—	1.21 (0–5.6)	0.43 (0–3.4)	0.88 (0–9.6)	—
EF	—	4.11 (0–6.9)	3.35 (0–7.6)	2.58 (0–6.5)	1.23 (0–3.1)
Early post-guard stage (3–9 Jan.)					
Direct observation record					
EF	—	—	0.69 (0–3.2)	0.4 (0–15)	0.78 (0–1.5)
Prey remains record					
H	0.49 (0–1.3)	0.36 (0–9)	0.44 (0–0.6)	0.18 (0–0.5)	0.28 (0–0.7)
EF	0.84 (0.3–1.8)	0.57 (0.3–1.3)	0.36 (0.1–0.7)	0.21 (0.1–0.5)	0.51 (0.1–1.0)
Late post-guard stage (10–19 Jan.)					
Prey remains record					
H	0.31 (0–0.7)	0.41 (0–0.6)	0.15 (0–0.3)	0.08 (0–0.6)	0.08 (0–0.4)
EF	0.64 (0–1.2)	0.61 (0–1.5)	0.19 (0–0.3)	0.23 (0–0.7)	0.11 (0–0.3)

Estimates of the average numbers of eggs and live chicks taken by the skuas on the H and EF study areas can be obtained for most years from observation logs and prey remains (Table 8.2). The estimates given for the incubation stage (entirely of eggs) and for the chick guard stage (eggs and small chicks) are from observation logs. The estimates of live chicks taken during the early post-guard stage are from direct observation and prey remains, and are recorded separately, while those for the late post-guard stage are calculated entirely from prey remains. EF skua pairs gained more eggs and chicks on average than H skuas for all stages in each year. This difference is most clearly marked during the guard stage where they took over three times as many eggs and young chicks on average as those on H Block.

There is a great amount of variation among the pairs in both study areas and this is shown by the ranges given. For example, on H Block in 1968–69 the 14 pairs averaged 0.88 eggs and chicks each day during the chick guard stage but all of these were taken by just two pairs, with one of these taking 10 eggs and chicks.

It was not possible to demonstrate differences among the years in the numbers taken by EF skuas for the incubation or chick guard stages but there were significant differences among them for the post-guard stages [1]. In the early stage these skuas took more prey during 1965–66 when there were the highest chick numbers on the colony than in 1968–69 when there were the fewest. In the late stage more were taken in the first two years than in the following two.

8.2.3   *A check on the observation and prey remains statistics: the average numbers of eggs and chicks lost each year from the breeding groups on H and EF blocks*

In each year breeding statistics were obtained for a number of penguin groups in different parts of the colony. These records provide key data on the amount and nature of the prey available to skuas on the colony and can also be used to show numbers of eggs and chicks being lost from the colony during the different stages of the breeding cycle. Virtually all eggs and chicks lost are eaten by skuas, as prey or carrion. Differences in numbers between the checking dates, which are usually at five-day intervals, represent numbers gained by skuas both through scavenging and through predation. This is not an exact measure of food gained by the study pairs for two reasons. First, not all eggs and chicks lost on the colony are immediately obtained by skuas. Eggs may be frozen or become embedded in the guano of the area and not retrieved until much later, or not at all. Chicks killed within the breeding group area may be trodden into the ground, becoming flattened and dried-out and ignored by skuas or if retrieved not eaten. Second, the selection of a small number of study penguin groups throughout the colony was to represent the colony as a whole; they less satisfactorily represent the special character of groups within the two study areas. An unbiassed analysis of their losses would require a separate study in which the different types of breeding group were proportionately represented. For the present analysis egg and chick loss figures are obtained in each year by selecting groups that were most like those of the study areas. Groups on the periphery of the colony are used to represent H Block; groups on B, D, E and F are used for EF Block.

The most informative data on skua foraging on the colony is of the predation on live eggs and chicks. The present analysis of losses from the groups, for symmetry, also focussed on this predation. In essence this means that the large numbers of infertile and addled eggs that show up

Table 8.3. *Numbers of eggs and chicks lost on average per skua pair each day from the breeding groups in the H and EF study areas during three life-cycle stages*

Data are numbers of eggs or chicks skua pair^{-1} day^{-1}.

	1965–66	1966–67	1967–68	1968–69	1969–70
Incubation stage (from 19 Nov.)					
H	0.27	0.67	0.09	0.45	0.13
EF	1.07	0.70	1.18	2.38	7.65
Chick guard stage (9 Dec.–2 Jan.)					
H	1.81	2.65	1.42	0.47	0.53
EF	5.44	8.99	5.68	3.36	4.37
Post-guard stage (3–19 Jan.)					
H	0.55	0.84	0.73	0.46	0.24
EF	3.86	4.44	1.56	1.36	0.81
Numbers of eggs laid in the breeding groups used for this estimation					
H Block	346	122	176	187	93
EF Block	642	266	430	234	290

progressively during the end of the chick guard stage (late December) and are retrieved during the post-guard stage as the colony structure breaks down, are ignored in these counts. Compared with penguin chicks their contribution to the food taken is small. Their importance will be considered further in section 8.3.

These figures of losses from the breeding groups provide an independent check on the estimates of the numbers of eggs and chicks taken by the skuas estimated from the observation logs and from prey remains. In Table 8.3 average egg and chick losses per skua pair are given for three stages: for the incubation stage (for its final 20 days), for the chick guard stage (25 days) and for the post-guard stage (for the first 17 days), which comprises the early stage of seven days and 10 days of the late stage, up to January 19). Beyond this date breeding groups tended to merge so that individual counts were no longer possible. For a general measure of eggs lost each day, for comparison with numbers estimated to have been taken each day by the skua pairs, the duration of the incubation stage needs to be shortened from its full 33 days (from first egg to first chick) used generally in this work. First eggs were laid c. 6 November but large numbers were not available to the skuas until well into this month (peak laying was about 11–15 November in this colony) and skuas themselves did not really show much interest in foraging for eggs until towards the end of the month. This lack of interest shows both in the observation logs and in the counts of egg shells on each skua territory. In 1966–67, the only year in which shells were carefully counted, the 14 pairs on H averaged only 1.7 eggs each day until 21 November but took over 5 each day on average in the first week in December. Similarly, 11 pairs on EF averaged 0.7 eggs each day until 28 November, but were taking up to 17 a day in the first week of December. For comparability with the estimates obtained from the observation logs, which were universally in late November and early December, the estimates of losses from the breeding groups should be based on the last 20 days, i.e., from 19 November.

The numbers of eggs and chicks lost during each stage from the penguin breeding groups have been used to determine numbers for the skuas in each area by a simple conversion based on the numbers of nests within territories. The figures shown in Table 8.3 are the average numbers of eggs and chicks lost from penguin breeding groups within the combined area of the skua territories in each study area. They do not, therefore, reflect the different interest or ability of the individual skua pairs, as was recorded in the observation logs.

These figures nevertheless point up very large differences in the numbers of eggs and chicks lost from the breeding groups from all causes during the different stages of the breeding cycle, with most being lost each day from the chick guard stage. The lower rates for the post-guard stage would have been even more pronounced if the estimates had been extended beyond this first period. Later in January and in early February even fewer are lost each day. In each year fewer eggs and chicks were distributed among the H Block skuas than those in EF Block in all three stages.

These calculations of numbers being lost from the breeding groups give good support for the estimates of prey taken made from the observation and prey remains records, providing independent confirmation of the sense of these estimates and for their use in estimating the amount of food taken.

8.3 *The amount of food taken by the skuas on the penguin colony and its energy content*

The preceding section has described the success of the skuas in obtaining eggs and chicks, concentrating on the numbers of food items obtained at different stages of the penguin breeding cycle. These figures can now be translated into biomass and energy content – the currency of the analysis of the bioenergetics of this system and the units needed to develop the equations of energy requirements of breeding skuas against the amount actually obtained at the penguin colony. Any deficiency must be made up by foraging at sea. This analysis continues within the same penguin life-cycle format established in the earlier section. This format has the added advantage of defining three types of diet: penguin eggs (incubation stage), penguin eggs and young chicks (chick guard stage) and penguin chicks (post-guard stage). As will be shown later the small proportion of eggs that survive into the post-guard stage and retrieved at this time by the skuas have such a small food content in comparison with chicks that they make a negligible contribution to diet, even though they represent a significant proportion of the *number* of prey items taken at this stage.

8.3.1   *Mass and energy content of penguin eggs and chicks*

The mass and energy content of each of the components of the skua diet on the penguin colony must be determined before the skua's intake within each stage can be estimated. This is most easily achieved for the eggs, which are a very uniform dietary item. By contrast, estimating intake from penguin chicks is very difficult: the size, mass and energy content of penguin chicks on any given date is variable and it changes rapidly through the season. Estimating the size of the prey when it is seen taken during the observation logs is itself a major hurdle. Moreover, the proportion of the prey consumed by the skuas varies for chicks of different ages. The following section outlines these difficulties and describes the methods used for estimating the biomass and energy content of the eggs and chicks.

Incubation stage: penguin eggs as dietary components

The penguin egg was described in Chapter 3. For the present calculations three figures are required: the mass of the egg contents, and the mass and energy content of yolk and albumen separately. Skuas spill much of the albumen when feeding on eggs and it is not until the embryo is well developed, incorporating the albumen, that the entire egg contents may be consumed. Herbert (1967) has shown that it is not until after 20 days incubation that there is a substantial embryo (of 45 mm length); this grows to near full size by 28 days. For estimating intake by the skuas only yolk mass and energy content should be used for the egg to 25 November (c. 20 days from the start of egglaying) but later the combined mass and energy content of yolk and albumen provide a more accurate estimate.

The variability in laying date and in incubation rate, and the varying skills of skuas eating the egg contents, with more or less loss of yolk and albumen, make the application of any finer division of egg quality and amount consumed by the skuas impractical. The estimates that follow are for the second half of incubation so that an energy content of 540 kJ (representing the entire egg) is appropriate. Only a small proportion of skuas are active in the colony earlier in the breeding season.

Chick guard and post-guard stages: eggs and chicks as dietary components

Estimates of the food taken by each of the skua pairs from the penguin colony is much more difficult in these stages than during incubation for two reasons. First, there are now four types of food: live and addled eggs and live and dead chicks. Second, the mass and energy content of chicks changes rapidly through this period, from being the exact equivalent of eggs at the start to being over ten times as much at the end.

There are six quite specific questions that must be resolved in order that the estimates of numbers of prey items taken may be converted to food volume and energy intake. These are, in order of discussion: 1. the significance of addled eggs in the food equation; 2. estimating the age and size of the chicks taken by the skuas as prey and carrion; 3. the amount of food taken from the carcass; 4. the significance of food sharing among pairs during the different stages; 5. whether chicks taken as carrion are to be included within the food equation; and 6. the energy content of penguin food.

1. The significance of addled eggs (Fig. 8.1)

The first question can be easily disposed of. Although infertile or addled eggs make up a significant proportion of the eggs on the colony, and an increasing proportion of the eggs taken by the skuas through the guard and post-guard stages, their food content is relatively unimportant compared with that of live chicks. This is because they are usually so addled and runny that the skuas lose much of the contents when breaking them open and because they make up a progressively less significant proportion of the total food on the colony as the chicks grow and mature. A single chick in late December is worth 10–20 addled eggs in energy content.

2. Estimating the age and size of penguin chicks taken as prey

Determining the size and age of the chicks taken by the skuas and their energy content is by far the most difficult part of deriving the food

Fig. 8.1. The problem of the addled eggs and their contribution to skua diet.

equation. Most chicks taken during the guard stage are small enough to be killed and swallowed entire by one of the pair, or torn apart and swallowed by the two birds, in a very few minutes so that these penguin chicks are usually only seen being killed during the formal, direct observation periods and no remains are left after feeding for estimating size. The need to maintain undisturbed conditions on the colony during the observation logs precluded retrieving these chicks from the skuas at the time of their capture for measurement.

From late December and through the post-guard stage to fledging the chicks mature in strength and coordination and the body becomes progressively toughened with an increasingly strong skeleton. These chicks take longer to subdue and the disturbance caused attracts and pinp ,ints observer attention. Many kills at this stage are witnessed. Moreover, the carcass takes much longer to consume and substantial remains mark its existence after feeding has finished. Because of the amount of disturbance caused by these attacks even low level monitoring of the study areas during other work, or for H Block from the field station overlooking it, can provide accurate records of the numbers taken and the opportunity for collecting prey data. At the very least this monitoring can corroborate absence of kills found in searches for prey remains. Negative records are as valuable as positive ones when assessing the sufficiency of penguin food for the skuas.

The mass of individual prey The ages and mass of chicks taken by skuas on each day were estimated in two ways. For most of the guard stage, when the chicks were consumed entirely, the mass of individual prey was related directly to the mean mass of chicks on the colony at that time. The colony data were obtained from the regular weighings of samples of chicks on selected breeding groups each year. The variation in chick mass on any date results in substantial error to these estimates of means in spite of the penguin's high breeding synchrony. For the post-guard stage when the chicks have matured sufficiently for remains to be left after feeding, estimates of prey size can be refined because these remains can be used as an index of age, and, consequently, probable mass. Thus, for the post-guard stage there are three options available for estimating the mass of chicks as prey or carrion. In order of increasing precision these are: using the mean mass of live chicks on the colony at the same date (as for the guard stage) as an indication of chick size; using the prey remains as an index of probable size from regressions of mass on foot length of live chicks on the colony; using the same regression of foot length on mass but of autopsied prey, measured before feeding had begun.

(There are alternatives to these methods of estimation. One is that such close watch is kept on the skuas that all prey can be measured when taken to provide exact records of the amount taken. There are, however, serious drawbacks to such an approach. It would cause continuing disturbance to both skuas and penguins, and this disturbance would certainly affect the performance of skuas. It would make them reluctant to leave the territory to forage at sea and through distracting penguins from chick protection enhance predation success. It would, in addition, be very labour intensive, requiring special teams to watch over each group of skuas throughout the 24-hour day. Alternatively, predation could be sampled for a set duration

Fig. 8.2. The first penguin remains found in the season after skua feeding – the linked limbs and synsacrum.

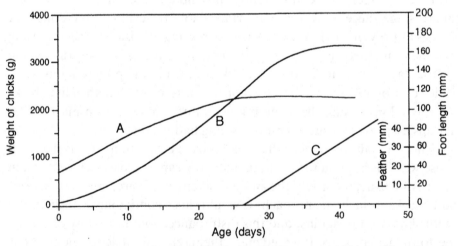

Fig. 8.3. The growth of individually marked known-age chicks on the colony in 1968–69. A, foot length; B, body mass; C, feather length from the measurement of feathers across the back. The regression equation describing the growth of these feathers is $y = -53.84 + 2.044x$ ($r^2 = 0.98$) where y is feather length in mm and x is age in days.

at intervals, but this approach becomes increasingly less productive later in the season as fewer and fewer chicks are captured.)

The few remains left after feeding severely restricts the options available for estimating live mass. In each season the first remains found after feeding are the linked legs, pelvic girdle and synsacrum (Fig. 8.2). As the chicks mature progressively more of the skeleton is left, together with the skin for the oldest chicks. Thus, attention focussed on the feet and feathering as the most useful indicators of age and size for routine autopsy of prey. Although it was soon readily apparent that both were related to overall size, a special study of their growth in known-age chicks was undertaken to understand this relationship more exactly, and especially to relate size to age for this colony.

In this study chick growth was recorded by regular measurements of a sample of individually marked, known-age chicks. The growth of the foot and the juvenile feathers is related to age and body mass in Fig. 8.3. Foot length is a good measure of body mass and age up to about 25 days when it reaches its maximum size. The feathers first show at *c.* 26 days and then grow at a constant rate. Taken together, foot length and feather length can determine age within a day or two for chicks up to about 45 days. Both of these growth rates appear insensitive to nutrition levels and the feathers appear at the same age, and grow at the same rate, in starving chicks and in well nourished ones. The same phenomenon of constant growth irrespective of condition was found earlier in the growth of the primary flight feathers of skuas (Young, 1963a) and may well be universal.

The study of known-age chicks has shown that foot length is a good predictor of chick mass throughout growth provided a series of regressions for different ages is used. It is in fact a much better predictor during early growth when both mass and size are increasing rapidly. At all times though, older chicks are heavier for the same foot length. For example, chicks of 110 mm foot length weighed on average 1390 g on 31 December, 1698 g on 5 January, 1980 g on 9 January, 2040 g on 14 January, and as much as 3000 g by the end of the month. Mean egglaying date for penguins at this colony is 14 November (Spurr, 1975c) with mean hatching date at *c.* 18 December, so that by 11 January, 25 days later, approximately half the chicks on the colony have reached their maximum foot length. From this time the regressions of mass on foot length become increasingly poorer predictors of body mass. A typical sequence of regressions for the chicks in one breeding group through the season is shown in Table 8.4.

Table 8.4. *Regression equations of mass on foot length of live chicks on penguin breeding groups*

Breeding group D29, 1968–69. Y = mass in grams; X = foot length in mm.

Date	Regression equation	r^2	n
28 Dec.	$Y = -778 + 18.87X$	0.88	40
2 Jan.	$Y = -1853 + 33.1X$	0.86	39
8 Jan.	$Y = -3299 + 46.8X$	0.59	39
14 Jan.	$Y = -10542 + 113.4X$	0.53	39
19 Jan.	$Y = -2893 + 51.5X$	0.29	30

The use of regressions of mass on foot length of autopsied chicks
Selection by the skuas of penguin chicks could take two forms. They might take younger or older chicks preferentially on any date. Alternatively, they might take larger or smaller chicks of the same age class. In the first case prey mass will be correctly determined from foot length regressions for the live chicks on the colony in each sampling period; in the second case these regressions will either over- or under-estimate mass, depending on which size was preferred. The degree of error is related to the degree of selection and becomes increasingly significant towards the end of the season as most chicks grow out of the skua's range for successful attack and only the smaller chicks can be tackled. At all stages, starving chicks which are light for their age (and foot length), attract skuas.

Because it was clear from comparing autopsied chicks with those on the colony at the same date that skuas were selecting lighter than average prey, it was decided to estimate the live mass of partly eaten chicks or skeletal remains found on the territories from regressions developed from the autopsy records of prey measured and weighed before the skuas had begun feeding on them. Estimates from live chicks on the colony would have significantly over-estimated prey mass [2]. The linear regression equations used for the estimation of mass of prey taken during the post-guard stage are shown in Table 8.5. Although it is usual to compare growth through log–log transformations, this made such small improvement to the fit or the homogeneity of the variance (assessed by the distribution of residuals) of any of the samples that this transformation was not considered necessary. The log–log regressions are, however, also given in the table.

Table 8.5. *The regression equations of mass on foot length of penguin chicks autopsied as prey taken before feeding by skuas had taken place*

Y = mass in grams; X = foot length in mm.

Date	Regression equation	r^2	n
A Untransformed			
3–9 Jan.	$Y = -2185 + 34.6X$	0.85	41
10–14 Jan.	$Y = -3667 + 49.3X$	0.63	51
15–19 Jan.	$Y = -6647 + 77.0X$	0.78	25
20–29 Jan.	$Y = -2790 + 41.5X$	0.62	17
B Transformed			
3–9 Jan.	$\log Y = -6.02 + 2.85 \log X$	0.85	41
10–14 Jan.	$\log Y = -8.15 + 3.31 \log X$	0.68	51
15–19 Jan.	$\log Y = -13.7 + 4.53 \log X$	0.77	25
20–29 Jan.	$\log Y = -6.64 + 2.99 \log X$	0.44	17

There were too few chicks available as prey late in the season to extend this series into February. All regressions were significant: by t test, $p < 0.01$.

3. *The amount of food taken from the carcass*

Skuas consume the entire body of small chicks but leave more and more of the skeleton and skin of maturing ones. There is a general pattern of feeding evident in skuas feeding on older chicks: the viscera are first eaten and only then is effort given to cleaning the flesh from the skeleton. The first remains found in each season from feeding skuas are the linked legs, pelvic girdle and synsacrum with part of the lower backbone. All the rest of the body, including the head and flippers will have been eaten. Later in the season the skuas leave completely cleaned skeletons, with or without the head. Later again much of the skin will be left with the skeleton. The amount left of the body depends on two factors: the age and maturity of the chick and the hunger of the feeding skuas. In practice a carcass scavenged by several pairs in turn will be cleaned more completely than one held by a successful pair able to get chicks more or less as needed. During field work each year chick remains were described and weighed to provide information on the proportion of the carcass eaten. As is to be expected from the variable behaviour of feeding skuas the amount left on a carcass in any week varies widely, the range for the first week in January, for example, is from 200 to 900 g, and in the second week, from 125 to

Table 8.6. *Mean mass of skin and skeleton of penguin chicks left after feeding by skuas at different times during the breeding season*

Date	Mean and SE (g)	Percentage of live mass	n
30 Dec.–2 Jan.	185 ± 17.6	15	8
3–9 Jan.	385 ± 42.7	29	25
10–14 Jan.	460 ± 30.5	30	30
15–19 Jan.	605 ± 52.6	32	22
20–24 Jan.	800 ± 40.6	36	7

1190 g. In view of this wide variation there is little point in taking the accuracy of this determination too far. For the estimations of food intake, therefore, the figures given in Table 8.6 have been used.

4.   *The significance of food sharing among the pairs in an area*

In measuring food intake of individual pairs from the direct observation of captures and, later in the season, from remains found on the territory, it is important to know how much of the carcass is eaten by the original pair. Although it is generally true that eggs and the small chicks of the guard stage are eaten exclusively by the capturing pair, larger chicks may not be able to be defended after a first feeding and may be stolen by neighbours so that a single large chick taken by one pair may feed several others over the following several hours. Viewed from another perspective a high degree of food sharing could mean that a small number of skilled and aggressive skuas could be supplying much of the food for the local skua group.

Different skuas were consistently more or less successful in holding their prey. Success depended on a range of factors including proximity of the prey to the boundary of the territory, distance from the nest area, the dominance hierarchy of the pairs in the immediate area and the hunger of neighbours at the time. Pair 25, for example, invariably lost their prey to either pair 24 or pair 26 after their first feeding, although they might be able to retrieve the carcass again later, but pair 13 on the opposite side of H4 penguin group invariably retained the few chicks they managed to take. The usual practice of skuas to bathe after feeding gives neighbours considerable advantage in scavenging. Not only are the original pair satiated, and less disposed to fight for the carcass, but only a single bird may be left on the territory at this critical time, and that one may well be tied to the nest or nest area through incubation or chick brooding. Overall, direct observation of skuas scavenging for carcasses on neighbouring territories and the movement of marked carcasses among territories indicate that although considerable scavenging occurs, the original owners, the pair that gained the chick by predation or scavenging, invariably took most of the food from it. For estimates of food intake this is an important conclusion in view of the high level of secondary scavenging exemplified for the post-guard stage in the predation record shown in Fig. 8.4. Of the 19 chick kills recorded in the figure, nine were subsequently lost to neighbours. Not until the carcass has been greatly reduced through feeding will it be light enough to be lifted in flight from the territory. Most carcasses, therefore, are scavenged by neighbours dragging them across the ground into their own territories. Secondary scavenging of this sort involves considerable risk of conflict and injury from fighting.

In the estimates of food intake it has been assumed that chick remains found on a territory belong to that pair and that all captures observed are attributed to the capturing pair. Inspection of the study territories several times each day reduced the likelihood that prey was being attributed to the wrong pair.

5.   *The status of dead chicks scavenged from the breeding groups*
      Undoubtedly, chick carrion makes up a significant part of the diet of some skuas and it would be incorrect to ignore this contribution entirely. The relative numbers of carrion and living chicks in the diet are shown in Table 8.7. The key determinants on whether these chicks should be included or excluded from this analysis are: 1. whether they are eaten or rejected by the skuas once retrieved, and 2. whether they are of much the

Table 8.7. *Numbers of dead and live chicks taken by skuas at different times of the breeding season in EF area*

In each year the number of dead chicks is given first.

	1965–66	1966–67	1967–68	1968–69	1969–70
Early post-guard					
3–9 Jan.	5, 48	9, 31	5, 22	7, 6	3, 27
Late post-guard					
10–14 Jan.	3, 31	5, 31	4, 14	2, 11	3, 5
15–19 Jan.	2, 20	8, 20	1, 4	3, 4	1, 2
20–24 Jan.	0, 3	6, 7	—	—	—

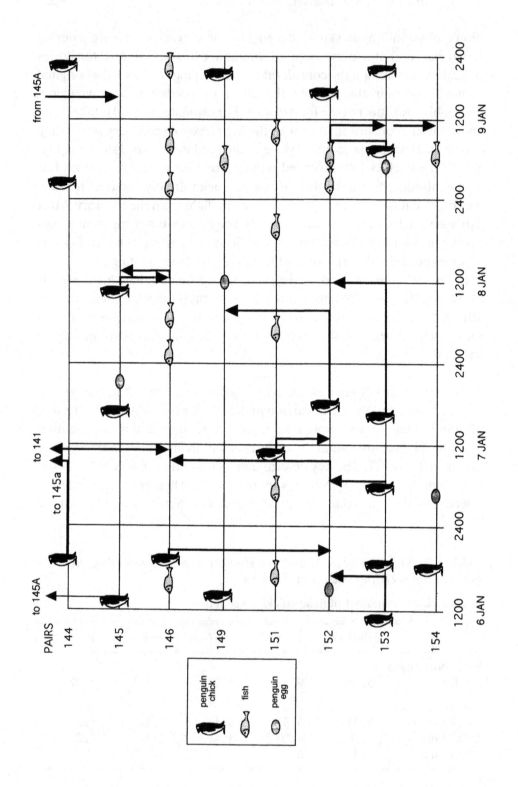

Fig. 8.5. A rare starving chick.

same size as the live chicks on the colony at the same date so that their mass can be fairly estimated. Surprisingly, most chicks retrieved when dead are eaten by the skuas. Only the most rotten ones or those flattened and dried from being embedded in the guano are rejected. Of 24 taken between 5 and 16 January for which there is information 18 were consumed normally, four were partly eaten and only two were rejected entirely. One objection to including dead chicks within each skua pair's food estimate is that chicks starved to death would be much lighter than the normally nourished ones on the colony. Although there is clearly a great difference in their mass there were so few starving chicks on this colony in any year that the point is of little importance. Starving chicks on crèches are very obvious: their hunched bodies with disproportionately long flippers are immediately apparent in even a cursory inspection (Fig. 8.5). So few chicks at this age were starved that they were treated as of exceptional interest by observers

Fig. 8.4. Food taken by 8 pairs of skuas on EF Block during the post-guard stage. Records are for the four days 6–9 January, 1970. The figure records both the small number of successful kills by these skuas and the subsequent movement of the carcasses among neighbours as secondary scavenging or kleptoparasitism. The importance of fishing by pairs 146 and 151 is also apparent.

and often weighed and measured. It must be concluded that most of the chicks taken as carrion during these five seasons died from injury, not from starvation.

The second condition needed to be fulfilled is their representativeness of chicks living on the colony. Although dead chicks certainly seemed to be more obvious, and more accessible to the skuas, once the nesting structure on the breeding areas began breaking down at the onset of crèching, a steady supply of dead chicks occurred throughout the season. Comparison between dead and living chicks during the post-guard stage using foot length as a measure of the age of chicks showed that they were not significantly different in age from live chicks during the early post-guard stage (3 to 9 January), but were on average between 5 and 10 days younger during each of the later five-day intervals (from 10 to 24 January). Too few occur after this date to have much importance in skua diet. It appears, therefore, that dead chicks are sufficiently attractive as food and sufficiently representative of the live chicks on the colony to be included in the totals.

6.   *Energy content of penguin chicks*

For this study all estimates of food intake have been made not of mass but of energy so that chick mass estimated from both the remains and the weighings of breeding groups representing the penguin colony as a whole requires conversion to energy content. Although the energy content of Adélie chicks of different ages does not seem to have been determined there is a detailed account of the way the caloric (=energy) density (in kJ g^{-1}) changes during chick growth in the closely related gentoo and chinstrap penguins (Myrcha and Kaminski, 1982). Of the two species chinstrap growth most closely resembles Adélie growth so that the regressions relating caloric density to chick mass for this species have been used. The figures for chinstrap penguins are, however, of the body minus the stomach contents. Direct application to entire Adélies could, therefore, engender significant error depending on the proportion of the total mass made up by the stomach contents and the difference between the energy values of the contents and body. Although the stomach may exceptionally attain 40% of the total mass it is usually a much lower proportion. The proportion of body mass contributed by the stomach contents (and incidentally the contribution to variance of mass) was determined by weighing the same chicks 24 hours apart after they had been held in enclosures without feeding. The change in mass was assumed to

Fig. 8.6. The apparent inability of penguin adults to recognise their chicks away from the nest site, or their apparent indifference to their plight. An inadvertent experiment from the study of overnight weight loss in chicks. Breeding group E5.

represent the loss of stomach content through digestion. Forty chicks weighed on 23 January averaged 2880 g and averaged 2466 g 24 hours later, an average loss of mass of 414 g $\pm$ 22.2 (SE) g, representing 14.5% of initial mass. The maximum loss from an individual chick was 870 g, with nine chicks losing over 500 g. A second sample from another breeding group at the same time lost on average 432 g $\pm$ 30.7 (SE) g, 13% of initial mass. These chicks were about 30 days old and when feed normally would gain c. 100 g each day.

(This enclosure experiment inadvertently illustrated a surprising feature of penguin biology. Returning parents of chicks settled at the nest and ignored their chicks collected together at the edge of the breeding area (Fig. 8.6.).)

Stomachs routinely weighed during autopsies of prey averaged c. 10% of body mass and many were empty. Even this proportion would, however, be significant if the energy values of the contents and the body were very different. The energy value of krill (4.5 kJ g^{-1} wet mass for *E. superba* (Clarke and Prince, 1980)) is similar to that of guarded and early post-guard chicks and only as the chicks mature through the late post-guard

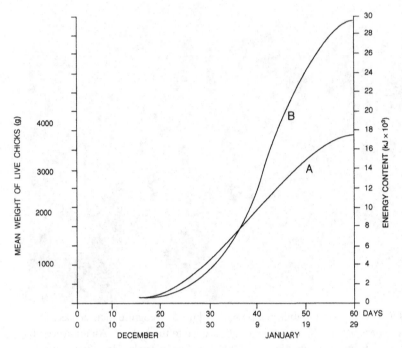

Fig. 8.7. The mean mass of chicks on the colony (A) and their energy content (B) through December and January. Energy content estimated from application of the figures for the energy densities of chicks in Myrcha and Kaminski (1982).

stage does the body value begin to diverge significantly from the contents' value. Overall, it is concluded that because of the small proportion of the body made up by the stomach and the similarity of the energy values, application of the chinstrap value to whole mass should not significantly detract from the accuracy of the equations.

The mean mass and energy content of chicks on the colony at the different stages of the breeding cycle are shown in Fig. 8.7. The scales used for the conversion of chick numbers to biomass and energy content are given in Table 8.8. These measures were used routinely for prey which were either eaten entirely, leaving no remains, or where remains occurred but were not measured.

As a general caution it should be noted that the figures given for the amount of food taken that are shown in the tables in Appendices 3 and 4 give a false impression of high accuracy and precision. They are, however, the product of a long series of estimations and extrapolations each with its associated error. There are large errors in the estimation of numbers of

Table 8.8. *The mean mass and energy content of penguin chicks on the colony at different dates*

	Mean mass of chicks (g)	Energy density[1] (kJ g^{-1})	Gross energy content (kJ)	Net energy content[2] (kJ)
Chick guard				
8–14 Dec.	100	3.40	340	340
15–19 Dec.	140	3.49	488	488
20–24 Dec.	240	3.71	890	890
25–29 Dec.	560	4.39	2462	2462
30 Dec.–3 Jan.	1050	5.45	5728	4868
Early post-guard stage				
3–9 Jan.	1550	6.54	10 128	7188
Late post-guard stage				
10–14 Jan.	2400	8.38	20 107	16 244
15–19 Jan.	2850	8.62	24 553	19 340
20–24 Jan.	3400	8.91	30 278	23 154
from 25 Jan.	3940	9.19	36 200	27 030

[1] From Myrcha and Kaminski (1982).
[2] The difference between the gross content of the entire chick and that of the remains left of the carcass after feeding. Up to 29 December it is assumed that all of the carcass is consumed by the skuas.

eggs and chicks taken, in the determination of the energy content of chicks, in the estimated mass of chicks, in the mass of the remains left after feeding, and in the proportion of the prey actually retained by the original pair. The errors are least for the incubation and chick guard stages, but are still substantial, and are greatest during the late post-guard stage where there appears to be greater selection of prey. Even if each of these steps could be estimated with a standard error of c. 10% of the mean (which would be a very acceptable level) the final figures of energy intake, the culmination of the sequences of estimates and guesses, would have an error exceeding the mean, with correspondingly wide confidence intervals. Unfortunately, each attempt to improve the precision of an estimate by refining the quality or quantity of the amount adds to the estimated error. The figures shown are the best estimates possible under the circumstances of high variability and uncertainty.

8.3.2   *Food taken by individual pairs during the different penguin life-cycle stages*

The incubation stage (see Appendix 3.1)

This stage runs from the date of the first egg being laid (c. 6 November) to the date of the first chick hatching (c. 8 December). The numbers of eggs taken by each of the study pairs, obtained from the observation logs, is the basis for this estimation of food intake. The skuas were slow each year to begin exploiting eggs and most showed little interest in the penguins until towards the end of November. Indeed some skua pairs were not even present on the territory until the end of the month. The estimates of numbers taken apply, therefore, more correctly to the second half of the incubation stage.

Records of the numbers of eggs taken for the second half of the penguin incubation stage are available for two years on H Block and three years for EF. These have been converted to energy intake by the skuas on the basis that the entire egg was consumed.

There are very big differences in the amounts gained by the individual skua pairs. Most pairs on H Block were not recorded taking eggs in either year and overall the skuas on this area gained less than those on EF. The few records available (from egg shell counts and from direct observation) of the early incubation period indicate a much lower interest in the penguins then and a much lower success overall. Skuas are able to defend and consume the entire contents of the eggs they gain (in contrast to the chicks they kill) so that eggs taken by one successful or skilled pair are not available generally to nourish neighbouring pairs.

The chick guard stage

The amounts taken by each of the pairs shown in Appendix 3.2 have been estimated from the numbers of eggs and chicks taken and the mean mass and energy content of chicks on the colony at the same time interval, together with the standard measure for the consumption of whole eggs. Dead chicks have been included in the calculations when they were eaten by the skuas. At this stage the entire chick is consumed by the skuas.

In general there are better records for this stage than the others, with more observation hours and greater numbers of prey taken, and this is reflected in the uniformity of the estimated intakes for the individual pairs. At the same time there is equal certainty of recording prey numbers in different years so that the lower rate shown for 1968–69 is probably a true indication that food was more difficult to get in this year when penguin numbers were at their lowest level.

The trends showing in the amounts taken in each interval on the basis of the constant average number of eggs and chicks for each pair are strikingly apparent: a constant success rate in taking chicks is much better rewarded at the end of the stage, when the chicks are bigger, than at the beginning. The different pairs were variably successful. Until they were removed from their territories pairs 146 and 148 were consistently the most successful pairs in EF area. Their replacements in 1968–69 (146 new pair) showed much less interest and were much less aggressive.

The post-guard stage

The estimates of food taken for this stage, shown in Appendix 3.3, are derived from the numbers of chicks taken during the early post-guard period (3 to 9 January) and for successive five-day intervals from 10 to 29 January. The numbers used here are those of chick remains and purport to be the actual totals taken, rather than estimated as in the other periods, and are recorded for each day. The accuracy of the estimates of food intake rely strongly on the accuracy of finding and assigning carcass remains correctly to capturing pairs. This accuracy depends in the main on the effort given to the inspection of territories for remains so that differences between years, or even within a year, could be an artefact of search effort. This possibility should be borne in mind when comparing intakes among different years. Comparisons among pairs is not invalidated by different search effort in the same way – once a search is undertaken all territories would be examined equally.

Eggs left on the colony at this date provide little food value and have been ignored in the estimates of food taken. As in the previous section, scavenged dead chicks were included in the food estimates where they were eaten and were roughly equivalent in size to living chicks at the time.

The numbers of chicks taken by the pairs declined through January, more than compensating for their increase in individual size, so that overall there was a steady fall through the month in the amount of food taken by each of the pairs.

Each prey item taken at this time is nevertheless a substantial food resource and can sustain a skua family for many hours. The irregularity in the pattern of food gain at this time (illustrated in Fig. 8.4) is in striking contrast to its regularity during the guard stage. Only 18 chicks were killed during these 84 hours, compared with 17 during 15 hours of records for the guard stage. The larger post-guard stage chicks are more difficult to capture, but at the same time are far more rewarding in terms of food content.

8.3.3   *Average intake levels for all pairs in the local area*

In the preceding sections the amounts for each pair have been calculated to give individual totals. Alternatively, the amounts taken by all the skuas in the local area might be averaged. This second estimate takes into account the movement of carcasses among the pairs and provides a measure of overall impact on the area occupied by these skuas. Differences in intake calculated in each way should only become significant during the post-guard stage. During the incubation and chick guard stages food taken by a skua pair is invariably consumed by them but during the post-guard stage the amount of food on a large chick is too much to be eaten at a single feeding and may become distributed among the local pairs. A number of attempts were made to estimate the significance of this secondary scavenging. The most complete account is for H Block skuas during January, 1966. In this month an estimated 167 kg of chicks were taken of which 19.7 kg (12%) was lost to neighbours after feeding. The amount lost was predominantly made up of the skeleton and skin. There were big differences in proportions lost. Pairs 12 and 13 took a total of 27 chicks and lost only one; pair 25 took eight and lost four. At the other extreme pair 23 took a single chick and gained a further six by scavenging from neighbours. Skuas demonstrate that for them this is not a trivial aspect of feeding: many birds drag carcasses away from boundaries during feeding and most will protect them strongly.

Because of the potential for quite significant movement of prey among pairs it is of interest to estimate on average how much could be made available to the local pairs from foraging by the group as a whole. For H Block the food taken during the incubation and chick guard stages is divided among the 14 pairs sharing the breeding groups. For the post-guard stage it is divided among 17 pairs with the addition of pairs 8, 10 and 17 whose territories ran with those of the foraging pairs. For EF Block the food has been divided among the eight foraging pairs.

The average intakes shown in Appendix 4 demonstrate the trends in food abundance through the year very clearly. Peak levels were attained in each year in late December and early January, falling away rather rapidly to small amounts through February.

STATISTICAL TESTS

[1] Statistical analyses comparing the rates of food capture (numbers of eggs and chicks) each year by the six to eight pairs in each year on EF Block for the four years 1966–67 to 1969–70.

1. Comparison of numbers taken each year during the incubation stage.
(1) pairs 146, 148 included. $F_{2,16} = 0.693$, $p = 0.506$.
(2) pairs 146, 148 excluded. $F_{2,12} = 1.314$, $p = 0.297$.
Means not significantly different among years.

2. Comparison of numbers taken each year during the chick guard stage.
(1) pairs 146, 148 included. $F_{3,22} = 1.98$, $p = 0.143$.
(2) pairs 146, 148 excluded. $F_{3,16} = 0.99$, $p = 0.418$.
Means not significantly different among years.

3. Comparison of numbers of live chicks taken each year during the early post-guard stage, pairs 146, 148 included. $F_{4,65} = 3.262$, $p = 0.017$.
By the Tukey Test, year 1965–66 is significantly different from 1968–69 but other year comparisons were not significant.

4. Comparison of numbers of live chicks taken each year during the late post-guard stage.
(1) pairs 146, 148 included. $F_{4,33} = 5.22$, $p = 0.002$.
(2) pairs 146, 148 excluded. $F_{4,25} = 3.01$, $p = 0.036$.
By the Tukey Test, the first two years show a significantly higher level than the others.

[2] Comparison between mean mass of live chicks on the colony and prey mass at the same date (Mean and SE)

	Live chicks on the colony (g)	Chicks autopsied as prey (g)
3–9 Jan. $t = 2.9$, df 70, $p < 0.01$	1574 ± 86.9	1240 ± 71.6
10–14 Jan. $t = 7.6$, df 90, $p < 0.001$	2478 ± 120.5	2341 ± 92.6
15–19 Jan. $t = 8.73$, df 63, $p < 0.001$	3243 ± 98.1	1872 ± 134
20–29 Jan. $t = 11.8$, df 55, $p < 0.001$	3595 ± 89	1744 ± 118

9

The costs and returns of foraging at the colony and at sea

9.1 *Introduction*

Although predators generally have a range of foods available to them and may have a diverse diet, these skuas are interesting in that they really have just two natural food sources and these require very different methods of foraging. At the penguin colony skuas can scavenge for eggs, chicks and food spilled by the penguins or attack adult penguins in order to be able to get at defended eggs and chicks. Once these prey are taken then the eggs must be opened or the living chicks killed. These all require very different techniques from foraging on small fish or plankton at sea. What then are the benefits of foraging on the colony compared with foraging at sea? Without doubt many skuas prefer foraging at the colony and only go to sea when forced to by lack of colony food. These skuas with colony food might, therefore, be considered advantaged over the majority of skuas in this region, which breed away from the penguin colonies and of necessity forage throughout the entire season at sea. Reaching this conclusion was the most significant result from the first work I did with these skuas at Cape Royds.

For the skuas breeding at Cape Bird, food is available at the penguin colony or at sea. Their daily energy needs during the demanding schedule of breeding activity and for the development of the chicks can be obtained only from these sources. Conceptually, the solution of the energy equation in this situation is straightforward: the energy requirements each day for the breeding unit can be estimated and is balanced by food energy obtained either at the colony or at sea by fishing. It is the demand for energy and nutrients for maintenance and growth that drives the foraging behaviour observed each day. Therefore, the bioenergetics of skua feeding should first be explicitly described. Whether or not the skuas forage among the penguins, and for how long, may then become explicable. Without an understanding of the bioenergetic demands the behaviour of these birds cannot sensibly be interpreted. It would merely allow description of the behaviour but without interpretation of the complex relations with the penguins. The overall aim of the account is to identify and interpret the points at which foraging decisions are made by the skuas. Key decision points shown in Fig. 1.3 are where skuas switch from foraging at the colony, with its apparent advantages of proximity to the nest and mate, to distant foraging at sea. An important question is whether these switches are exclusively energy mediated.

Because more than one food type is being taken and compared it is important to distinguish between gross food and metabolised food. At the broadest level the difference between the mass of a whole egg is clearly different from the amount actually taken up by the skua during feeding, when not only is the heavy shell discarded but much of the albumen may be spilled and lost. But even after ingestion not all the food is digested and metabolised. Some may be lost as regurgitated pellets and a proportion will be lost as faecal egesta. The comparable measure in all cases is the metabolisable energy (ME), that part assimilated. This proportion is the metabolisable energy coefficient (MEC) (=assimilation efficiency) and in birds varies between 20 and 99% depending on the nature of the food and the species being tested (Wolf and Hainsworth, 1978). The sequence in energy refinement of the prey for skuas feeding on the colony on eggs and chicks can be shown as a series of steps in which proportions of the food are lost (Table 9.1).

This chapter begins the evaluation of the two different foraging systems as the first stage in an attempt to assess the advantages of holding territories on the penguin colony. The first requirement is to evaluate the energy equations of food gathering at each place, comparing scavenging and predation at the colony with foraging at sea. This comparison depends crucially on the estimation of the energy costs of foraging on the one side and of the metabolisable energy content of the prey on the other.

Table 9.1. *Energy refinement of the prey*

Refinement stages	Materials lost
Gross energy of prey (GEP)	
$\downarrow$	Egg shells and spilled yolk and albumen Skeleton and skin of chicks
Net energy of prey (NEP) (=Gross energy intake (GEI))	
$\downarrow$	Faecal energy (faecal egesta) Ejected food pellets
Digestible energy (DE)	
$\downarrow$	Urinary waste (excreta)
Metabolisable energy (ME) (The assimilated energy for metab- olism, activity, growth and reproduction)	

Unfortunately, both measures are still clouded with uncertainty. This will become very apparent in the calculations that follow.

9.2   Foraging at sea

9.2.1   *Obtaining records of foraging skuas*

It is worth recalling that when the modern ecological studies of skuas in Antarctica began it was generally accepted that they depended entirely on the local colonies of breeding penguins and petrels and scavenging from seals. Since the 1960s (Young, 1963b) foraging at sea has become well documented from studies relying on the direct observation of undisturbed birds (Neilson, 1983; Pietz, 1987; Peter *et al.*, 1990) and from natural experiments of foraging disruption (Parmelee *et al.*, 1978). Even so, universal acceptance of the importance of sea foraging was not easily achieved. There seem to be two main reasons for this reluctance. First, skuas appear to have an enormous and easily obtained food source at the penguin colonies and their predation and scavenging there are all too obvious to even casual observation. Second, feeding at sea occurs well away from most land-based observers and can only be detected from systematic long-term observation of breeding groups, while the feeding of mates or chicks, the most obvious indicator of sea foraging, occurs very rapidly, within a few seconds of return to the territory, and is suppressed and delayed by disturbance. It is in fact virtually never seen during routine work or movement among breeding birds.

There are few data for both the foraging duration *and* the amount of food taken during individual foraging events. In observational studies of skua activity duration is recorded but the numbers and quantity of fish captured can only be estimated from the few second's sighting as it is fed to the waiting, often frantically hungry, birds on the territory. Good data on duration appear in Pietz (1987) for example, but are not matched by information on amount. Any attempt to gather detailed information on amount during these watches would quickly backfire as the skuas would become so distressed by the disturbance of rushing on to the nest area to retrieve the food before it could be re-ingested that they would become reluctant to leave the territory at all. Alternative approaches to this one have been employed. Spellerberg (1966) for example, obtained amounts from the regurgitates of chicks, but these data were not matched by foraging duration. Overall, there is now much better information on the duration of sea-foraging events than on the identity and amount of food taken.

The difficulty in getting information on sea foraging is exacerbated by the need to carry out long watches of a group of skuas. The few foraging flights that any pair carries out in a day coupled with their long duration means that it is very expensive in time to watch a single pair. To watch a group of breeding skuas, however, one needs to be some distance off to have them all in clear view and the disturbance caused in retrieving the food at feeding then affects the whole group. Overall, the difficulties in obtaining good data routinely seem unsurmountable with present methods. Clearly we need some new approaches or some new technology for this work.

9.2.2 *The energy returns from fishing*

The energy gained by skuas in foraging at sea depends on the amount taken and the species caught, each with its specific mass energy content and its metabolisable energy coefficient (MEC). These need to be determined before foraging gain can be estimated.

In McMurdo Sound the Antarctic silverfish *Pleuragramma antarcticum* (Fig. 9.1) is by far the most important prey of skuas foraging at sea and estimations of energy content of fish prey are derived from this species. The measure of energy density used for this account is 6.3 kJ g^{-1} wet weight (R. W. Davis, pers commun.).

There does not seem to be any information on MEC for this species. Silverfish compensate for the absence of a swim bladder by gaining neutral buoyancy through lipid sacs and by reduction of skeletal ossification

(a)

(b)

(DeVries and Eastman, 1978). (Ash constitutes between 0.3–0.5% of the total body mass compared with the 2% usual for fishes.) Both features increase the energy density of the species and the weak ossification should also increase MEC. Metabolisable energy coefficients for piscivores are c. 75–80%. Furness (1990) uses 80% for great skuas feeding on sandeel (*Ammodytes marinus*) and Gabrielsen *et al.* (1987) determined an efficiency of 76% for kittiwakes (*Rissa tridactyla*) fed capelin (*Mallotus villosus*). Because of the weak skeletal ossification of silverfish the assimilation efficiency is no doubt greater than for these species, thus for this work an MEC of 85% is assumed, giving a metabolisable energy content of 5.35 kJ g^{-1} fresh weight.

There are few data on the amounts of fish taken during each foraging event. Young (1963b) measured the amounts taken by one skua during three successive flights as 75 g, 105 g, and 140 g. This was the mass of fish disgorged on to the ground in chick feeding on return to the territory. Spellerberg (1966) estimated a mean fish mass of 106 g (range 95–118 g) from chick regurgitation after feeding.

These seem small amounts for such a large bird, about 8% of body mass, but are about the same as calculated for the great skua in the Shetland Islands (Furness and Hislop, 1981) from weighing chicks before and after feedings in which all the food regurgitated by the parent was eaten. The measurements for the great skua gave a mean mass fed to chicks of 91 g (range 50–125 g). These authors consider that the maximum amount that can be carried is 'not much more than 125 g' noting that although the otoliths of heavier fish, up to 200 g, sometimes occur these are rare.

9.2.3 Energy costs of foraging at sea

The energy needed in foraging is a product of the flight metabolism and the duration of the flight. Skuas foraged long distances out to sea from the colony so that almost all the energy costs of sea foraging are from flight to and from the foraging zone.

Fig. 9.1. (*a*) Antarctic silverfish as disgorged onto the ground from a foraging skua. The fish to the left of the pliers has the head already digested from its carrying position in the crop. The other two are in exceptionally good condition, suggesting a short foraging distance. (*b*) Antarctic silverfish being feed to chicks by the two birds of a pair of skuas.

The energy costs of skua flight

There are no direct measurements for skuas of the power required for flight and this must be estimated from research on similar species or from general equations. Four avenues are available. These are: 1. use of a fixed multiple of basal or existence metabolic rate; 2. use of aeronautical principles (Pennycuick, 1989); 3. use of allometric equations using empirical values, for example those of Kendeigh *et al.*, 1977; and 4. use of regression equations of measured power and mass determined by the doubly labelled water technique (Birt-Friesen *et al.*, 1989). The accuracy of any of these approaches for estimating skua flight metabolism depends on similarity in body size and proportions and on flight performance. Flapping birds with stubby wings reluctantly lurching into the air cannot be expected to fly with the same economy as the oceanic petrels or albatrosses. Determination of the flying mode is the first decision to be made in looking for comparative species. Although skuas wheel and glide fairly well about the colony, up to moderate wind speeds, their flights to and from the foraging areas at sea are characterised by steady wing beating; they are at these times flapping birds rather than gliders. In high winds their flight is markedly affected. They compare poorly with snow petrels *Pagodroma nivea*, for example, under these conditions when the two birds are flying together across the colony. They clearly lack the facility to exploit turbulence and ground effects that marks the boisterous passage of petrel flocks. The effect of strong winds on flight performance may be as significant as that found by Gabrielson *et al.* (1987) for kittiwakes, where energy use increased 1.5 times during flights in strong winds.

For the estimations of the power needed for foraging flights skuas are considered flapping birds, flying too high in the air column to gain much benefit from the turbulence and lift of winds at the sea surface.

Range of estimates of metabolic costs of flight

In this series of estimates of the metabolic costs of flight, skua mass (M) is taken to be 1.315 kg (the mean found for breeding birds at Cape Bird, section 5.4). Power and energy use units are watts (W) and kJ h^{-1} (kcal times 4.1868) respectively. The estimates are of the total metabolism during flight, the sum of basal and specific flight metabolism. These values may then be compared directly with the energy use of birds foraging on the penguin colony, equating flight with scavenging and attack activities.

1. Comparisons by Pennycuick (1989) and Flint and Nagy (1984) show such wide variation in the ratios of basal to flight metabolic rates for different species that the method of using a fixed multiple of the basal rate

to estimate flight rate has little merit unless used for closely similar species. It is not considered further.

2. The flight power model of Pennycuick (1989) utilises mechanical components of horizontal flight together with the estimated basal metabolic rate (from the Lasiewski and Dawson (1967) regression equation) and an 'overhead' for the increased blood flow and respiratory physiology associated with flight. This total, the mechanical power, is converted to chemical power, the usual measure based on fuel or oxygen consumption by the use of an energy efficiency conversion ratio of 0.23.

For a skua of mass 1.315 kg and wing span 1.39 m (Spellerberg, 1966) the model computes values of minimum power speed (V_{mp}) of 10.5 m s^{-1} (37.8 km h^{-1}) and power of 88.7 W (319 kJ h^{-1}) and maximum range speed (V_{mr}) 17.2 m s^{-1} (61.9 km h^{-1}) and power of 111.7 W (402 kJ h^{-1}). For skuas undertaking lengthy foraging flights the maximum range speed, the speed giving the most economic flight for distance covered, is the more appropriate estimate.

3. Equation 5.43 of Kendeigh *et al.* (1977) for birds in sustained flight is $E = 0.3157M^{0.6980}$ where E is in kcal h^{-1} and M is in g. For a skua of mass 1315 g this equation gives an energy use value of 198.7 kJ h^{-1} (with a power of 55.1 W). The additional energy cost of flight over the basal rate (equation 5.45) is 171.4 kJ h^{-1}.

4. The approach of deriving regression equations from empirical data has been utilised by Birt-Friesen *et al.* (1989) to produce a general relationship for the metabolic rates of birds during flight or pursuit diving. The data for this equation are from rates measured directly using doubly labelled water in six species in conjunction with time budget data to record the amount of time flying. Their equation is: log $E = 1.86 + 0.748$ log M (or $E = 72.4 \, M^{0.748}$) where E (existence plus flight energy) is in kJ h^{-1} when M is in kg. This regression, admittedly tentative from such a small sample, gives the flight metabolism for skuas weighing 1.315 kg as 88.9 kJ h^{-1} (with a power of 24.7 W).

It is difficult to know how to assess this range of measures and select the one most likely to represent true skua foraging costs. The highest values are from the Pennycuick (1989) model (402 kJ h^{-1}), and these are over four times as much as those calculated from the Birt-Friesen *et al.* equation (88.9 kJ h^{-1}). The estimates from Kendeigh *et al.* (1977) fall midway between these two. These individual estimates, however, may be compared with results of recent work on species with similar ecologies. In the kittiwake for example, a sub-Arctic seabird, the field metabolic rate (FMR) of foraging birds away at sea for a day at a time was 3.16 times the basal rate for this species and although individual FMRs in the sample exceeded five times the basal rate overall this is a rather modest increase for flight (Gabrielsen *et al.*, 1987). Similarly modest rates were found also in the sample of flying birds considered by Birt-Friesen *et al.* (1989). One of the best studies for comparison is also one of the earliest. Baudinette and Schmidt-Nielsen (1974) trained wild-trapped herring gulls (*Larus argentatus*) to fly in a wind tunnel so that direct measurements of oxygen uptake could be made. When gliding, two birds had metabolic rates of 12.5 W and 15.4 W ($=45$ kJ h^{-1} and 55.4 kJ h^{-1}), respectively 1.9 and 2.4 times resting rates. Unfortunately, the rates for these birds during flapping flight were not determined but would have been significantly higher.

The basal metabolic rate for skuas is between 500 kJ day^{-1} (R. W. Furness and D. Bryant, pers commun., 1991 for great skuas) and 800 kJ day^{-1} (south polar skuas, Ricklefs and Mathew, 1983). The lower rate is equivalent to c. 21 kJ h^{-1} and the higher one 25.8 kJ h^{-1}. These rates in relation to the flight rates shown above suggest that the Pennycuick model estimate is much too high, being 12–19 times BMR and even that from the Kendeigh *et al.* equation (7–9.5 times BMR) also seems high. It appears therefore, that the rate from Birt-Friesen *et al.* (c. 90 kJ h^{-1}, 3.5–4.3 times BMR) is the best approximation. All these rates will be related later to the foraging duration in determining a cost–benefit equation for foraging. Unrealistically high metabolic rates will stand out there when compared to the likely return from foraging.

The duration of foraging flights

The skuas at Cape Bird invariably flew towards the north when foraging, flying sometimes tight about the glacier to the immediate north of the colony (Fig. 9.2), sometimes further to the west, heading out directly into McMurdo Sound. No flights were recorded to the south. Most skuas at Cape Royds also fished towards the north of that colony (Young, 1963b), although some flights from there also went south. The clear separation in flight direction between foraging flights towards the northern quadrant and southerly flights along the beach towards the bathing pools facilitated record keeping. These two flight patterns were distinguished also by their altitude. Birds heading out to sea steadily increased height, those to the bathing ponds sped along the shore line, flying just beyond the edges of the coastal territories at low altitude, no more than a few metres above ground.

Birds leaving or returning to the territories were invariably tracked to and from the area by eye, often by field glasses as well. On days with good

Fig. 9.2. The ice cliffs to the north of the Northern Colony around which the skuas fly when foraging to the north of Ross Island. Beaufort Island shows on the horizon.

visibility over the sea, field glasses kept skuas in sight for up to eight minutes, to about 7 km, before they began merging with the broken background of the mountains on the far side of the Sound or their track became confused with other birds in the area. These trackings were important for establishing unequivocally that the birds were heading off on foraging flights and were not simply going out to the nearest iceflow or clear water pool for sea-water bathing. Skuas beginning a foraging flight were also marked by characteristically strong flight, heading in a set direction and seldom deviating. At the end of the flight discrimination between local and foraging flights was just as clear. Birds returning to the territory from foraging appeared even more purposeful than when leaving, flying hard and directly on to the territory from high altitude. Until the end of incubation, while the male was still feeding the female, his imminent arrival was often indicated by the behaviour of the mate on the nest. Even when the male still appeared as nothing more than a small dot on the horizon the females became excited, and abandoned the nest to run to the area where feeding usually occurred. They also seemed to be able to discriminate whether the male was carrying food or had been bathing. Returns from bathing were never met with an expectation of feeding. The ability to identify mates was quite uncanny. In the numerous records of female response to a returning bird incorrect identification was never observed. Even where a single bird was returning towards close-packed nests invariably just one female responded and this was invariably the mate. In the great majority of returns the male immediately fed the female or chicks and it was often possible to see the food being fed. This event was a clinching point in the certain determination of foraging behaviour. In a small proportion the return was not followed by any feeding, sometimes none was expected, in others the female begged strongly. Older chicks always begged, often so fiercely that the returning bird had to fly away from them briefly to gain some respite from harassment. Once the chicks had hatched both males and females foraged for them and the parents may then be reluctant to feed each other so that chick feeding is delayed until they can be fed alone. If the eggs or chicks are lost the female returns to begging from the male. Males are at first reluctant to feed their mates at this time and again feeding may be delayed, even avoided, on the return to the territory. On a small number of occasions, however, it was hard to avoid the conclusion that the failure to feed was because foraging was unsuccessful. Overall, though, the overwhelming majority of foraging flights were immediately terminated by feeding the mate or the chicks. The small number of exceptions were so unusual that they provoked comment in the daily record sheet.

There is another good indicator of sea foraging in early and late summer. When air temperatures are very low, returning birds are masked with ice, frozen onto the breast and head feathers. In exceptionally severe conditions the face was so heavily armoured that vision seemed obscured.

There is no information at all on the distribution or abundance of silverfish in the Ross Sea or locally within McMurdo Sound. However, what is apparent from the foraging of skuas is that there were few if any accessible shoals close to shore. Only once during the season's study at Cape Royds were skuas seen fishing near the shore and none has been seen fishing close in during many hundreds of hours observation at Cape Bird.

When assessing whether the skuas on the penguin colony could be supported exclusively by the penguins in their territories, evidence of sea foraging, from flight direction or flight behaviour or from seeing fish being fed to chicks or mates, was adequate. Determining the duration of foraging, however, as distinct from merely recording its occurrence, requires that the times of leaving and return are monitored. In watches of 6–8 hours, the usual spells in this study, many foraging duration records were inevitably lost as one or other of these points fell outside the observation period. The numbers of records that could be included in the analysis of foraging duration would be increased (from 122 to 138) by using

Table 9.2. *Mean foraging times for skuas foraging at sea*

The records are in minutes (mean and sample standard deviation). Number of records on each observation log occasion are shown in brackets.

	1966–67	1967–68	1968–69	1969–70
15–30 Nov.		41 ± 8 (11)	32 ± 16 (9)	—
Dec.	200 ± 86 (5)	352 ± 74 (2)	39 ± 14 (3)	263 ± 137 (11)
		52 ± 3 (2)	46 ± 21 (5)	112 ± 51 (5)
		93 ± 32 (3)	94 ± 66 (4)	124 ± 64 (14)
		94 ± 25 (5)	71 ± 19 (3)	
			63 ± 37 (3)	
			85 (1)	
			73 ± 28 (6)	
1–20 Jan.		87 ± 32 (2)	53 ± 17 (7)	223 ± 105 (18)
				58 ± 36 (5)

the median, rather than the mean, duration but while allowing the inclusion of lengthy flights going beyond the observation period this change would bias the records in the other direction by under-estimating the number of short duration flights beginning just before observations began. The records shown (Table 9.2) are the mean durations for 124 completed foraging flights.

There is a very large range of foraging times recorded in these figures, with mean durations ranging from 32 minutes to 352 minutes – from under an hour to six hours. Nor is there a consistent pattern with stage in the season; both the shortest and longest times occur early in the year. On the whole, consistently longer durations occurred during late December than earlier or later. Comparisons within years shows that the time needed could quickly change, for example, from 42 minutes on 25 November, 1967 to over 300 minutes on 5 December; from 46 minutes on 4 December, 1968 to 106 minutes on 10 December. Such variability was clearly unrelated to overall seasonal changes, for instance to the progressive loss of ice cover in the Ross Sea through summer. On the contrary, the variability suggests dependence on rapidly changing and extremely variable environmental conditions. No doubt this wide variability imposes very different energy costs on the skuas from day to day.

There are two general factors that combine to determine the success of experienced foragers. These are the movement and density of the pack-ice, determining access and visibility, and the availability of silverfish at the surface which can be captured by plunge diving. Subsidiary factors are wind force and general visibility. Although both can affect flight speed and navigation they were much less significant than ice cover at Cape Bird.

Until about mid to late January, when the Ross Sea is essentially free of pack-ice, the amount of pack-ice in McMurdo Sound, and within the southern sector of the Ross Sea within the flight range of the Cape Bird skuas, is determined by the action of two opposing factors. The coastal current flowing about Cape Bird and past the colony to the south carries pack-ice into McMurdo Sound from the Ross Sea. Countering this steady build-up are the periodic southerly gales which blow the ice out towards the north. These opposing trends have been well described by Ainley and LeResche (1973) for the ice cover along the northern side of Ross Island at Cape Crozier. The Cape Bird colonies, and more particularly the Northern Colony, are protected from the main force of this wind, often lying within a light north east wind, the back eddy from around Ross Island, while the southerly storm rages past a short distance off shore. Until well into January, when most of the ice has dissolved throughout the Ross Sea as a whole, intervals of calm or light winds see the Sound blocking steadily with ice. This packing is very rapid. Within a day of a two-day southerly storm which has pushed the ice beyond the horizon it is back again. Similarly, the storm effect is equally rapid. Southerly winds overnight can clear the Sound. Some of these ice conditions are shown in Fig. 9.3.

The wide variability in foraging times recorded for these skuas makes it very difficult to provide useful estimates of the foraging costs without relating these to foraging conditions. The two environmental factors of ice cover and weather conditions stand out as being the most important elements for foraging skuas so that in reviewing the foraging costs of the skuas it is important first to consider times achieved by the skuas under various sea-ice and wind conditions (see Table 9.3). The broad categories considered are:

- open water, light winds or calm conditions;
- open water, storm conditions;
- moderately closed pack-ice, from 40% to 80% cover, calm or light winds;
- dense pack-ice, over 80% cover, calm or light winds.

(It is not possible, as symmetry might suggest, to have a category 'southerly storms with pack-ice'; these winds soon clear the Sound of pack-ice. Moreover, there were very few times at Cape Bird when visibility was

Table 9.3. *Foraging times under different conditions of ice cover and wind strength*

Data are mean duration and sample standard deviation.

Date	Foraging times (mins)	*n*
1. Open water, calm or light variable winds		
25–27 Nov.	41 ± 8	11
27 Nov. 1968	32 ± 16	9
16 Jan. 1969	53 ± 17	7
19 Jan. 1970	58 ± 36	5
Overall mean	43 ± 17	32
2. Open water, foraging in southerly storm conditions		
29 Dec. 1966	200 ± 86	5[1]
8 Dec. 1967	93 ± 32	3
13 Dec. 1967	94 ± 25	5
10 Dec. 1968	94 ± 66	4[2]
11 Dec. 1968	85	1[3]
29 Dec. 1968	71 ± 19	3
12 Dec. 1969	two records >173, >260	
Overall mean	117 ± 70	21
3. Moderately heavy pack-ice, light variable winds or calm conditions		
15 Dec. 1967	52 ± 3	2
1 Dec. 1968	39 ± 14	3
4 Dec. 1968	46 ± 21	5
17 Dec. 1968	73 ± 28	6[4]
20 Dec. 1968	63 ± 37	3[5]
18 Dec. 1969	124 ± 64	14
Overall mean	86 ± 56	33
4. Dense pack-ice, light variable winds or calm conditions		
5 Dec. 1967	352 ± 74	2[6]
9 Dec. 1969	263 ± 137	11
29 Dec. 1969	112 ± 51	5
6–9 Jan. 1970	223 ± 105	18
Overall mean	227 ± 120	36

[1] plus one, >240 mins; [2] plus two, >77 mins, >66 mins; [3] plus one, >105 mins; [4] plus three, >32 mins, >63 mins, >155 mins; [5] plus one, >210 mins; [6] plus six, >115 mins, >120 mins, >65 mins, >85 mins, >390 mins, >300 mins.

(a)

(b)

(c)

Fig. 9.3. The difference between smooth fast ice at the start of the season (*a*), and jumbled pack-ice later in the year (*b*). The sound clears of pack-ice late in summer (*c*). Small amounts of ice remain attached on the beach of B Block. View from the Northern Colony towards the west.

so affected by dense fog or low cloud that skuas were prevented from foraging, so this factor can also be ignored in this general review.)

The first conditions should offer the most favourable foraging conditions for skuas, with silverfish both easily visible and accessible, the last should preclude foraging, by cutting off access to the sea. Times taken in foraging during the storms, when the ice has been swept away, test the skuas' ability to find and catch fish in broken water. For most of the summer the Sound is moderately ice-covered, so this category is probably the most representative of foraging conditions in this region.

It is possible that a mosaic of iceflows and open water may provide the best conditions of all, providing still water ponds among the iceflows that should enhance visibility into the water.

It must be appreciated, however, that the assessment of pack-ice density had to be made from the land and that the local conditions recorded may not faithfully mirror the conditions of the skua foraging areas far out in the Sound to the west or northwest, or along the northern coast of Ross Island.

Nevertheless, there was clearly a very general correspondence, and the ice conditions seen locally in McMurdo Sound corresponded with those being recorded by Ainley and LeResche (1973) at the same time from Cape Crozier 100 km away.

The foraging durations (Table 9.3) record a very marked difference in the skuas' ability to forage under different ice and wind conditions. The shortest times recorded for skuas at Cape Bird were three flights of 17–18 minutes each on 27 November, 1968. These times were unusually short and about 30 minutes is a more realistic minimum duration for these birds. It is not possible to record the longest durations so readily as skuas may well simply give up foraging under the worst conditions, returning to the colony after hours or days. The longest certain record of a skua returning with fish back to the territory was 414 minutes (nearly seven hours) on 8 January, 1970, when all flights on that day were protracted and the 18 monitored averaged 223 minutes.

Overall, the durations recorded here for the different ice conditions show that skuas took twice as long to forage in moderate pack-ice than in open water and five times as long in dense pack-ice. These are predictable responses taking the impact of the ice cover into account and knowing that in severe ice conditions the birds were apparently flying around Cape Bird into more open water to the northeast and east of Ross Island. Even so some birds managed to exploit pockets of open water even when the ice pack appeared solid to the horizon. For example, on 9 December, 1969 the male of pair 149a completed two successful foraging flights of just 41 and 54 minutes respectively when the other skuas in this area were taking on average 311 ± 96 minutes. The second flight occurred at an unusually short interval after the first, as though the skua appreciated how fortunate it was. Usually, however, the foraging times on any occasion were rather uniform. On 16 January, 1969, for example, the six flights (each by a different bird) took 50, 30, 49, 42, 42 and 80 minutes. This uniformity suggests that most of the time was taken up with flying to and from the foraging area, and little time had to be spent in searching widely for prey once foraging began.

The skuas' ability to forage in rough water during the southerly gales, testified by a number of very good foraging times in these conditions, was more surprising. Even on the worst days of the storms the skuas headed out directly into the Sound to disappear from sight into the low boiling clouds and furiously broken white water that accompanied the high winds. As far as could be determined from following them with field glasses they

Table 9.4. *Metabolisable energy gain by skuas from foraging at sea under different ice and wind conditions*

Wind and pack-ice conditions	Flight duration (mins)	Energy gain (kJ min^{-1})
Open pack-ice	17 (minimum)	39.4
Open pack-ice	43 (average)	15.6
Moderate pack-ice	86 (average)	7.8
Dense pack-ice	227 (average)	2.9
Southerly storm conditions	117 (average)	5.7

For this example an energy load of 670 kJ is assumed from a wet mass of 125 g of fish, with an energy density of 6.3 kJ g^{-1} and a MEC of 0.85. The flight durations are from Table 9.3. The minimum time shown (17 minutes) is the shortest time recorded for foraging by these skuas.

made no attempt to avoid the winds and rough sea by attempting to forage in the sheltered water near the penguin colony (the rough seas began 1–2 km offshore running to the north from the tip of McDonald Beach) or fly around the Cape to more sheltered water along the north coast of Ross Island. Their ability to forage in the storm conditions on these records is about the same as for foraging in moderate pack-ice. Apparently silverfish are at the surface during these storm conditions and skuas can both see them and catch them there.

The mean figures for foraging under different conditions have been used in Table 9.4 to show the metabolisable energy gain for each from an average load of 125 g fish. These gains range from 39.4 kJ min^{-1} for the most favourable foraging conditions to 5.7 kJ min^{-1} for the least. The costs and gains for skuas foraging at sea have been brought together in Fig. 9.4 comparing the cumulative costs of flights against the possible energy returns from the load of fish brought back to the territory. The wide differences among the estimates of flight energetics derived from the different methods is immediately apparent. The highest estimated rate of energy expenditure (from the Pennycuick, 1989) model seems to preclude net gain in all but the most favourable conditions, whereas the lowest rates still give a positive energy return for flights during dense pack-ice (which have a mean duration of 220 minutes).

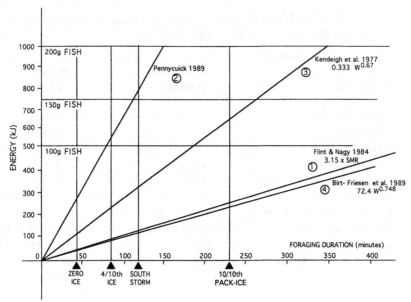

Fig. 9.4. The relationship between estimated energy costs of skua flight and energy return from food gained during foraging at sea. The four estimated rates are from 1. Flint and Nagy (1984); 2. Pennycuick (1989); 3. Kendeigh *et al.* (1977); and 4. Birt-Friesen *et al.* (1989). The energy value of the fish taken in each flight was estimated for three loads (100 g, 150 g and 200 g) of *P. antarcticum* (having an energy density of 6.3 kJ g^{-1} and MEC of 0.85).

9.3   *The energy costs and gains from foraging on the penguin colony*

9.3.1   *Simplifying the variables*

The uncertainties evident in the previous section on foraging at sea appear models of exactitude compared with the difficulties of developing energy cost and return equations for foraging on the colony. Where sea foraging exploits a single rather standard prey, skuas on the colony exploit an array of prey, of differing types and energy content, varying through the season, but at times available together. Whereas sea foraging has a major energy cost in the flying component, which can be measured, albeit roughly, and which has an observable start and end to give an elapsed time for each individual event, foraging on the colony has instead a subtle intermix of resting and intense activity, of sitting and watching, of flight, of aggressively active encounters. Nor is there generally a precise start and end to a foraging cycle, the activities drift throughout the 24-hour day punctuated by the successful obtaining of prey, the necessities of brooding and incubation spells and the imperatives of territorial defence. How then is the duration of colony foraging to be recorded? The time in hours and minutes of the observation watch? The time the skuas were engaged in scavenging or predatory behaviour, no matter how tentative? Or the time they were actively engaged in flight or attacks requiring the use of energy in excess of the resting state?

The calculations are simpler in the early part of the season where the prey at least is standardised as eggs. But these may be gained in a few seconds when thrown out of the nest by the penguins or gained after only the most demanding and long-lasting attack sequences, taking hours rather than minutes. For the skua the return in energy is identical, but the cost in time, energy use and risk of injury is very different. In part this difference in effort reflects the availability of prey, the numbers of nests within the territory. This variability provides an additional factor not so obviously evident for the skua foraging at sea, where a common prey source is available to all birds.

It is necessary also to acknowledge that there are great differences in the interest and ability of skuas in exploiting the penguin food within their territories. Some obviously work hard and continuously to gain eggs and chicks, others act merely as scavengers, others show almost no interest. Which of these is to be used as the norm for working out returns for effort? The first gives the more realistic measure of what might be achieved; the second gives the most favourable return, as scavengers can pick up food with almost no effort involved; the third acts as if holding a territory away from the colony. Should this last group be included within the colony-wide calculations of effort and return? Indeed, how can this mess of variables be rationalised so that meaningful figures of the cost and return from foraging on the colony might be determined?

The first requirement for rationalisation seems to be to exclude 1. pairs which show little interest in the penguins and 2. pairs which appear to be using colony food only as a backup for sea foraging. The remainder are then to be considered dedicated predators and scavengers on the penguins which only take up sea foraging when forced by hunger to do so. The switch from one foraging place to the other should in these skuas be quite mechanistically predicted.

A second simplification would be to calculate effort and return separately for the various stages of the penguin breeding cycle. Essentially, this means a return to the stages used so far throughout this analysis. The uniformity of the egg as prey, either predated or scavenged, offers the best likelihood of reaching an accurate measure of gains at the colony comparable with the measures for foraging at sea. A second advantage of measurements at this stage is that this stage precedes skua egg hatching so that the foraging units uniformly comprise two adult breeding birds.

A third rationalisation would be to measure the effort when it can be fairly well assumed that the skuas are working as hard as is possible. One way to be assured of this is to measure their effort when it is known that sea foraging is both difficult and time consuming, when the option of getting fish has become energetically expensive. Skuas faced with repeated flights of over four hours a time to forage at sea might well look more attentively at the penguin food closer to the nest on the territory (but see [1]). There is good evidence for this view in the many examples in the observation records of skuas with hungry chicks suddenly carrying out quite dangerous attacks on penguins, quite out of character with their usual timidity. Similarly, others prevented from going to sea at all by enveloping fog have taken to cannibalistic attacks on chicks of neighbouring pairs, a very dangerous activity indeed for skua predators.

Nothing has been said so far about how to include the wide variation in penguin nest numbers within the different territories. The solution to this problem is surely to examine effort and return for different ranges of numbers of nests. Even the most capable skua predator cannot survive on a few nests, whereas large numbers provide a 'cushion' for those less able.

The exclusion of many of the skuas on the colony from these calculations does not dispense with them altogether. Overall, when assessing the benefits of nesting with or away from penguins, the value of the penguins as a local food source, as an alternative to sea foraging, or as an insurance against its loss, will again be considered. For the moment the concern is to measure the costs in energy terms of living entirely, or as far as possible, on penguin food for comparison with the costs of foraging at sea.

9.3.2   *The energy costs of foraging on the penguin colony*

The several assumptions and simplifications documented in the previous section have been taken into account for the specific requirements of calculating the energy costs and gains from foraging on the penguin colony. The most important has been in the way in which pairs with similar numbers of penguin nests in the territory have been grouped together, irrespective of whether the territories were on H or EF areas. This decision recognises the cardinal significance of prey availability to the skuas. It also allows means to be determined for each prey-number group bringing together both successful and less successful pairs. The second simplification adopted has been to group records into five different stages of the penguin breeding cycle – the egg stage (from 20 November to 9

December) early and late guard stages (from 10 to 20 December and 21 December to 2 January respectively) and early and late post-guard stages (3–9 January and 10–20 January respectively). The first stage is of predation of eggs, the second two of eggs and young guarded chicks, and the post-guard stage of large chicks. By grouping the data in this way the variability in numbers of records can be smoothed to provide generally adequate measures for the skuas at each stage of the penguin breeding cycle, with its sequence of very different types of prey.

One of the difficulties in attempting to calculate rates of food gathering when the focus is on individual birds and pairs is to account for records of activity that failed to obtain prey. By summing the activities of a group of skuas these records can be incorporated in an overall measure of efficiency. Unsuccessful records should not, however, be lost sight of as they are an important measure of uncertainty in foraging at the colony. The perceptions of probable foraging success of skuas that consistently obtain food on the colony must surely be different from those that are consistently unsuccessful, irrespective of the amount of effort expended.

The third simplification has been to convert all activity records to minutes of activity and convert the different types of food taken to standard units of energy (kJ). In the end what is required of the analysis is a set of estimates that can be used to compare directly the costs and benefits of foraging on the colony with foraging at sea. The fourth simplification has been to equate the rate of energy expenditure during active preying and scavenging on the colony with that of flight. For this reason only the more vigorous behavioural attack and scavenging categories (categories 4–38; see Table 7.1) are used for the compilations of activity and energy use. This restriction has been necessary because most records of skuas interacting with penguins are of alert skuas watching the penguins (behavioural categories 2 and 3), activity little different from resting. Finally, although foraging at sea is an individual activity, and foraging on the colony often a shared activity of the pair, to allow comparison the costs and benefits of each should be determined on an individual skua basis. In order to demonstrate how effective foraging can be on the colony the individual records for the two most successful birds, the males of pairs 25 and 153, are also shown.

Organisation of the data set

The calculations that follow derive from the records made of skua interaction with penguins during the observation logs, specifically of the more vigorous interactions involving both flight and ground searches and attacks.

The skua pairs were grouped on the basis of numbers of nesting penguins within their territory.

These were:

I   More than 250 nests – pairs 144, 146, 146 new pair, 148, 152, 153.
II   Between 100 and 249 nests – pairs 145, 154, 9, 25.
III Between 50 and 99 nests – pairs 11, 12, 13, 27.
IV   Between 20 and 49 nests – pairs 149, 151, 22, 24.
V   Between 1 and 19 nests – pairs 145a, 150, 152a, 8, 14, 19, 20, 21, 26.

This banding is sufficiently wide to allow for the variation in penguin nest numbers that occurred in the various years of the study. Pairs 8–27 were on H Block, the remainder were on EF Block.

Energy content of the prey

In tabulating the records of energy gain only those prey that were eaten by the skuas have been included within the totals: eggs or chicks scavenged by them and discarded without being eaten have been excluded. At some stages of the year carrion occupied a very large part of the skuas' interest in the penguins, and were as vigorously worked for as living prey, but if not eaten they cannot be included within the energy equation as a gain for effort.

Eggs

As in the earlier sections the energy content of eggs at all stages of incubation is assumed as 540 kJ.

Penguin chicks

There do not appear to be any published determinations for assimilation coefficients for skuas for any of their diverse diets. It is necessary, therefore, to use records for other species. For carnivorous birds as a whole the MEC is between 68% and 87% (Wolf and Hainsworth, 1978). Recent publications of birds feeding on terrestrial prey, for example, give 75% for kestrels (*Falco tinnunculus*) feeding on birds (Masman *et al.*, 1986); an average of 77% for raptors (Bozinovic and Medel, 1988); 61–79% for owls (*Asio otus*) feeding on mice (Wijnandts, 1984); 71% for *Bubo virginianus* feeding on turkey poults (Duke *et al.*, 1975). In the absence of more specific information on skuas or closely

Table 9.5. *Average rate of metabolisable energy gain from penguin eggs and chicks by skuas with different numbers of penguin nests in the territory*

Units are kJ min^{-1} for individual skuas.

| Skua group | Incubation | Guard stage | | Post-guard stage | |
		Early	Late	Early	Late
I >250 nests	8.1	5.7	11.9	30.6[1] 10.8[2]	9.0
II 100–249 nests	11.7	1.9	39.4	17.2[1]	—
III 50–99 nests	14.0	1.7	11.4	—	—
IV 20–49 nests	14.2	10.1	5.1	52.0[1]	32.3
V 1–19 nests	0[3]	0[3]	0[3]	—	—
Skua pair × hours	1236	904	1242	1280[1] 328[2]	312

Activity categories 4–38, active searching from flight or ground, scavenging and predatory behaviours.
[1] Records over 80 h continuous log of behaviour 6–9 Jan. 1970.
[2] Records from a series of logs totalling between 13 h and 21 h respectively during 1967–68 and 1968–69.
[3] No food taken by any skuas.

related species an MEC of 80% (Furness, 1978) is used for skua feeding on penguin chicks and 90% for feeding on eggs. This latter figure represents the mean of an assumed 100% efficiency for fresh eggs and 80% for those in which the chick is well developed.

For the guard stage the energy density of the chicks is averaged as 3.5 kJ g^{-1} during the early stage and 5 kJ g^{-1} during the late stage. Chicks of this age are almost completely ingested.

During the post-guard stage the chicks are only partly eaten, with skin and much of the skeleton left. For these chicks the MEC is the product of the proportions ingested and assimilated. For the early stage on average 80% of a chick is eaten, for the late stage 75%. With an assumed assimilation of 80% the overall MEC is 64% of gross mass for the early stage and 60% for the late stage. The energy density of these chicks averages 6.5 kJ g^{-1} for the early stage and 8.5 kJ g^{-1} for the late stage.

The records shown in Table 9.5 summarise the average rates of energy gain during active foraging by birds having different numbers of penguin nests within the territory. The most consistent rate occurs predictably for skuas in group I (those skuas with the largest numbers of penguin nests), ranging for individual birds from 5.7 kJ min^{-1} during the early guard stage to 30.6 kJ min^{-1} during the early post-guard stage in 1970 (but 10.8 kJ min^{-1} for a shorter sequence of records in the two previous years). Overall, the most reliable record is for the incubation stage where each prey (egg) taken provides only a modest gain so that the skuas must maintain a high rate of interest and activity. Later in the season, and especially during the post-guard stage, each penguin chick taken can provide food for long periods and the levels of skua activity recorded depend very much on when the behaviour logs were made in relation to when prey were taken. High rates of activity, with or without success, would be recorded for hungry birds, low, even nil, rates would be recorded for skuas that had killed and fed from chicks within the previous few hours. The records for groups II and III during the early guard stage seem anomalously low compared with the others at this time and reflect the variability inherent in the activity and success of skuas reliant on small numbers of penguin nests.

As noted earlier, chance plays a major part in the success of skuas with small numbers of penguin nests. With few exceptions these skuas did not work consistently so that food gained easily (e.g., an egg or chick found outside the breeding group requiring almost no effort on the skua's part) gives a high return for the amount of effort expended. This is the explanation for the high gain rates seen in the group IV skuas, none of which were dedicated to feeding on penguins. The poorest record is for the group V pairs. These pairs showed little interest in the penguins within their territories and none managed to obtain any food there during these logs. This uniform indifference has occurred by chance in this sample: there were other pairs on the colony with just as few penguins that were very active.

The average rates of energy gain from foraging on the colony incorporate records from a range of abilities and variable interest in the penguins. All the skuas in group I attacked penguins with various success, none in group V carried out sustained attacks, whereas there were both aggressive and indifferent skua pairs in the other groups. The means, nevertheless, mask a wide range of aptitude and success. Two skua pairs, 153 on EF Block and 25 on H Block, were consistently over all the years and at all times of the year the most successful predators in their respective areas. Their performance can be used to set a standard for food gain from the colony by committed and able predators with exclusive access to high and

Table 9.6. *Average rate of energy gain for the males of the two most committed predator pairs within the study group*

Rates shown are the mean intake of metabolisable energy (kJ min^{-1}) from active foraging for each male bird of these pairs. Effort recorded as minutes of activity within categories 4–38.

	Pair 25	Pair 153
Number of penguin nests at 10 Dec. each year	54–105	425–585
Incubation stage	21.9	25.8
Early guard stage	21.6	22.4
Late guard stage	172.0	80.0
Early post-guard stage		
Long watch (80 h)	—	301.2
Other logs (38.5 h)	—	135.1
Late post-guard stage	—	85.4

medium penguin numbers respectively. Neither pair needed to forage at sea for much of the penguin breeding season. The success of each pair depended largely on that of the male bird. The female of pair 25 showed little interest in the penguins, the female of pair 153 actively scavenged for penguin food but had little success as a predator. The high food gains shown in Table 9.6 reflect the skill of these two predators, which took many eggs and chicks following only the briefest attack. For 153 the abundant nests in the territory offered great scope for opportunistic attacks. This bird was rarely seen in long-sustained attacks on single nests. More likely was a fast sortie from the nest area, even from the nest itself, at an exposed egg or chick in a nest or at an egg or chick out of the penguin breeding group.

Even so the great flush of readily obtained food in the early post-guard stage was soon over. Both birds were recorded fishing on 16 and 17 January, 1969.

Both these very active pairs had fair numbers of nests in contrast to the situation of pair 151, which held a very small territory on the upper edge of EF Block and had a half share of F10 – no more than 20 nests. Although also extremely active, their poor record of success resulted from having few opportunities to take prey.

At the other extreme to these pairs were skua pairs which evinced great interest in their penguins but carried out few actual attacks and gained little food. Pair 146 new pair is a good example. During the early and late

guard stages during 1968 they spent 12% and 10% of their time respectively interacting with penguins but gained no food. Their failure was not
through lack of opportunity: they had 350 penguin nests in the territory.

Another group of skuas was those that had access to penguins but took
almost no interest in them. Different pairs reacted very differently through
the season. Some, such as 144, for example, did not really begin to take
any interest in their penguins or to behave aggressively towards them until
the first chicks hatched. Pair 144 had access to a large number of penguins,
but most pairs in this category had few penguins. The long hours of
observation failed to record more than a few tentative, timid pokings along
the margin of H2 by any skuas of pairs 19, 21 or 22 at any time, even though
pair 9, their closest neighbour, could be seen making a good living off the
same breeding group.

9.4   *The energy equation of foraging at sea versus foraging on the
        penguin colony*

The two sets of measures of foraging costs and benefits for the two
locations can now be compared. This is of necessity rather simplistic as we
have seen in the preceding sections the extent of the variation and
uncertainty. A tentative comparison is possible, however, for the performance of each of the five groups of skuas (based on numbers of penguin nests
in the territory) with the performance of skuas foraging at sea under the
different ice and wind conditions. Each of these sets of foraging records
has been calculated as energy gained per minute of activity, equating
activity on the penguin colony with flight.

The figures for energy gained from foraging under the different ice and
wind conditions provide the key measures for assessing the benefits of
foraging at the penguin colony, in terms of pure energy economics.
(Aspects of injury risk and breeding failure associated with foraging
absence will be considered in Chapter 10.) Comparison of Tables 9.4 and
9.5 demonstrates that for skuas with more than say 50 penguin nests the
return from foraging on the colony during the incubation and early guard
stages is about the same as for foraging at sea under light pack-ice
conditions – both return about 10 kJ min^{-1} of activity, of flight or attack
behaviour respectively. Later in the season foraging on the colony gives a
better return than for all sea foraging except in the most favourable sea and
ice conditions. Similar conclusions can be drawn from the comparison with
pairs 25 and 153 alone (Table 9.6): not until the late guard stage in late
December was foraging on the colony appreciably more rewarding than
foraging at sea under good conditions. However, as foraging conditions at

Table 9.7. *Sea-ice cover in McMurdo Sound each year in relation to southerly storms through the breeding season from 10 November until 20 January*

Number of days (percentage of time).

Year	Number of storms	Sea-ice cover			
		0% (with storms)	<40% Light pack-ice	40–80% Moderate pack-ice	>80% Dense pack-ice
1966–67	6	10 (14.1)	12 (16.9)	15 (21.1)	34 (47.9)
1967–68	6	15 (21.1)	8 (11.3)	24 (33.8)	24 (33.8)
1968–69	5	12 (16.9)	6 (8.5)	20 (28.2)	33 (46.5)
1969–70	5	15 (21.1)	7 (9.8)	20 (28.2)	29 (40.8)
Mean percentage of time		18.3	11.5	27.7	42.2
Foraging duration at ice level		117 min	43 min	86 min	227 min

After 20 January the Sound was generally clear of ice in any wind until freezing began in early winter. Light pack-ice occurred within the first four days from storm abatement, moderate pack-ice from four to five days and dense pack-ice after five days without strong southerly winds.

sea deteriorate, either because of storm conditions, or more commonly because of pack-ice blocking into McMurdo Sound, the benefits of penguin ownership and colony foraging become increasingly advantageous. Under the most severe sea-foraging conditions recorded (but short of total failure when the skuas would have deserted the territory altogether) penguin foraging was from 3–15 times more advantageous. To underline this advantage even further it should be recalled that the fish load set here for this comparison was 125 g, and that this is a generous figure. The average loading brought by these skuas from each foraging event is probably less than 100 g.

This comparison between the two foraging efficiencies has highlighted the significance of foraging conditions at sea as the most important factor determining whether foraging at sea or on the colony is advantaged. It is important, therefore, to determine what proportion of each season offered favourable conditions for sea foraging. In each year the date and duration of storms were recorded together with the movement and density of pack-ice. It is possible from these records to estimate the extent of ice cover by combining the storm records with measures on the rate the ice packed back into the Sound once the storm abated. In Table 9.7 the Sound was assumed to be clear of pack-ice during the storm and for one to two days afterwards, depending on the duration of the storm. Thereafter the Sound became progressively more tightly packed with ice reaching dense pack conditions by five days, which was retained until the next storm swept the Sound clear again. From about mid January there is so little pack in the reservoir to the east of Ross Island that even under continuous northerly winds little pack enters the Sound. The table shows the ice conditions for 71 days from 10 November to 20 January.

Although there is some variablility in the number and duration of storms over summer the ice conditions were remarkably uniform, with on average 18.8% of the time free of ice during storms, 11.5% with zero or light pack in calms or light winds, 27.7% with moderately dense pack and 42.2% with dense pack. That is, for little more than a tenth of the early and mid season did the skuas encounter the most favourable foraging conditions while for nearly half of this time they met very difficult conditions with such dense pack that foraging would have been precluded within the Sound itself. This high proportion of time in which foraging would have been difficult and time consuming needs to be kept clearly in mind when assessing the advantages of owning territories with penguins. The distribution of storms and the density of pack-ice through the season is shown in Fig. 9.5.

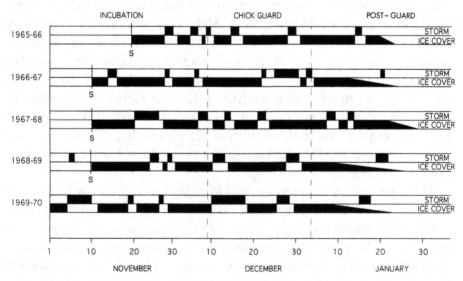

Fig. 9.5. The distribution and duration of southerly storms and the density of pack-ice in McMurdo Sound over summer during four years. S = start of records.

9.5   *Why don't all the skuas with penguins exploit them strongly?*

All the information presented so far demonstrates the advantage of foraging on the penguin colony under most wind and sea conditions for much of the breeding season. On energy terms all skuas should work to exploit this local food source. In fact only a few pairs were dedicated predators. In view of the seeming advantage of foraging at the colony why did not all the skuas attempt to exploit this food source fully?

The simple answer is that many skuas did not have the determination and perseverance needed to get eggs and chicks from defended nests or defending adults, or to retrieve eggs and chicks from among penguins as scavengers. (In most circumstances the same skills and determination are needed to scavenge free eggs and chicks from *within* the breeding group as are needed in predation.) In time skills can be learned by trial and error – and the natural aptitudes and attributes of such a large, highly mobile predatory bird gives them an enormous advantage over the sedentary penguins from the very beginning. But this learning process can only occur if the skuas first attempt to scavenge and prey on the penguins. This means more that just sitting nearby waiting for eggs or chicks to roll out of the breeding groups. The most successful skuas *cause* food to become available – by disturbing the penguins so that eggs or chicks are thrown out of the nest, by obtaining food from the nest by speed or subterfuge, or by intimidation and aggression. The single factor that links all these approaches is that they all involve risk of injury. It is a fact of life that timid skuas obtain little food. It is also a fact that the most successful skuas worked the hardest. For the incubation stage, for skuas taking eggs there was a significant direct relationship between amount of activity and reward in food capture [2]. It was not possible to establish the same relationship for the guard and post-guard stages where chicks provided extremely variable rewards.

Skuas may be inherently timid or pugnacious, and this can be seen in the generally more aggressive activity of the males of pairs compared with the females, but it is likely also that their experience on the penguin colony is an important factor in how they interact with the penguins.

This conditioning can begin even before obtaining a territory. At Cape Bird each year the non-breeding skuas settled into a clear area among large groups of penguins on A Block. (In the first season they were for a time on high ground above H Block well away from penguins.) On A Block they had ready access to breeding penguins and if they wanted to they could scavenge and prey upon them, either as individuals or as pairs and groups. The behaviour of these predominantly young birds clearly underscores the importance of experience. Although as many as 60 skuas were on the non-breeding club at any given time they took very few eggs or chicks, most of which were scavenged, and their tentative attacks on penguin nests on the margins of colonies were laughably inept.

What is the risk to skuas holding territory and foraging on the colony? The risk of nesting within or close to the colony arises from interference by penguins and from the risk of enhanced brood predation by other skuas, arising from the high density of breeding birds and from the attraction of outside skuas to the penguins. This general consequence of nesting near penguins will be considered in the following chapter and can be left for the moment. The specific risk of foraging among penguins is from the danger of being caught and held by the penguins and being injured or killed. This risk is lowest during mid-incubation in the penguin cycle when there are few unattached penguins on the colony. At all times of the season the risk of being caught is most acute during strong winds when the skuas have difficulty controlling the part-opened wings during ground attacks. The very large difference in size and strength between the smaller predator and larger prey should not be underestimated. Nor is this a one-sided contest between forceful predator and timorous prey: the two species are equally pugnacious. Indeed in many encounters the penguin is far more pugnacious. Birds of both species can fight to the death for nests and mates, and are equally belligerent in their defence of offspring.

There are two constraints to obtaining information needed to refute the hypothesis that the history of interactions between a skua and the penguins has no influence on their later predatory behaviour. The first is that it is simply not possible to be certain that any individual skua may or may not have been hurt by penguins, even seriously injured, on some earlier occasion in its life. A single event causing injury could occupy no more than a minute or two. Second, it is difficult to see when a skua has been pecked or even caught momentarily during the routine observation logs. From the usual observation points at some distance from the penguins only serious events will be noticed – events in which the skua is obviously being held (as in Fig. 9.6) or is being attacked in the breeding group. Occasions in which the skua is chased off by pecking or a blow from a flipper would not be visible, especially when the skua has crashed into the penguin in a jump-attack or is reaching under or around it to get at eggs or chicks in a nest.

The first point to establish is how commonly skuas are caught by penguins. The second is how serious these events are, and what risk they

Fig. 9.6. Demonstrating that skuas may well be caught by defending penguins when preying or scavenging on eggs.

offer to skua life. The best record of the frequency of being caught is from 1966–67 when a written record was made of skuas interacting with penguins. (In subsequent years much greater emphasis was given to providing a coded record of the details of behaviour and only the most dramatic events would have been noted.) In this year, during 44 hours observations of eight pairs of skuas on EF Block between 6 December and 3 January (during the penguins' late incubation and guard stages) nine skuas were recorded as being caught and held by penguins when attempting to scavenge or prey on nests. The probability of being caught is a significant 0.013 bird^{-1} hour^{-1}. Clearly, being caught by the penguins seriously enough so that it can be seen from a fair distance is not uncommon. At this rate each of these skuas could be expected to be caught on average once every 80 hours or so, or about once every four hours of active attacking behaviour. Of course, not all skuas are equally fallible or foolhardy. Although three of these records were of captures of the male of 149 and two were of 148, none of pairs 152 and 153 was seen caught during this season (although as will be seen later 152 male was well caught in a later year).

Some skuas are either very lucky or very skilled at avoiding capture. The male of pair 25 was easily the most aggressive skua working H Block penguins but was never seen caught seriously during many hours of observation from a clear vantage point. The detailed activity records of this bird in December, 1969 show that it was caught momentarily by the wing only once in 920 minutes of forceful predation, in which the favourite technique was to attempt to swivel or pull nesting penguins away from the nest by the tail (attack category 12). The speed with which this bird jumped and flew from nest to nest seemed to enable it to avoid even the most aroused and alert penguins. At the time of these observations though, there were only a handful of unattached penguins, and areas where they ran out to meet the skua as it landed on the margin were soon avoided.

It is important to appreciate how potentially dangerous are any catches by penguins. Skuas may well buffet their way clear from being caught by one penguin by the wing or tail and unless thrown into a neighbouring penguin (which will now also be furiously alerted) it will be able to lift quickly into the safety of flight. The surprising feature of these captures is that the penguins let go very quickly compared with the way they hang on grimly to opponents in their intra-specific battles. Only rarely are skuas held by individual penguins for more than a few seconds before being released or tumbled away. But if caught in a colony with free penguins not tied to nests even this limited capture is potentially fatal. If firmly caught they end up within a plunging, fighting group of penguins. These events are a maelstrom of anger and furious activity as penguins rush into the mass of fighting birds from all over the breeding group. The skuas cannot fly out of this mass of penguins, being held by at least one or more at any time, so they can only fight and struggle their way through the nesting penguins to reach the margin. Once there the fury subsides as the unattached penguins give up and return to their own territories. Two of the nine records listed for 1966–67 were of this sort. One was the result of genuine misadventure. Male 145 was caught by the wing by a sitting penguin when attempting to retrieve a deserted egg in the middle of E16. Before it could break clear it was then attacked by five other penguins which rushed into the fracas. The second event was through foolhardy behaviour. The female of 149 flew over a row of margin nests to grab an undefended 250 g chick and then attempted to drag it back to the outside through the penguins (it was too heavy in these circumstances to carry in flight). The skua was caught almost immediately and although attacked by four unattached penguins managed to escape – succeeding against all odds in dragging the chick out with it! It should not be thought that the penguins combine forces in some group defence activity in which their own territoriality is submerged momentarily. Much of the fracas occurs because the penguins attack each other as well as they tumble and fight across the birds on territories. Without this heightened disturbance, confusion and divided interest it is unlikely that any skua would ever be able to fight clear.

What evidence is there that skua capture is potentially lethal, given the evidence that most seem able to break free and that although the frequency of capture is high no skuas were actually recorded as being killed by penguins directly on this colony over the five years of observations? No dead or mortally wounded skuas were ever found among the penguins or near the margins of breeding groups. The greatest risk to skuas would seem to lie in having their wings broken from flipper blows, as the wings are always opened and are beating in these attacks. But surprisingly none has been seen, although broken wings have occurred in attacks by skuas on each other. The very thick feathering that characterises these birds no doubt protects them from physical injury as well as from cold. The certainty that skuas can sustain very serious injury has come from fortuitous observations.

The most convincing of these are:

1. The male of pair 133 was caught by a penguin on a margin nest and held by the neck in the beak while being beaten by the flippers. It was flung clear after no more than half a minute to roll down the slope away from the nest and was then unable to stand or move. This bird survived.

2. The male of pair 152, a very experienced predator on EF Block, was seen to stoop into the centre of F13. Although held and attacked by up to 10 penguins at a time it was able to fight its way across 7 m and through 11 nests to reach the margin 51 seconds after first being caught. This bird could neither fly nor stand after reaching the relative safety of open ground outside the margin. Unfortunately, the stoop was into 153 territory so that it had to rescued by hand to save it from being killed by this pair. Even if the stoop had been successful the reward was going to be derisory – 120 g of very dead, rotten penguin chick.

What evidence is there in these records that capture and beating by penguins causes skuas to be more cautious? In the long term this cannot be assessed and the cumulative impact of getting caught may well deter skuas from preying on the colony. But for some birds at least this does not seem to happen. By chance all of the really committed scavengers and predators will be caught several to many times a season yet there was no evidence that they were becoming more cautious over the years of this study. Quite the contrary. Both 25 and 153 males seemed to get even more daring and capable. Nor did male 152 change greatly after the horrific episode of being nearly killed in F13 group. Next season its behaviour was apparently unchanged. All these are of course pretty tough and experienced skuas, an equivalent imperturbability may not exist in young, inexperienced skuas in their first breeding year on the colony.

Neither did capture seem to deter skuas in the short term. In some attack sequences skuas were seen to be caught several times yet continued to attack forcefully. For example, male 148 spent four minutes in a frontal attack to get an egg out from beneath a lunging penguin even though caught by the neck momentarily several times. Two minutes later it resumed the attack on the same penguin to carry off the second egg. Clearly this skua had not been greatly disconcerted by being caught. In conclusion of this part it appears that once these skuas developed the interest and facility of preying and scavenging on the colony they were not deterred later through being caught or beaten. It is not possible, however, to conclude that those skuas that were weak or timid predators had earlier been deterred. There are no records to test this possibility. The balance of evidence, especially of the behaviour of the non-breeding birds on the colony and of birds newly taking up territories, is that all of this behaviour is perfected through practice and trial and error.

Much of the argument advanced in this chapter on the balance of advantage between foraging on the colony or at sea seems pretty well summed up in the following incident – it contrasts the certainty of sea foraging with the chance of easy gain on the colony. On 22 December, 1966 the male of pair 149 left to go fishing at 1608 h, flying hard around the glacier cliffs to the north of the colony towards open sea north of Ross Island. The female was then alone at the nest but abandoned it abruptly for a five second jump attack at a poorly defended penguin nest to grab a 300 g chick. She had gulped most of this chick down before the male returned at 1845 h (after 157 minutes hard flying) with about a 100 g of fish.

9.6   *Assessing the food sufficiency of pairs foraging on the penguin colony*

In the preceding section the figures for the amounts of penguin food ingested by the skuas at different times of the breeding season have been brought together for the individual pairs and on average for the birds of the local area. The interesting question that now must be addressed is – was it enough?

There are three ways that this question might be answered. First, the behaviour of the mates and chicks should clearly demonstrate whether they were well fed or hungry. Hungry females harass mates on the territory for food, begging and pecking at the breast and face, and chicks do the same at both parents. Moreover hungry chicks use the whistling call with monotonous regularity. Second, in pairs usually feeding from the colony the observation of foraging at sea is a clear signal of hunger and of shortage of colony food. Third, the food intake can be measured against food requirement estimated from allometric equations, determined from direct measurement of metabolic rate, or from experiments recording the amount of food ingested by captive or free-ranging pairs.

In practice all these methods can be brought together to give the most secure assessment of sufficiency. Observations of behaviour will certainly indicate that the birds are hungry at that time but although they cannot measure the food deficit, they nevertheless highlight short-term deprivation not easily recorded by other methods.

9.6.1 *Daily energy requirements*

Experimental determination from feeding free-ranging skuas

Müller-Schwarze and Müller-Schwarze (1977) provide a number of measures of the amount of food eaten by free-ranging pairs of skuas at Cape Crozier. The skuas were fed by hand *ad libitum* several times each day with either eggs or penguin flesh. Feeding on eggs. Pairs took on average 10.6 eggs each day, equivalent to 5724 kJ day^{-1} (10.6 times 540 kJ) if they managed to retrieve the whole egg content. Feeding on penguin chick flesh. Three pairs with small chicks took on average 1279 ± 163 g (SE) a day ($n = 10$ days), 640 g for each adult as the chicks would have consumed little of the food taken at this time. This amount is equivalent to 4180 kJ day^{-1} of ingested energy for each adult (640 g times 6.54 kJ g^{-1}) and ($\times 80\%$) 3344 kJ metabolised. The authors give the lower figure of 471 g per day for each adult for food ingested over the entire study. These are, however, minimum estimates as the skuas were free to take other food away from the territory.

Metabolic rate estimations

Evaluating food sufficiency from the penguin colony requires comparison between the food available and the field metabolic rate (FMR), which is the basal metabolic rate (BMR) with the additional costs of living in the local environment (the costs of thermoregulation, locomotion, feeding, alertness, posture, digestion, reproduction and growth (Nagy, 1987)).

Historically the FMR was estimated from the BMR determined from oxygen utilisation of resting birds in an enclosure or calculated from allometric equations based on mass. The FMR was then estimated by using a multiplicand of the BMR to take all the activities of the free-living bird into account. Setting the scale of the multiplicand was critical to the accuracy of the final estimate.

During the last decade the use of labelled isotope methods on free-ranging birds has revolutionised the study of bird energetics, and in conjunction with traditional techniques of BMR determination based on oxygen consumption has provided important information on the relationship of the two measures of bird metabolism. There have so far, however, been few determinations of sea birds and none of skuas. The most relevant for skua studies are those on black-legged kittiwakes (*Rissa tridactyla*) (Gabrielsen *et al.*, 1987), sooty terns (*Sterna fuscata*) (Flint and Nagy, 1984) and northern gannets (*Sula bassanus*) (Birt-Friesen *et al.*, 1989) which are flapping birds foraging at sea for fish in cold temperate environments, but there are others on penguins, (Davis *et al.* 1989) and albatrosses (Costa and Prince, 1987; Pettit *et al.*, 1988) which are also useful for comparison.

There are two determinations of BMR for skuas from oxygen consumption. For south polar skuas, Ricklefs and Mathew (1983) determined a mass specific rate of $130 \, 1 \, O_2 \, kg^{-1} \, h^{-1}$ for skuas resting within a closed chamber. Using the equivalent $0.0201 \, kJ/ml \, O_2$ this oxygen uptake corresponds to a metabolic rate of $627 \, kJ \, kg^{-1} \, day^{-1}$ for skuas weighing 1315 g.

A second record is for great skuas by R. W. Furness and D. Bryant (pers. commun., 1991). They determined a rate in a through-flow apparatus of $16.9 \, J \, g^{-1} \, h^{-1}$, equivalent to $533 \, kJ \, day^{-1}$ for skuas averaging 1315 g. This rate is appreciably less than that recorded by Ricklefs and Mathew. It is, however, within the range estimated for birds of this mass from the equations of Kendeigh *et al.* (1977) (equation 5.5, $0.522 \, W^{0.734}$), the general equation for non-passerines which gives $425 \, kJ \, day^{-1}$ and the equation for non-passerines of Lasiewski and Dawson (1967) which gives a rate of $399 \, kJ \, day^{-1}$.

From these comparisons the rate determined by Ricklefs and Mathew seems high. For the present study the lower rate of $533 \, kJ \, day^{-1}$ from Furness and Bryant is accepted as the most likely estimate of the true BMR. (The basal metabolic rate is defined in practice (Kendeigh *et al.*,

1977) as the 'rate of energy utilization of fasting, inactive animals in the zone of thermo-neutrality, or at least at high, relatively unstimulated ambient temperature'.) Provided the techniques are comparably exact the consistently lowest record of a series of determinations is the most likely to approximate the BMR. Any movement or restlessness in the birds or changes in temperature causing stress will automatically raise the metabolic rate, and consequently the oxygen uptake.

The estimated BMR can be converted to an FMR from the relationship between the two rates determined in studies where they have been simultaneously determined. Comparisons for 18 species made by Birt-Friesen *et al.* (1989) gave a mean ratio of FMR/BMR of 3.29 ± 0.29 SE. Using this ratio, the FMR for skua adults is *c.* 1750 kJ day^{-1}, for pairs 3500 kJ day^{-1}. For pairs with a single chick (almost none of the pairs raised a pair of chicks) this rate would need to be adjusted upwards progressively from early January, when the first chicks hatch to their fledging and post-fledging dates of mid February and early March respectively. By these dates the chicks will be consuming at least as much food as the parents – giving a total for the family of *c.* 5000 kJ day^{-1}. But to a degree this estimate is of academic interest only; from mid January there is so little food to be gained from the colony that the skuas are almost universally foraging at sea. The FMR estimated here is significantly less than that determined by Müller-Schwarze and Müller-Schwarze (1977) from the feeding experiments. For comparisons between between energy need and energy return from foraging on the colony it has the advantage of providing the most favourable comparison. If the skuas cannot meet this lower estimate they would certainly be unable to meet the higher one.

9.6.2 *Comparing the field metabolic rate with the estimated amounts gained each day from the penguin colony*

Appendices 3.1–3.3 show the amounts of food it was estimated that the skuas obtained each year from the colony. These amounts are of metabolisable energy and may be directly compared with the amounts needed for the pair each day (3500 kJ day^{-1}).

Direct comparison indicates that none of the skua pairs studied gained sufficient food during the incubation stage to meet their assessed energy needs. In general it is not until 20–24 December each year that they reached this level of food volume. The groups of skuas on EF and H blocks may be usefully contrasted at this time. For the skuas on EF Block, which had the most favourable prey ratio, although none attained sufficiency during 8–14 December, some few reached sufficiency during 15–19

December and most exceeded requirements by the end of December. By way of contrast only pair 25 on H Block gained enough food from the colony before the end of December.

In general all skuas found foraging on the colony easier for a time during the post-guard stage. It is a short-lived abundance however. The increasing size of the chicks provides at first an increasing resource but they become too hard to kill later in January, and though even more valuable, can seldom be taken. The majority of skuas on EF Block during the first two years, when there were the highest numbers of breeding penguins, were able to obtain enough food from the colony between 10 and 19 January, but fewer achieved this during 20–24 January and sufficiency fell away quickly from this time so that only one or two pairs achieved this level at the end of January. During the 1968–69 season, when fewest penguins were breeding, no pairs were supported entirely from the colony during the first week of January and only two of eight pairs were able to get enough food there during its maximum food availability in mid January.

The food taken by the skuas on H Block showed the same pattern of abundance and sufficiency as on EF Block, with the majority of pairs meeting their requirements from the colony in the first two years but having much less success in the following ones, when only one or two pairs gained sufficient food from the colony at any time during January. Again, the most favourable time for skua foraging was between 10 and 19 January.

> *The evidence from foraging flights at sea that the skuas were not able to gain sufficient food from the penguin colony at all times during the breeding season*

In Table 9.8 the number of foraging flights seen have been brought together for pairs with different numbers of penguins within the territory for 10 day intervals over most of the penguin breeding season. There were no observations during the early part of the season (before mid November) when few eggs were being taken, and from the end of January. Irrespective of numbers of penguins within the territory at least some of the skuas in every group were foraging at sea during all parts of the season. There were too few records, however, to be able to show clear patterns of foraging incidence, with peaks during periods when less food was available and higher rates in pairs with smaller numbers of penguins. What is certain, nevertheless, is that the possession of even substantial numbers of penguins did not confer an easy availability of colony food.

Table 9.8. *Incidence of foraging flights at sea for pairs with different numbers of penguin nests within the territory*

Data are foraging flights pair^{-1} day^{-1}.

Pairs with:	10–19 Nov.	20–29 Nov.	30–9 Dec.	10–19 Dec.	20–29 Dec.	30–9 Jan.	10–19 Jan.
>250 nests	—	5.64	1.08	1.80	1.58	4.17	0.79
100–249 nests	—	1.29	0.91	0.38	0.50	0.30	0.91
50–99 nests	1.84	0.55	0.64	1.24	0.50	0.33	1.20
1–49 nests	—	1.42	1.66	0.88	1.27	0.19	1.43

Weighted means taking into account the numbers of pairs in each study area with the different nest numbers and the numbers of hours observation. There are five or six pairs in each nest number category.

The single exception to the need to forage at sea routinely during these weeks was pair 153 on EF Block. This pair had more than 500 nests and the male was a sufficiently skilled predator to be able to feed entirely from them. Only one foraging flight was recorded by them (on 25 November, during the incubation stage) in 317 h of observation. For comparison, pair 145 with more than 250 nests within the territory undertook 13 flights during the same interval. The success of pair 153 was matched by pair 25, the male of which was the most able predator on H Block. This pair was also largely able to feed exclusively at the colony, with only six flights noted during 115 h of observation, but this was an even more meritorious achievement than for 153 as it had few penguins on which to forage.

STATISTICAL TESTS

[1] Comparison of the levels of activity of EF skuas foraging on the penguin colony at times when sea foraging was easy and difficult. 'Easy' and 'hard' foraging conditions were inferred from the average time that the skuas required to obtain fish. Activity levels were measured as number of 2.5 min active intervals each hour for the individual skuas of nine pairs. Analysis was confined to records made of behaviour during the penguin late incubation stage. (Mean and SE.)

'Easy' sea-foraging conditions
4 December, 1968, 46 min.
27 November, 1968, 41 min.
29 November, 1967, 32 min.
5 December, 1969, less than 2 hours.

'Hard' sea-foraging conditions
4 December, 1967, 352 min.
8 December, 1967, 93 min.
9 December, 1969, 263 min.

1. Comparing all activity (categories 2–38)
During 'easy' conditions, mean rate = 2.366 ± 0.50.
During 'hard' conditions, mean rate = 2.622 ± 0.86.
$t = 0.255$, df 16, $p = 0.80$, ns.
2. Comparing vigorous activity levels (categories 4–38)
During 'easy' conditions, mean rate = 1.376 ± 0.54.
During 'hard' conditions, mean rate = 0.90 ± 0.31.
$t = 0.76$, df 16, $p = 0.45$, ns.

[2] Regression of food gain against time spent in active foraging and scavenging (categories 4–38) during the incubation period for 15 pairs. $Y = -0.235 + 18.93X$, where Y is energy gain in kJ h^{-1} and X is number of 2.5 min intervals of foraging activity ($p < 0.001$, $r^2 = 0.902$).

10

Immediate impact of the skuas on penguin breeding

10.1 *Introduction*

In each year large numbers of eggs and chicks were lost from the penguin colony (between 30% and 42% in each year of all eggs laid) and most of these sooner or later were eaten by the skuas. For this large colony at Cape Bird these losses represent a substantial food resource (Table 10.1)

What must now be determined is the skua's contribution to penguin breeding failure. Essentially, the problem is to determine how much of this enormous loss of eggs and chicks is directly attributable to skua predation and disturbance compared with that resulting from the behaviour and mismanagement of the penguins themselves and from environmental factors. In the first situation the skuas are acting as predators, in the second merely as scavengers. This is a crucial distinction bearing on both short-term ecological and behavioural adjustment between the two species and on long-term evolutionary modifications to breeding distribution and breeding cycle seasonality and synchrony.

Even the most cursory observation of the attacks by the males of pairs 25 or 153 demonstrated undeniably the skua's potential as a predator. On the

Table 10.1. *Estimated numbers of eggs and chicks lost from the Northern Colony each year*

	Estimated numbers of eggs laid[1]	Percentage loss to fledging[2]	Numbers of eggs and chicks lost
1965–66	54 270	33.4	18 130
1966–67	54 000	36.2	19 550
1967–68	46 700	29.5	13 780
1968–69	32 340	42.3	13 680
1969–70	42 890	35.8	15 350

[1] Total number of eggs laid was estimated from the number of nests in early December, corrected for losses to this date, each with on average 1.9 eggs.
[2] The percentage loss to fledging of eggs laid was derived from the records of breeding success each year in the study groups. In each year the survival of between 610 and 2012 eggs was monitored.

other hand there are any number of observations of timid skuas acting merely as scavengers, and nervously at that when close to the penguins. Overall can skua predation be identified as a significant impact on the reproductive success of this colony? Much can be achieved in resolving this problem by direct observation, which records both predatory and scavenging impacts, and which may lead to an assessment of their significance. But direct observation alone cannot measure the overall impact on the penguins across the wide area of this colony. To achieve this aim the prey itself must be studied, specifically by comparing places and breeding groups where predation could be predicted to play a major role with places where it might have less impact.

The interaction of species – as predator and prey – has both immediate and long-term influence on each contestant. Immediate consequences are, for example, the reduction of reproductive output of the prey through the loss of eggs and chicks. Of no less interest in such an association are the longer-term impacts on the overall breeding behaviour, on its timing and synchrony, on the aggregations of both predator and prey and on the evolution of agonistic and defensive behaviours.

It is of some importance, therefore, to establish the nature of the association between these skuas and penguins and to assess the relative impact of predation as a factor in the breeding behaviour and success of the penguins. This is attempted in the following sections. The implications of this association in the long term will be explored in the next chapter.

10.1.1 *Are skuas predators?*

The short answer to this question is yes and no. Some skuas attack penguins at nests in order to get their eggs and young chicks and many skuas will attempt to kill penguin chicks away from the breeding groups. Other skuas seem almost indifferent to the penguins within their territories. Moreover, in spite of the apparently abundant food on the colony all skuas foraged at sea for at least part of the breeding season. The real question is not whether they are predators or not, but how strongly the threat of predation is perceived by the penguins (which threat, depending on degree, may or may not affect their behaviour), and how predation compares with other life-threatening factors during breeding?

Minor predation threat or impact might well induce a scarcely recognisable selective response; strong threat and impact would provoke strong anti-predator selection.

Up to now the status of skuas as predators on the colony has been assessed from the records of their predatory and scavenging behaviour and from prey remains recorded on the territories. It is now time to examine the other side of the association, to measure impact on the prey.

At first sight the most obvious and simplest way to measure the level of predation is to record the loss of eggs and chicks from identified nests. There have been numerous studies of this sort. Their conclusions on the role of skuas are, however, variable and generally equivocal. They are, nevertheless, a great advance on the earlier accounts which unequivocally concluded that skuas were ferociously successful predators (e.g. E. A. Wilson, 1907; Ponting, 1921; Levick, 1914; Siple and Lindsey, 1937). Skuas have been dogged by this reputation ever since, clouding the perceptions of both visitors and scientists.

10.2 *Methods used in measuring skua impact on breeding penguins*

There are a large number of variables to be taken into account in any study attempting to measure impact directly. For skuas these are variability in skill and interest, the cycles of hunger and satiation, differences in numbers of penguin nests within the territory, the ease of foraging at sea, the previous experience of penguin attacks and the recency of being injured. For the penguins the variables are nest location, experience as breeding birds and with skua attacks, weather and sea variables, nutrient condition and time into incubation fasting. To complicate comparison with earlier studies there has also been uncertainty and variation in how the results of breeding success studies are to be presented. After all these years of penguin study there is no agreed methodology for this work nor are the defined categories or life-cycle stages applied uniformly. Some of these uncertainties will be confronted in the attempts later in the chapter to review the earlier work. Moreover, implicit in all work is the assumption that the penguins were little disturbed, even undisturbed, by the study and that the record obtained represented the natural situation. Lishman (1985b) is refreshingly honest in recording significant impact, providing more than the usual statement that the birds were undisturbed or that the success of study groups was not significantly lower than that of selected controls. Only recently with the advent of research designed specifically to measure the impact of disturbance by recording physiological responses (Wilson *et al.*, 1989; Culik *et al.*, 1990) has the real impact of research contact begun to be appreciated. Like most before me, I will state formally that the study penguins were watched over carefully and were not significantly disturbed. To do otherwise would be to negate the value of the results.

All of this uncertainty in determining the methods used and in reporting results underlines the need for better guidelines and universal measures on how studies on these birds are to be undertaken. Not only would this foster conservation aims it would also provide assurance that results are comparable. This last factor is of great significance for a species being so widely studied across such a large geographical range.

The impact of predation by skuas on the penguins at Cape Bird has been studied using variants of the usual methodologies of recording penguin breeding success by following the fate of individual nests and through direct observation. The methods were, however, predicated by the goal of determining skua impact rather than penguin breeding success itself. Therefore, the selection of study groups and the precise methodology was different, focussing on potentially different levels of impact and on distinguishing between scavenging and predation.

In assessing skua impact there seem to be two overriding factors that must be untangled. These are broadly the factors of place and of prey characteristics. Place refers both to the location of the breeding groups of penguins on the colony, and to the place of nests within each breeding group. It is concerned importantly with skua access to the prey. The first aspect has received little study to date (but has been touched on by Müller-Schwarze and Müller-Schwarze, 1973; Müller-Schwarze *et al.*, 1975 and Trillmich, 1978). The second has been considered in all accounts of breeding success of penguins since the suggestion by Sladen (1958) that less experienced breeders were more likely to be found on the peripheries of breeding groups than in the centres, and were likely to be poorer breeders. The factor of prey characteristics is concerned more directly with selection of prey types – whether skuas preferentially select living or dead prey, or chicks of a certain size, for example. Responses of penguins to these factors have the potential to modify the breeding behaviour of the penguin, both within the short term through learning and in the long term by selection, but the two factors would have different effects. For instance, the first could affect the selection of nest site within the colony and subsequent changes with season, the second could affect breeding synchrony.

Three different comparisons were attempted: 1. comparisons of a series of breeding groups chosen to reflect potentially different predator impacts; 2. direct comparison of peripheral nests, which were more exposed to predation, with central nests of large breeding groups; and 3. comparison of groups protected from skua predation with unprotected groups. This last experiment, essentially a predator exclosure experiment, was designed specifically to distinguish between scavenging and predation. Each of these comparisons was run across several years to test whether population density and seasonal factors, especially local weather and pack-ice conditions, were significant in modifying or enhancing the levels of predation by precluding or curtailing sea foraging.

Each breeding performance comparison depended on mapping the nests of breeding groups and in following the fate of the eggs and chicks. Records were made at four to five day intervals through the season, beginning in mid November. Nests were checked from the margin except for the central blocks of nests of the central versus peripheral comparison in which penguins were lifted momentarily by the tail. None of these birds deserted but it cannot be said that they were undisturbed. (This experiment was not repeated in the last year because of recognition of the great amount of disturbance caused in the ten minutes or so needed to check the nests and through an understandable reluctance to work among these furiously angry birds. I would certainly not attempt to repeat this sort of comparison now that we have a much better understanding of the effect of research disturbance on penguins.)

Breeding success has been variously defined in different studies of penguins and may be given as eggs and chicks per nest or as eggs and chicks surviving of eggs laid, which may again be divided to distinguish between one- and two-egg clutches, and again split among pairs with different laying dates – using the actual date or working in relation to mean date. There are differences also in whether pairs holding territories but failing to lay eggs are included in the calculations. The measure of success also varies, ranging from survival to departure at sea at one extreme to survival to the beginning of the post-guard stage at the other (Penny, 1968). A common measure is 'number fledged' but this needs to be qualified by a clear statement about the criterion used for fledging. Success may also be given in terms of numbers or proportions of pairs raising at least one chick. As in advertising, one needs to read the small print very carefully in making comparisons across different studies, and even with a clear understanding of what was done, comparison may still not be possible unless all the parameters are given. The more usual case is to present the results as a summary table.

Comparisons of breeding success in this study have been made on the basis of survival of the eggs and chicks to three benchmarks: 1. to mid December, the peak of chick hatching, 2. to c. 3 January, the beginning of the post-guard stage and 3. to mid–late January, marking the end of the stage when penguin chicks were vulnerable to skua attack. This last mark is closely equivalent to fledging success – the numbers of chicks going to sea from the numbers of eggs laid. The use of survival to mid December as an equivalent to the more usually employed hatching success was deliberate to overcome the problem of the variable number of non-viable eggs in each group. This study is trying to find out whether skuas have a significant impact on the *survival* of eggs and chicks, that is, on whether these are predated. It is their survival as units within the nest, and later in the post-guard stage, which is important, not whether they did or did not hatch. As will be shown later it is unlikely that disturbance of nesting birds by skuas influences egg viability. The majority of the non-viable eggs are found to have undergone very little development. For example, of 51 eggs still present on 26 January, 1969, 45 (88%) had no apparent development. The remaining eggs contained well-developed but dead embryos. None was chipping.

In summary, the first mark used to compare breeding success is survival of eggs and chicks to mid December, whether the eggs were viable or not, whereas the second and third marks record the more usual breeding success statistics of survival of chicks to the post-guard stage and to fledging respectively, both from the total numbers of eggs laid. These last two figures, therefore, incorporate non-viable eggs within losses. The alternative to this approach would be to delete non-viable eggs from the original total of numbers laid, but they are not always recognisable even late in the hatching period unless the colonies are checked very regularly. This approach anyway is more likely to lead to even greater uncertainty (replacing the variability of non-viable egg number by the uncertainty of their identification and origin) and was not thought appropriate for the large-scale counts being made for this work. Simply incorporating their numbers within losses brings this study more into line with others, which have also accepted this error, and reduces the amount of disturbance to the breeding group through not having to check these eggs in the hand regularly.

10.3 *Place of breeding as a factor in the breeding success of penguins*

10.3.1 *Location of breeding group within the penguin colony*
Consideration of Fig. 10.1 showing breeding success in different colony sites suggests two possible reasons why groups on the periphery of the colony might suffer heavier predation than those lying within the body of the colony. The first reason is that the skua:penguin ratios at these are generally higher. The second is because as these groups were more isolated from neighbours they allowed greater freedom of movement for skuas and once penguin chicks had been captured they were less likely to escape back into a group or be rescued by adults passing near them.

There are two data sets bearing on location. The first compares the numbers of chicks surviving to mid January with the numbers of nests in early December for breeding groups on the colony for the two seasons, 1968–69 and 1969–70. All groups were counted in the first year, most were counted in the second. Essentially, this is a measure of fledging success (recorded as chicks per nest), as few eggs were lost prior to the December count and few chicks were lost later in January. There is a small error for some contiguous groups where the crèches had merged. Unless there was any prior evidence for differential predation in these groups their chick numbers were assigned in proportion to the earlier nest numbers. The timing of these counts of fledging chicks needs to be carefully set: if it is too early some of the mortality will be missed (and the estimates of success rate will be inflated); too late and so much movement will have taken place among the groups that false records for individual groups will invariably be produced. The chick counts recorded here occurred at the first signs of chick movement away from peripheral or small colonies and were carried out during 15–20 January. The second data set is from the study breeding groups.

Counts of chicks compared with earlier nest numbers throughout the colony
In 1968–69 the mean breeding success for the 125 breeding groups was 0.98 ± 0.16 (mean $\pm$ SD) chicks per nest on 9 December (minimum occupation number). Figure 10.1 has been constructed to show how the

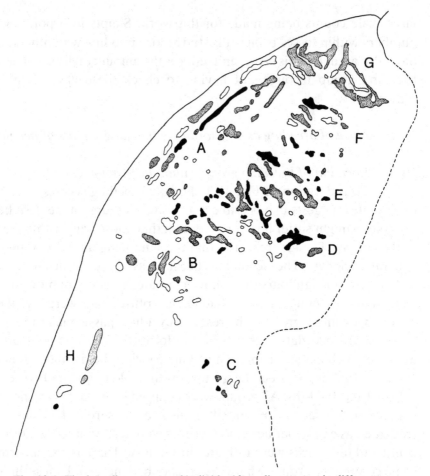

Fig. 10.1. The breeding success of individual breeding groups in different locations on the colony in 1968–69. Mean success to fledging per nest based on nest counts on 8–11 December for the whole colony was 1.25 ± 0.29 (mean and SD) for this year. Unshaded groups: Low success, below 0.5 SD from mean, i.e., <0.92 fledglings per nest. Shaded groups: Average success, ±0.5 SD about the mean, i.e., 0.93–1.06 fledglings per nest. Groups shown in black: High success, greater than 0.5 SD above the mean, i.e., >1.06 fledglings per nest.

individual groups varied about this mean by establishing three categories of breeding success: low success (below 0.5 SD from the mean), mean success (from −0.5 to +0.5 SD about the mean), and high success (more than 0.5 SD above the mean) – respectively <0.92, from 0.93 to 1.06, and >1.06 chicks per nest. Groups with different breeding success appear distributed throughout the colony with not much evidence of an edge effect. This effect could, however, be established statistically by comparing peripheral groups with inner ones. (Groups were judged peripheral if there were no others between them and the skua nesting areas that ring the colony.) There was a significant difference in breeding success overall between central and peripheral groups and this difference was evident irrespective of whether the comparison was of groups with more than 40 nests (moderate to large groups) or confined to groups with less than 50 nests (small to moderate groups) [1]. These comparisons excluded the outside groups in block A along the sea edge, which are really part of the inner colony area in lacking a skua nesting margin, and those in G Block, which sprawl and merge over a series of ridges and cannot be readily assigned to these two categories.

Although the overall breeding success was higher in the following year (1969–70) (1.28 ± 0.24 sd chicks per nest) the distinction between central and peripheral breeding groups was maintained in comparisons including both small and moderate groups but not for moderate groups (with more than 40 nests) alone.

On the whole, therefore, it appears likely that the position of the breeding group within the colony relative to skua nesting areas does bear on likely breeding success, with peripheral groups having lower success, but the difference is small on average and there is wide variation among individual breeding groups.

Breeding success of the study groups in different parts of the colony

The general conclusions from the colony-wide estimations of breeding success, which were based on a single count of nests in early December and a single count of chicks in mid January, have been amplified by the detailed analysis of breeding statistics derived from the regular monitoring of the selected study groups.

In each year between 600 and 1000 nests (2.4–4.0% of the colony total) were followed in 12 to 20 breeding groups. Two criteria determined the selection of groups: 1. that they were sufficiently isolated that the post-guard chicks would remain on the group area and be recognisable as a population unit until leaving for sea; and 2. that they were representative of breeding locations and skua–penguin situations on this colony. The first condition was the most constraining as it precluded the use of most penguins on A and G blocks, and the large, closely-sited groups on the upper colony (blocks B–F). Most groups selected were thus small to medium-sized groups, containing 20 to 100 nests at peak occupation. They could all be mapped and recorded readily from the margin. In the light of the findings by Ainley *et al.* (1983) that the oldest, and most experienced, penguins shun the centres of large groups this was a fortuitous choice of

Table 10.2. *Comparative breeding success of central and peripheral breeding groups of penguins on the Northern Colony*

Proportion of chicks fledging in mid January of eggs laid in study groups. Numbers of eggs laid shown in brackets.

	Central groups	Peripheral groups
1965–66	0.69 (1392)	0.52 (620)
1966–67	0.59 (705)	0.36 (236)
1967–68	0.75 (667)	0.62 (321)
1968–69	0.57 (364)	0.58 (224)
1969–70	0.78 (596)	0.73 (279)

group size as these were most likely to comprise mixed (unstratified) age classes and be more representative of the colony as a whole than the largest groups.

In each year groups were selected on the perimeter and in the centre of the colony, the former representing what was considered the most favourable siting for skua predation with access by higher skua numbers and with more space about the margin in which post-guard chicks could be tackled. There were, however, few of these breeding groups on the colony so that choice was limited and the range of different topographic situations, for example, could not be explored. Of course the breeding groups do not fall neatly into two categories anyway but form a continuum ranging from large groups within the territories of weakly predatory skuas at the one extreme to small, isolated groups apparently 'besieged' by several aggressive pairs at the other.

The comparative breeding success of central and peripheral groups (using the same criterion as before) varied markedly in the different years depending on the groups monitored (Table 10.2), with peripheral groups being less successful in the first three years (and markedly so in 1966–67) but performing equally well in the last two years. By chance all the peripheral groups used in 1966–67 had low success, each being the least successful group within their local area. The difference in breeding success shown is not because the peripheral groups were smaller – in each year the peripheral groups produced from 49 to 155 eggs on average, with the best survival relative to central groups (in 1968–69) coming from the lowest average number.

The breeding success of very small breeding groups (fewer than 25 nests in early December)

There were a small number of breeding groups dotted throughout the colony that contained only a few breeding pairs. Judged on the basis of the substantial depths of guano built-up none was a very new group. (In the five years of this study only one completely new breeding group was established.)

The smallest of these groups, with fewer than 10 nests, invariably had low breeding success, and two of them (A19 and A29) disappeared during the course of this study. However, a number of the slightly larger groups performed surprisingly well, even when on the colony periphery.

Overall, these groups averaged slightly below the colony average in both 1968–69 and 1969–70. In the first year 24 small groups with a total of 338 nests (mean 14.9 ± 7.0 SD nests) fledged 320 chicks, on average 0.95 chicks per nest (compared with the colony average in that year of 0.98 chicks per nest).

In the following year 21 groups with 326 nests (mean 15.5 ± 6.8 SD nests) fledged 370 chicks, 1.13 chicks per nest compared with the colony average for the year of 1.28 chicks per nest. However, if the selection is of groups with <20 nests these averaged only 0.96 chicks per nest and for the five groups with <10 nests the average was 0.18 chicks with three producing no chicks at all.

10.3.2 *Nest location* within *the breeding group*

For most of the breeding cycle the penguin breeding groups are so tightly compacted that skuas can only reach the eggs and chicks from the margin or from flight over the group. Skuas are not able to land or forage within the breeding area. All observations show that most skua activity is in fact directed at the margins of groups. This constraint on foraging is especially severe once penguin chicks have grown too large to be lifted from flight.

This general appreciation that skuas tend to do most of their foraging about the margins of groups has long been recognised and has led to the almost universal separate analysis of penguin breeding success for central and peripheral nests as a measure of skua predation. Not only are the penguins on the margin more exposed to predation they are on average less aggressive than are the birds on central nests (Spurr, 1972). At first sight this seems a straightforward comparison. Peripheral nests are exposed to skuas, central nests are not. The analysis is, however, confounded by a number of other factors, including possible differences in age and breeding experience of penguins in the centre and margin of breeding groups, differing levels of disturbance caused by the reoccupying penguins in late incubation and chick hatching, and differences in egglaying synchrony. In practice this complex interrelation between nest position and penguin factors makes the interpretation of differences found in breeding success extremely difficult. This difficulty becomes especially acute in the absence of banded known-age penguins. With a population of these birds it would be possible to separate out the effects of age and experience from the effects of location in the breeding group, as has been achieved by Ainley *et al.* (1983).

The first formal definitions of central and peripheral nests are those of Penny (1968, p. 96) who considered a nest peripheral 'if on the colony edge' and central 'if it had at least one nest between it and the edge'. However, definitions need to take into account the very big changes that occur to the groups through the season as nests classified as peripheral during minimum nest numbers will probably be bounded by one to several rows during the peak numbers of occupation and reoccupation. For the assessment of skua disturbance and attack on breeding success, nests described as peripheral need to be open to the margin throughout the breeding season. Ideally, they should be selected from peripheral nests at the peak of occupation and egglaying, selecting more than needed for the work in the expectation that a proportion will be barren or lose eggs early. Through selection at this time one can be certain that these nests will be exposed on the margin at least until reoccupation is well advanced. With luck many will remain fully exposed to skuas even beyond this life-cycle stage right through breeding until the group structure collapses during the post-guard stage. This is essentially the approach taken by Spurr (1975c) whose definitions of central and peripheral refer to exposure to the edge of the breeding area from egglaying. This is a more specific definition than that of either Penny (1968) or Tenaza (1971); the latter is used by Davis and McCaffrey (1986). (Tenaza's study of nesting behaviour took place from 6–11 December, during minimum occupation, so it may be assumed that his nest selection applied to this date with the study birds selected from successful breeders left after failed nesters had departed. These birds were not, therefore, likely to typify peripheral birds of the occupation stage.) Authors need to state very clearly how they selected peripheral nests to ensure that different studies may be compared.

There is generally less ambiguity in the selection of central nests, only if the outside rows were dramatically eroded would original selections be exposed on the margin.

Records made across the season at Cape Bird indicate the degree of uncertainty that can occur when classification changes. Ninety-three nests on the margin of four breeding groups were marked on 14 November, during peak egglaying and occupation. On 6 December at minimum occupation 37 had been deserted exposing 18 nests from inner rows to the margin. During the following three weeks as reoccupying birds returned to the colony the margin again began to be covered by new birds so that by 30 December 23 (25%) of the original set were now well protected and should thus more properly be classed as central nests. Results from this set of nests on the impact of skuas on peripheral nests that did not take these changes into account would clearly be misleading. The extent of change in any group depends on egglaying and retention of eggs, on the scale of the reoccupation event and on how the new nests are distributed about the breeding group. It depends as well on the terrain covered by the group, which may foster or deter new arrivals. Of recent authors, only Spurr (1975c) has indicated when his study nests were selected and none say what they have done with the records of nests changing classification during the time of the research.

Attempts were made to compare the breeding success of pairs in the centre and on the periphery of large breeding groups in four years (see Table 10.3). These groups were so big that the central block of nests was usually separated from the edge by several rows of nests. All peripheral nests could be approached by skuas from the ground throughout the season and those protected later by reoccupying birds were discarded from the analysis. Between 35 and 50 nests were monitored in each part of three or four breeding groups each year. These nests were selected during or immediately after peak egglaying. They could only be monitored until the beginning of the post-guard stage in early January when the breeding group structure collapses and chick identity is lost (the cut-off point for Penny also). Neither adults nor chicks were banded for these experiments: the disturbance caused at the time would no doubt have invalidated the results. The nests were mapped and most records of their progress were obtained from the margin.

There was considerable variation in the relative successes of the two sets of penguins in the different breeding groups in each year and in different years. Overall, pairs on central nests outperformed those on the margin but the discrepancy between the two was not as great as might have been expected from watching skuas working around these breeding groups. Survival to early January was, however, universally very good in these trials, with the lowest survival, at just 50% of eggs laid, occurring in 1968–

Table 10.3. *Breeding success of pairs in the centre and on the margins of large breeding groups*

Survival of chicks recorded to 5 January (early post-guard stage).

	Number of nests with eggs	Eggs laid	Eggs hatching	Chicks fledging	Percentage fledging of those laid	Percentage nests with chicks[1]
1965–66						
A31 centre	38	74	—	61	82	—
periphery	52	90	—	70	78	—
E2 centre	48	90	—	76	84	—
periphery	30	59	—	42	71	—
F15 centre	42	82	—	68	83	—
periphery	24	45	—	27	60	—
1966–67						
A31 centre	32	63	58	52	82	75
periphery	39	66	48	42	64	74
E2 centre	35	61	50	40	66	82
periphery	30	58	47	42	72	93
F15 centre	32	64	53	47	73	84
periphery	25	44	32	28	64	80
1967–68						
A31 centre	43	82	68	60	73	81
periphery	36	63	45	38	60	72
E2 centre	36	68	58	56	82	89
periphery	27	49	44	39	80	85
F15 centre	40	67	57	55	82	81
periphery	24	46	31	29	63	79
B3 centre	42	81	68	63	78	93
periphery	33	57	46	46	81	84
1968–69						
E2 centre	43	49	35	32	65	75
periphery	34	32	24	24	75	74
F15 centre	30	46	28	23	50	59
periphery	35	44	34	26	59	65
B3 centre	44	71	63	52	73	74
periphery	42	45	39	31	69	67

[1] Percentage of nests with eggs producing at least one chick.

69, in which all breeding groups on the colony had a depressed breeding success. Overall, the success in these groups compares well with that recorded in other reports. The criterion used by Penny (1968) was survival of at least one chick of nests with eggs (as in the last column of Table 10.3). On this basis he found 77% of central pairs ($n = 383$ nests) and 70% of those on the margin ($n = 298$ nests) were successful. In all years except 1968–69 the penguins at Cape Bird were more successful than this, with up to 93% of nests being successful.

Breeding success recorded here is also higher than that reported by Ainley *et al.* (1983) at Cape Crozier, overall 68% of eggs in central nests ($n = 408$ eggs) and 60% in peripheral nests ($n = 435$ eggs) and very much higher than that recorded by Davis and McCaffrey (1986) – 51% for eggs in central nests ($n = 357$ eggs) and an unusually low 27.1% for eggs in peripheral nests ($n = 501$ eggs). This latter study was done on H Block at Cape Bird in 1977–78. Such low success suggests, as indicated by Lishman (1985b) and Culik *et al.* (1990), that these birds were being adversely affected by observer disturbance. Alternatively one must conclude that this was an even harder year for penguins than in 1968–69; but there is no evidence for this.

Only Ainley *et al.* (1983) have been able to distinguish between location and breeding experience of the penguins. The significant difference they recorded overall between pairs in the centre and periphery of groups at Cape Crozier arose from a significant discrepancy among three- and four-year-old birds. Other age classes bred with equal success in each place. This result confirms the common sense notion that the more experienced penguins would have less difficulty in countering skua disturbance and predation.

What evidence is there at Cape Bird that the lower success of pairs on the group periphery might also result, at least in part, from less breeding experience? Two comparisons suggest this. First, the pairs on margins laid fewer eggs on average, indicating that they were younger birds. This difference shows up whether it is measured as numbers of eggs laid per pair in nests with eggs (1.78 and 1.62 on average for central and margin pairs respectively) or as numbers of eggs laid per occupied nest, not just those with eggs, with centre pairs producing 1.65 eggs and margin pairs 1.33. Second, pairs on the margin were also on average later breeders, also indicative of less breeding experience. For example, in 1967–68 65% of the central pairs had laid by 14 November compared with only 23% of pairs on the margin.

Although it has been possible to show overall that pairs on the margins of groups bred less successfully than those in central nests the difference is not as clear cut as expected and different breeding groups provided quite contrary results. In this series alone although central pairs were more successful than ones on the margin in nine groups, in four others the margin pairs were more successful and in one other the two sets were equally successful. Nor is this variability related to differences among seasons. Depending on the breeding group, pairs on the margin were more successful than those in the centre in both 1966–67 and 1968–69. In the final year (1968–69, Table 10.3) in two of three groups the margin pairs were more successful than those in the centre. Clearly, one must be very cautious about reaching firm convictions about the relationship between nest site in the breeding group and breeding success unless extensive trials have been carried out.

In these trials at Cape Bird large breeding groups were used so that pairs on the margin were never fully isolated from neighbours at any stage and this contiguity gave them protection from skua disturbance, with attacks possible only from their front. It would be very easy to select study areas in which pairs on the margin became isolated. In this situation even an averagely skilled skua can cause significant loss.

In summary of this section it is apparently better for penguins to nest within the breeding group rather than on its periphery and marginally better to be in a breeding group in the centre of the colony than in one on its edge. Certainly they should avoid very small groups on the colony periphery. Oelke (1975) is the only person to have investigated the behaviour of penguins in different parts of colony. In his transects across Cape Crozier he also found wide variability in the success of different breeding groups with the ones on the outer edge, within the territories of skuas, faring less well. These were all rather more isolated breeding groups than occurred at Cape Bird. But there is such wide variation among even apparently similar groups located in much the same sorts of places within the colony and such wide variation between the success of central and peripheral nests in different groups that the impact of skua predation and disturbance is clearly the most significant variable. Where weakly predatory skuas defend large numbers of penguins there will be little difference in the breeding success of different groups and between central and peripheral nests. Where few penguins nest within the territory of an aggressive skua pair then the impact of all factors impinging on success – age and experience and nest position – will be exaggerated.

10.4   *The role of disturbance in skua predation and scavenging*

In common with many other predators that need to attack the parent in order to reach offspring prey, predatory and scavenging roles are blurred. For example, is obtaining eggs from a deserted nest that resulted from the continuous disturbance of the incubating penguin to be classed as predation or scavenging? The final act of taking the deserted eggs is done as a scavenger, but the long series of attacks that preceded this, culminating in the penguin deserting the nest, were those of a predator. The strong defence of the eggs by the parent is unequivocal evidence for this.

The distinction between these two roles is not clearly apparent in the literature of similar situations of a predator acting against a larger colonial prey even though in most of these the defence of the brood by parents is a common theme.

10.4.1 *Amount and form of disturbance to breeding penguins*

Disturbance of incubating and brooding penguins has three effects. First, any disturbance in the breeding group immediately causes penguins locally to become alert and active, raising their metabolic rate, and consequently reducing the time they can stay at the nest while fasting. Confirmation of this impact has come from the studies of Culik *et al.* (1989, 1990), which have described the changes in heart beat rate (assumed to be directly proportional to metabolic rate) of penguins standing at the nest or disturbed while incubating. Contrary to earlier appreciations the Adélie was not found to be particularly well insulated and any activity that disturbed the penguins from the crouched, incubating stance on the nest was costly in terms of heat loss, compensated for by increased metabolism. A heart beat rate of 86 beats per minute (bpm) was raised to 127 bpm when standing and even moderate disturbance that provoked little more than head movements raised the rate to 99 bpm. Second, prolonged disturbance, especially continual attack, appears to reduce the 'self-confidence' of many penguins in their defence of the nest. These two effects of disturbance act in concert – a starving penguin is less able to repel skua attacks. Third, attacks and surveillance that provoke rapid adjustments on the nest, or the excited aggregation of free adults, may cause eggs and chicks to be thrown from the nest or be trampled. Displaced and frightened chicks running through the breeding group and unable to find a resting place in its centre, either on a vacant nest site or under another adult, are forced to the margin, and are exposed to skua attack. Up to this point none of the activity of the skuas in these situations is strictly predatory. Indeed, only when taking and killing chicks that are able to survive away from the nest is true predation involved.

The distinction between the two behaviours is not easily drawn and definitions such as in Curio (1976) distinguish between them on the basis of whether the prey is living or dead when gained. The problem of how the prey became available is not addressed. Curio's definitions are:

> Predation – 'a process by which an animal spends some effort to locate a live prey and, in addition, spends another effort to mutilate or kill it'. (The consumption of the prey following capture has been intentionally omitted from this definition as animals may be disposed of without being eaten.)
>
> In scavenging, animals 'feed on carcasses or on scraps and offal left by other animals'.

The definition of predation given here is similar to the first one in R. J. Taylor (1984): 'predation occurs when one organism kills another for food', but note that it is defined here in relation to obtaining food – a narrower focus than used by Curio. Even so, neither definition really considers how to distinguish *direct predation*, for example, of skuas on chicks out of colonies, from *indirect predation*, of skuas on guarding parents to obtain eggs or young chicks, or from *delayed predation*, when attacks 'force' the parents to desert the nest so that the eggs and chicks become accessible to attack or to scavenging. There are also two circumstances of nest desertion: either it might never have occurred without the skua attacks, or alternatively the attacks might merely have brought it forward in time. There is the complication as well that the desertion may occur hours or days after the last attacks occurred. If desertion always occurred during an attack there would of course be little ambiguity in recording a predatory cause. In part, solution to these difficulties resides on the level of surveillance and knowledge of the interaction between the two contestants. Without prior knowledge of attacks, eggs or chicks gained by skuas from an abandoned nest would be recorded as scavenging; with knowledge of persistent attacks it would more properly be recorded as predation.

But there does not seem to be any sensible precedent in the literature on this point. Scavenging within the formal definitions occurs on deserted eggs and young chicks (unable to survive away from the nest), predation occurs when eggs or chicks are taken from defended nests or chicks are killed during the post-guard stage. Although it might just conceivably be reasonable to consider attacks on young living chicks predation, attacks on abandoned eggs can scarcely be included within this definition. In point of fact once eggs and chicks are abandoned they should immediately be considered carrion, and thus are scavenged.

In Chapter 7 the amount of time spent by skuas in flight surveillance of the penguin colony was described from the point of view of the skua, of the proportion of time spent in this activity and on its energy need. It is now appropriate to consider the same activity from the point of view of the penguins as a disturbing and disrupting activity on the colony.

At Cape Bird the skuas were largely confined to their own territories in their searching flights above the penguins. Their height above the penguins depended both on the size of the territory, being generally higher in large territories and lower in small ones (pair 151 with a territory scarcely larger than the F10 penguin group it contained never reached more than a metre or so above the penguins) and on wind conditions. Strong winds brought the skuas down nearly to penguin head height. Most flights during light winds were at about 4–5 metres above the penguins in large territories, but were higher for skuas hunting outside the territory. Low flights within another skua's territory are invariably met with strong defence. The heights given by Müller-Schwarze and Müller-Schwarze (1977) for flights at Cape Crozier are a good description as well for those at Cape Bird: most flights there were below 6 m with mean height of all flights up to 10 m being 4.1 m.

All flights about the territory were followed to some degree by the penguins. Müller-Schwarze and Müller-Schwarze (1977) found that at Cape Crozier penguins were alert to skua flight up to 12 m. The proportion of penguins reacting to skuas above them depends, however, on the amount of disturbance immediately prior to the flight, because once penguins have been aroused they tend to react more strongly later, and to the flight characteristic. Direct flights across the group may elicit little response; slow wheeling flights, and especially hovering, provoke a stronger reaction; hovering just out of head reach provokes an angry response by all the penguins in the immediate area. Penguins not attached to the nest or brood rush towards hovering skuas, causing even greater disturbance as they provoke fierce, territorial defence behaviour from nesting penguins.

There are two measures of the degree of surveillance of the penguins from the records of skua flight paths. The first is the number of times each breeding group is contacted by flying skuas – the number of passes over

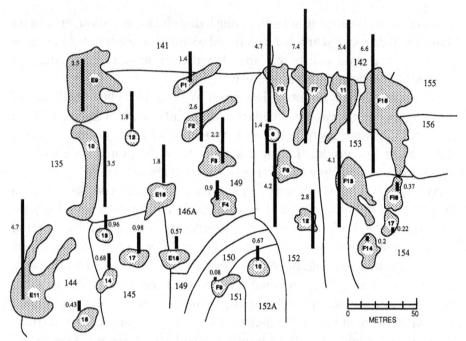

Fig. 10.2. Relative levels of disturbance on breeding groups from skua flight about EF study area. Number of flight contacts hour^{-1} on each breeding group. Shaded areas = penguin breeding groups; unshaded areas = skua territories. The level of disturbance on each breeding group is indicated by the length of the solid bar.

them. This is measure of the frequency of contact and disturbance. The second record is the length of the flight path above the penguins. This gives a rough index of the numbers of penguins affected.

The number of passes above the penguin breeding groups is shown in Fig. 10.2 as mean number of contacts each hour for the breeding groups in EF Block. The number of contacts varied from a minimum of 0.08 h^{-1} for F9 to a maximum of 7.4 h^{-1} for F7. There is a general trend of lower values for the small breeding groups in the small territories along the margin of the colony and higher values for the large breeding groups towards the centre of the colony away from the skua nest sites. Groups in pair 154 territory recorded exceptionally low rates (0.2 h^{-1} for F14 and F17) even though this pair had substantial territory on the colony; this reflected the low interest in the penguins by this pair.

Although these records of flights across the breeding groups indicate a high order of surveillance and disturbance this is much less than recorded in other studies. Müller-Schwarze and Müller-Schwarze (1977) recorded

Table 10.4. *Records of mean flight path per hour for skuas on EF study area*

Pair	Mean flight length (m h^{-1}) mean ± SE	Efficiency (percentage of flight over breeding groups)
144	178 ± 53.4	54.1
145	232 ± 41.2	36.0
146 new pair	186 ± 47.7	29.6
149	40 ± 249	23.7
152	601 ± 106.6	42.4
153	482 ± 141.6	65.7

Records are mean flight path from eight observation logs between 11 December and 22 January, 1968–69.

Flight length was traced directly from the maps uses a planimeter calibrated to map scale.

25 overflights per hour in a central breeding group (in common ground open to all skuas) at Cape Crozier and 34 per hour over a group within a territory. High rates were recorded also for brown skuas by Trivelpiece *et al.* (1980): 2.1 overflights per hour for territorial skuas but a surprising 11.1 for foraging intruders.

This is, however, a crude measure of impact. A more realistic measure of surveillance (and possible impact through disturbance) can be gained from the records of the distance flown above the penguins. Flights about the territory are much slower on average than flights out to sea and can be followed and mapped readily enough while in progress. (In one sample of 17 flights each between 400 m and 1700 m in length on EF the skuas averaged 5.53 ± 1.56 m s^{-1} (mean and SD) range 3.9–6.7 m s^{-1}, with the fastest flight speed occurring in pair 153, which had the largest territory. These speeds are less than half the usual flight speeds for this skua of c. 14 m s^{-1}.)

Data for 51 flight tracks are recorded for these skuas in Table 10.4. These skuas averaged between 40 m and 600 m of flight each hour about the penguins in the territory, with the longest flight lengths occurring in pairs with the largest territories. This record discloses a substantial coverage over a total ground area measuring 150 m by 115 m, with flight patterns reaching high efficiencies in flying over the breeding groups. Conservative estimates of the numbers of penguins affected by these flights lie between 3820 and 8870 penguins each hour (from a nest spacing of 0.7 m and assuming that a track five nests wide is affected). As there were only c. 1600 nests in total within these territories in EF these estimates indicate that each penguin was overflown on average between 2.4 and 5.6 times each hour.

Two factors influence this flight pattern: the size of the breeding group, related to the need for flight surveillance in larger groups, and the number and distribution of the breeding groups within skua territories. A single, small group close to the nest can be watched from a single vantage point whereas several breeding groups some distance from the nest can be watched most effectively from flight. In many pairs there is a rather constant pattern of long sweeping flights about the territory. This pattern of surveillance showed clearly for pairs 152 and 153. Both undertook repeated sweeps over their breeding groups touching the same groups time after time, the order depending on flight direction determined by wind direction. Most flights by 153 were from F13 to F15 across F11 and back to F13. Those of 152 circled across F12, F11 and F7 to F5 and F6 before returning again to F12. These pairs with long, narrow territories running down into the colony and containing large numbers of penguins have high flight levels. Individual preference by skuas is also significant: pair 154 had a moderately large territory with three breeding groups yet had exceptionally low flight levels with most surveillance being from the ground.

It needs to be remembered that this is not the total surveillance exercised on these penguins. An even greater level occurred from the ground by skuas walking about or sitting close to the margins of breeding groups.

10.4.2 *The possible role of disturbance in production of carrion on the colony*

Disturbance by skuas and penguin egg viability

The often vigorous movements needed at the nest by penguins meeting attacks by skuas and in shuffling about to face disturbance or attacks on neighbours in the breeding group could conceivably damage the embryo and contribute to the high proportion of eggs that fail to hatch in this species. This possibility can be examined by comparing hatching rates in the protected breeding groups (protected by wire exclosures from skua disturbance) with those in control groups. (Without egg predation or scavenging the rates found within the protected groups more certainly reflect true rates. In groups open to skua attack, however, some non-viable eggs will have been taken by skuas before becoming apparent but provided normal and non-viable eggs are taken equally, and provided the groups are monitored sufficiently closely to detect these eggs, the effect of skua predation on the proportion non-viable should be of small signifi- cance. Most non-viable eggs found are over-age, the laying date being known, or are dislodged from nests by the growing sibling.)

There is no evidence in this comparison between the protected and control groups for skua impact on egg viability. In protected groups in 1966–67, 21 (7.6%) of 277 eggs failed to hatch, in 1968–69, 28 (6.1%) of 462 failed, in 1969–70, 5 (4.3%) of 116 failed; in all seasons this was a total of 54 (6.3%) of 855 eggs. Identical rates were found in groups open to disturbance. In 1969–70, for example, 60 (6.3%) of 953 eggs in the study groups failed to hatch.

Similar rates have been recorded by Davis and McCaffrey (1986) (6.6%) and Spurr (1975c) (6.7%) for the same colony in other years. From research carried out at Cape Crozier (Ainley *et al.*, 1983) it is likely that most of these eggs are from younger birds. At Crozier the infertility rate of eggs of three- and four-year-old birds was 25%. The mere existence of these non-viable eggs among breeding groups protected from the skuas suggests that skua disturbance is unlikely to be a major factor. The similarity between the rates in the protected and control breeding groups goes further to indicate that disturbance may not be implicated at all. The figures for proportions of non-viable eggs determined so far, by several studies, are sufficiently accurate for population studies. On ethical grounds there is no justification for taking an egg sample to determine fertility rates directly.

Disturbance by skuas and egg and young chick mortality

The possible role of skua disturbance in egg and young chick mortality arises from its effect in causing incubating penguins to shuffle about on the nest to face or meet skua activity about the breeding group, from the rush and violence of free penguins chasing skuas hovering above the breeding group or landing near the margin, from the continual disturbance of incubating penguins aroused into alert attention by skuas from a resting incubation posture and physiology, or from the way penguins on the margin are simply 'worn down' by the need to meet and parry skua attacks time and time again. Penguins on the periphery of breeding groups seen standing off the nests, or backing away to the far side of the nest during a skua attack show all the symptoms of exhaustion and failing commitment with head drawn back, eyes rolled and slow wing beating.

In none of these events is the skua preying directly on the eggs or chicks; both become free as carrion when displaced from the nest or deserted there if the penguin leaves the breeding group. The only way egg and chick loss caused in this way can be distinguished from direct predation is by close monitoring of the interactions between the skuas and individual penguins and from direct observation of the final phases as the penguin defence collapses. Under normal breeding conditions, however, it is rare to see penguins actually abandoning the nest, i.e., walking off the eggs and going to sea. Although there were many records of penguins doing this in 1968–69 these were of starving birds that had been sitting through a long-extended incubation spell. This failure was in response to exceptional pack-ice conditions rather than skua disturbance. Mostly all one finds in the breeding group is an empty nest, more rarely a nest with eggs or young chicks and this often after attention has been drawn to it by foraging skuas.

10.4.3 *Possible role of disturbance on the timing of the onset of the post-guard stage*

There is a second possible role for skua disturbance in chick loss in the timing of the onset of the post-guard stage and its implications for the survival of the very first chicks to be left unguarded. At the very start of this stage, there is little open space for these chicks away from territorial adults so that a high proportion of the first unguarded chicks end up on the breeding group margin – where they have a heightened exposure to skua attack. A significant skua role would occur if adults left before their chicks were able to withstand skua attacks because the skuas chased them away.

The very difficult situation encountered by these chicks is evident in the notes made of one such encounter. On 27 December a chick standing with adults on the margin was caught by pair 148 and dragged 8 m away from the breeding group before escaping. Although it immediately rushed back into the breeding group it was soon chased out again by the nesting penguins to the margin – and was promptly attacked again by the same skuas. Similar occurrences were commonplace at this time.

The switch in penguin behaviour from guarding chicks on the nest to leaving them on their own is a critical time for the young chicks, matched in risk only by the movement to sea late in the season. This change takes place when the chicks are about 20 days old and allows both parent to forage for food. It occurs at Cape Bird during the first weeks of January but varies among breeding groups and between seasons. There are two quite independent aspects of the biology of this stage that need to be kept distinct in any analyses. The first is the change in the behaviour of the parents from guarding the chicks to leaving them on the breeding area. The second is the behaviour of the chicks after they have been left on their own. Although these chicks are usually termed crèche chicks in the literature (and the stage of the life-cycle the crèche stage), they should more correctly be termed unguarded chicks and the term crèche reserved for when the chicks are left together on the breeding group area in the virtual absence of any parents and more specifically when they are aggregating closely (see Chapter 1). The timing of the changeover in the parents' behaviour from guarding to leaving the chicks is thought to be influenced by such factors as the amount of food demanded by the chick or chicks compared with how readily it can be obtained. This determination is, therefore, ultimately physiological and ecological. (Although comparisons between single and pair broods would provide an obvious test of the 'food required' hypothesis this does not seem to have been carried out. Although Davis (1982) compares single and pair broods these were of small samples, not controlled for other factors influencing the changeover.) However, within the ten or so days in the season during which the switch occurs other factors such as weather, social stimulation and the activity of other pairs, and disturbance and attacks by other adult penguins and skuas may also be significant. As Taylor (1962a) found, these factors may advance or delay the transition significantly with different chicks being first left unguarded between 17 and 32 days old. Interestingly, Taylor suggested a significant role for skuas by noting that parents could not leave the more isolated nests 'until the chicks had reached a size when they were relatively safe from skua predation'. The minimum age when chicks can be left unguarded on the breeding group is determined ultimately by their maturity. Lishman (1985b) reports that efficient thermoregulation does not occur until they are c. 15 days old and that chicks cannot recognise parents or re-locate their nests until c. 17 days. It is unlikely that chicks could survive on their own earlier than this.

Most accounts of the behaviour of chicks at this stage are, however, of the behaviour of the unguarded chicks in relation to disturbance and predation by skuas and to huddling when cold, rather than to the transition from being guarded on the nest to unguarded. The role of adults on the breeding areas in protecting unguarded chicks from skua attack is now well recognised in the literature. It has been adduced by Yeates (1968, 1975), for example, to account for the absence of crèching at Cape Royds during two years in which there was an unusually high number of reoccupying penguins. The behaviour of the chicks during this stage is, however, a quite separate problem. The main focus of interest has been to determine why chicks come together in tight groups. Davis (1982) concluded that the adult/chick ratio in breeding groups was the most important factor in determining the onset and pervasiveness of the crèching he observed at Cape Bird in 1978. Because of the restricted definition of crèching used in his study ('a minimum of three chicks in close association') this work relates more specifically to the behaviour of the unguarded chicks rather than to the transition from the guarded to unguarded stage.

In different circumstances skua disturbance and predation could act to hasten or delay the parent's departure. It would delay the parent's departure if the parent was reluctant to leave its chicks exposed to predation. Conversely, a history of attacks by skuas might precipitate an early departure. The role of skuas in the tight crèching of chicks on breeding groups is not in question: attacks and disturbance by skuas of chicks at this stage invariably causes them to group together.

The changeover from the guarded to unguarded chick stage was routinely recorded in each of the study groups each year. Although it is a straightforward process to count chick numbers, care must be taken with the counts of adults on the breeding groups at this stage. Too brisk an approach to a breeding group causes tentatively-attached adults to drift away, so that counts comprise only those with well established territories. At this time of the season it is very easy to lose 10–20% of adults from a small group through approaching too closely or precipitously. Moreover, counts of adults need to distinguish between those attached to the area and those in passage or exploring. In the following account precautions were taken during the surveys to ensure that all adults attached to the groups were included in the counts.

In the 1969–70 season 1855 chicks in 24 breeding groups were checked each day from the appearance of the first unguarded chicks on 30 December. These chicks provided a first insight into how unguarded chicks arose at this date. Without exception all of them had been displaced from nests during territorial fights between resident parents and reoccupying birds. Of the 22 chicks that had been left by parents at this date, 10 were on the periphery of the group, 10 were alone on central spaces or nests and two were tucked in with other chicks beneath adults. (These latter chicks were only recognisable because they joined a two-chick brood.) Thus, the very beginning of the transition in this year was provoked by penguin conflict, not skua aggression, and occurred when displaced chicks were sufficiently large and mobile to find an open place in the breeding group to live.

What evidence was there at Cape Bird that skua activity was implicated in the timing of the onset of the post-guard stage?

The changeover from guard to post-guard stage proceeds very rapidly. In 1969–70 the first chicks were seen on 30 December, by 3 January 15.7% of chicks were unguarded, by 5 January 29.7% and by 10 January fewer than 1% of chicks were still being guarded by parents. Adult/chick ratios ranged from 0.72:1 to 1.51:1 in the different groups but were independent of the proportions of unguarded chicks in the colonies on either 3 or 5 January [2]. Nor was there any obvious pattern evident in the onset of the unguarded stage and location of the breeding group on the colony. On 5 January similar ratios of unguarded to guarded chicks occurred on A block, groups with low skua activity, and on the colony margin along D, E and F blocks with high activity. This analysis suggests that at least for this year with its moderate adult/chick ratios skua activity was not a significant factor in the date parents left their chicks. However, in the previous year there had been much higher adult/chick ratios, ranging from 1.46:1 to 3.8:1 in the different groups, and unguarded chicks occurred much later in the year so that there were few unguarded chicks anywhere on the colony on 4 January and only c. 20% by 10 January. It is nevertheless too simplistic to conclude from this comparison that high ratios deter skuas, and thus defer the time parents leave. It might be argued instead that with high numbers of reoccupying adults coming onto the breeding area parents are under great pressure to defend their nesting territories from usurpation and this cannot be achieved if both parents are absent for long periods foraging.

The evidence shown above for a small role for skuas in the onset of the post-guard stage is supported by the few records made at this time on the protected breeding groups. Two groups on A block in 1969–70 divided between protected and open sections had equal ratios of guarded and unguarded chicks on 5 January: 49% in the protected areas and 42% in the areas open to skua attack. In a second comparison of groups on C Block in 1969–70 when the open group had 22% of 63 chicks unguarded there was only one unguarded chick of 97 in the protected groups. This appeared at first sight to indicate a significant skua role but from inspection of the changes in pairs on the open group and the high numbers of nests destroyed there through fighting the result is more properly interpreted as reflecting differences between the penguin groups rather than skua activity.

In summary, the role of skua disturbance and activity among the penguins in the timing of the transition from the chick guarded to unguarded stages is equivocal. Although skuas potentially could have a major impact on the parental behaviour at this time, either advancing or deferring leaving depending on circumstances, overall it has not been possible to demonstrate unambiguously a relationship between time of leaving and either adult/chick ratios, and inferred access by skuas to the chicks, or different levels of disturbance dependent on location on the colony. A closer study involving both nutrition and penguin–skua interactions is clearly necessary.

What is clear nevertheless from this study is that the first days of this transitional period offer great opportunity to the skuas for predation on chicks displaced from nests. Until the structure of the breeding groups breaks down substantially there are few places where these chicks can rest within the group and a high proportion end up on the margin where they are often readily accessible to skuas. Also, there are many instances in the field records of repeated attacks on the last isolated guarded nests while the crèche chicks are virtually ignored. Many of these attacks are successful so that it would be much safer for these chicks to be abandoned into the crèche. These observations suggest that late chicks are at just as much at risk as the very early ones and that laying synchrony is an important aspect of penguin brood defence.

10.5   *Selection of prey on the penguin colony*

This section uses the records of foods obtained by skuas on the colony as an indication of the nature of the skuas' interaction with the penguins as scavengers or predators. In principle if the skuas forage exclusively for spilled food, or take dead or abandoned eggs or chicks without a history of disturbance, intimidation or attack, then they should be classified as scavengers. Conversely, if their food consists entirely of living prey obtained by immediate and direct attack then they should be classified as predators. It is already known, however, that most skuas obtain both sorts of food, subsisting on a mix of live and carrion eggs and chicks obtained by scavenging and preying. The uncertainties that exist are on the relative importance of these food classes and on the degree that all food gained, whether dead and alive, can be attributed to their disturbance and attacks. If this high impact is the reality of their association with the penguins then their role is that of predator.

Three broad penguin mortality factors have been identified: those of the environment (ice cover, flooding, snow and cold), those resulting from penguin competition and mismanagement, and skua predation and disturbance. However, these factors act synergistically and assigning a cause for any single mortality event is usually hazardous.

In many instances it is not at all obvious how an attack is to be catalogued, because to do this precisely needs an understanding of the psychological and physical state of the defending parent and of the circumstances immediately preceding and during the attack. Is, for example, an attack by skuas on penguins fighting over a nest containing eggs scavenging or predation? It would seem to be scavenging if the nest contents were already doomed, but predation if they were successfully retained by the parent on the nest. The role of the skua itself in these situations is also a complicating factor. To what extent does its interference during a penguin territorial fight affect the outcome of the fight, and the nest content's survival. If the parent had not been distracted by the attacking skua it may have been able to defend its territory more easily. Or without the skua attack the eggs and chicks may have been able to have been shifted just far enough away to allow both pairs to occupy territories, and for the eggs and chicks to survive.

There are two quite distinct aspects of selection of prey by skuas. The first is the relative preference for scavenging versus predation; the preference for undefended foods – of krill, and dead or displaced eggs and chicks. This activity poses lesser risk to skuas than predation on defended prey. The second is in the selection of defended prey, because of constraints on the predator to tackle big chicks, so that they take smaller ones late in the season, or because of ease of capture, for example, of small chicks rather than eggs from under a nesting parent. But it should be realised that selection by the skuas during both the incubation and chick guard stages and partly as well during the post-guard stage is selection of the defending parents, not of the prey – the eggs or chicks – directly. Only when the skuas are attacking unguarded chicks in the crèche in the absence of parents, or are attacking chicks outside the crèche, is direct selection on prey size and prey condition occurring.

10.5.1 *Records of prey taken and prey selection*

Although there are essentially only three foods available to the skuas on the colony – spilled krill or fish from chick feeding and penguin eggs and chicks – this occurs in diverse forms. Eggs are dead or living, defended or displaced, chicks might be guarded or displaced, large enough

Table 10.5. *Food taken by skuas on EF Block during the penguin chick guard and post-guard stages*

Stage	Deserted eggs	Defended eggs on nests	Dead chicks (scavenged)	Live chicks (defended)
Early guard stage (9–19 Dec.)	4	13	0	17
Late guard stage (20 Dec.–2 Jan.)	6	4	7	31
Early post-guard stage (3–9 Jan.)	9	0	8	3

to crèche and put up some defence when attacked or, conversely, weakened through starvation, or are carrion.

For much of the season these foods occur together so that skuas have the opportunity to select from among them – to show preferences for specific foods or preferences for scavenging or predation. The widest diversity occurs during late December and early January (during the late chick guard and early post-guard stages). At this time there occur living and dead eggs, and living and carrion chicks. Earlier in the season only eggs occur; later, only large chicks. It is important to reiterate that there was no food at all left on this colony (or in the Middle or Southern colonies) in any year at the end of the season, and consequently none was present when the skuas first arrived in spring.

All the foods present on the colony are exploited. The records of food taken by EF skuas on 27 December, 1969, for example, illustrate this foraging diversity. All skuas picked up krill spilled during chick feeding. Two rotten eggs were taken after rolling out of breeding areas. Five dead chicks were picked up from among the nesting penguins and three live chicks were killed. Two of these were taken from defended nests, the third had been displaced onto the margin of a breeding group. Even the most skilled predator of these skuas did not spurn carrion: pair 153 took one rotten egg and two dead chicks in addition to killing a defended 400 g chick.

The most detailed records of the precise nature of the foods taken from direct observation of skua foraging are from the 1966–67 season (Table 10.5). In this year these skuas were seen to take a preponderance of defended eggs early in the year but more deserted and rotten eggs later. These latter became increasingly available to them as the group breeding structure broke down and nests were deserted. There was a similar pattern for the chicks: live chicks made up most of the prey until the post-guard stage.

The significant comparison for considering whether skuas were preying preferentially on any of these foods is with the proportion of each available on the colony at each time. This is a straightforward comparison if considering simply eggs or chicks; preference, the proportions of food taken compared with food available, can be realistically assessed because both prey forms are similarly protected and similarly available. Displaced or abandoned eggs and chicks form a different class of foods. Although they comprise only a minute proportion of the total numbers occurring at any time they are especially attractive for skuas through being undefended. Because of their inequality of *availability*, as distinct from *occurrence*, direct comparison as a measure of preference is meaningless: any comparison made would overwhelmingly demonstrate strong preference. It is much easier, and safer, for skuas to pick up abandoned eggs and chicks than to get these prey out from under a defending parent.

The record of foraging by the nine skua pairs on EF Block on 20 November, 1968 indicate how much the skuas relied on deserted eggs in this year. Over 7.5 hours 12 eggs were seen taken; nine came from deserted nests and three had rolled out of breeding groups and were picked up in the open. By way of contrast there was only a single predation attack on a defended nest.

Four year's records of the comparative numbers of eggs and chicks taken during the guard stage (Table 10.6) demonstrate an early preference for chicks compared with eggs, reversing towards the end of the stage [3]. The preference for chicks is interpreted as showing that young chicks are easier to catch hold of than eggs in an attack on a defended nest; the apparent preference for eggs later simply reflects the increased number of rotten eggs becoming easily accessible to skuas as they are lost from nests. Many are displaced from nests by growing chicks.

The numbers taken, and the short time eggs remain on the breeding groups, demonstrate clearly that skuas scavenge assiduously for carrion, even though it is mostly of lower food content, and often inedible. Skuas gain little from a rotten or frozen egg, little or nothing from a putrefying or dried-out chick carcass. The availability at any time of this prey for skuas can be measured by regular searches of the colony. To appreciate the volume of this food becoming available, and to be able to deduce its origin, it is best to use breeding group exclosures where abandoned and displaced eggs and chicks accumulate in the absence of skua scavenging. There is a surprising volume of this food.

Table 10.6. *Numbers of eggs and chicks taken by skuas during four intervals during the guard stage in relation to proportion of each occurring on the colony at the time*

	Numbers taken	Proportion in colony (%)
Dec. 11–15		
Eggs	36	85
Chicks	7	15
Dec. 16–20		
Eggs	3	50
Chicks	15	50
Dec. 21–25		
Eggs	8	30
Chicks	28	70
Dec. 26–30		
Eggs	9	10
Chicks	18	90

10.5.2 *Carrion present on the colony: deserted and displaced eggs, displaced and dead chicks*

The numbers of eggs seen about the colony during the incubation stage reflects a complex interaction comprising: 1. size of breeding group and its shape and topography; 2. the amount of disturbance among the incubating penguins; 3. the weather and foraging conditions (determining nutrition of the penguin parents); and 4. the level of skua foraging activity. Factors 1 and 3 can be controlled for by observations in different seasons, factor 4 by the use of exclosures.

The contrasting numbers in the 1968–69 and 1969–70 seasons (Table 10.7) shows the impact of the harsh pack-ice conditions of the first year on the incubating penguins. Through failure of mates to return from foraging many penguins deserted their eggs. The high numbers of birds seen standing off the eggs at the nest, or standing near the nest, was an extraordinary feature of this year. Such behaviour had not been noticeable in any of the other years. Indeed, it was seeing this behaviour in mid November that first alerted observers to the possibility that unusually high egg desertion might occur in this season. The majority of free eggs seen on the colony in this year came from genuine desertions, rather than from fighting or the displacement of parents from nests by later arrivals. By early December ice conditions had improved dramatically allowing nor-

Table 10.7. *The numbers of deserted and abandoned eggs and parents standing off nests counted during repeated spot checks of the same breeding groups on the colony during the incubation periods over the two seasons 1968–69 and 1969–70*

Date	A Block (1)	(2)	(3)	B Block (1)	(2)	(3)	F Block (1)	(2)	(3)
1968–69									
20 Nov.	26	7	18	1	2	2	—	—	—
21 Nov.	39	10	21	5	2	3	—	—	—
22 Nov.	38	13	25	7	2	4	—	—	—
24 Nov.	18	8	10	2	6	2	12	1	8
27 Nov.	38	8	12	9	3	1	7	1	3
30 Nov.	39	4	10	2	3	1	2	6	3
3 Dec.	5	4	5	3	2	0	1	0	0
6 Dec.	2	0	1	0	0	0	0	0	0
Number of nests	3480			1190			1300		
1969–70									
20 Nov.	0	0	3	0	0	0	1	0	0
23 Nov.	0	0	0	0	0	0	0	0	3
28 Nov.	0	0	0	0	0	0	0	0	3
Number of nests	5070			1700			1810		

(1) Eggs displaced from nests.
(2) Eggs deserted on nests.
(3) Incubating parent standing off nest.

mal foraging behaviour and from this date egg and chick survival was very high. There were good foraging conditions throughout incubation of the following year and few deserted eggs could be found in searches. This is the normal situation for this colony.

The numbers of deserted eggs occurring on the colony at any time is dependent also on the rate they are scavenged by skuas. On average this rate is dependent on the numbers of nests in each skua territory. In 1968–69 the 15 skua pairs on A block were on average each associated with c. 230 penguins, in B and F blocks each with c. 85 penguins. Except for the record on 24 November there were relatively more eggs present at any time on A than on either B or F. Peak numbers were counted on 22 November when there were 14.6 deserted and displaced eggs per 1000 nests on A but 7.6 eggs on B. These numbers represent very large volumes of food for skuas scavenging on the colony, especially when it is remembered that it is not a cumulative amount, but the difference between those becoming available and those taken by the skuas to that time.

The amount of carrion on the exclosure plots
In both 1966–67 and 1968–69 five breeding groups and in 1969–70 two groups were protected from the skuas by wire screens. The prime aim of these experiments was to measure breeding success in the absence of skua disturbance and predation. A secondary aim was to record the volume and origin of carrion arising on breeding groups. In the absence of scavenging, with its attendant disturbance, it was expected that it would be easier to deduce how this food was produced – from desertion, from displacement of parents, from fighting leading to the displacement of eggs and chicks from surrounding nests, from chicks being crushed on the nest and from starvation. The amounts of carrion occurring on the protected groups in these years are shown in Table 10.8. As expected there were very large differences among the years with a high proportion (30%) of eggs displaced or abandoned in 1968–69 but with similar rates for carrion chicks in the three years (between 5% and 8%). All of this food may be listed unequivocally as carrion and could be obtained by scavenging as no attacks and only minimal disturbance from skuas flying high above the breeding group could have occurred. As such these figures represent the losses caused by the penguins themselves and by environmental factors. Differences from other breeding groups indicate the extent of disturbance and predation factors. This difference will be considered again in the accounts of breeding success of these breeding groups.

Late in the season there were few chicks still small enough to be attacked and the skuas very clearly began to search more diligently and actively for carrion. Any chick lying sprawled on the breeding area during a warm afternoon could be expected to be prodded or grabbed by a skua to be tested for life. During the course of a sultry afternoon all the chicks of a moderately-sized group could be tested. Their indignation in being grabbed like this and woken from abandoned sleep provided enlivening viewing. The hasty way they scrambled to their feet, squawking, stretching and wing beating, marked them out very clearly from the others on the area, and indicated the passage of the skuas as they searched their way through the group.

In each year records were kept during the watches of the numbers of each different food taken, to test for the importance of carrion in the diet of birds with different penguin numbers. It was expected that skuas with large numbers of penguins might be able to gain higher proportions of

Table 10.8. *Food becoming available as carrion on protected breeding groups*

	1966–67	1968–69	1969–70
Total number of eggs laid	518	225	116
Eggs available as carrion			
On deserted nests	27	67	10
Percentage of all eggs laid	5.2	30	8.6
Away from nests	2	2	1
Carrion chicks			
On deserted nests	4	0	0
On defended nests[1]	33	10	4
Away from nests	2	2	1
Total	37	12	5
Percentage of all chicks hatching	7.6	7.6	4.7

[1] Dead, starving or displaced chicks with parents or other adults.

their food as carrion than those with few penguins. There are two good reasons why this might not necessarily apply. First, these skuas were not confined to penguin food and could fish at sea during shortage, rather than being 'forced' to become more and more active and aggressive on the colony through hunger. Second, skuas as a group, or some of the skuas, might simply prefer to capture live food, eschewing the carrion within their areas. The results of this study seem to bear out these reservations on the way skuas might act. As shown in Table 10.9 the proportions of the different foods taken by skuas on EF and H blocks were similar, especially so for the proportions of dead and live chicks taken. It would appear that H Block skuas foraged at sea during food shortage on the colony and did not step up their predation efforts to the extent that live chicks became the predominant food taken. Or alternatively, even if foraging more keenly, they were unable to gain further chicks due to the anti-predator behaviour of the penguin adults.

Scavenging for spilled krill and fish on the colony
Krill and small fish were commonly seen as dribbles and small heaps on the generally rock-hard surface of the breeding groups in late summer. Although it was not possible to measure the amounts of this food picked up by the skuas their intense interest in penguin chicks being fed

Table 10.9. *The different foods taken by the skuas on H and EF blocks during the guard stage*

Data are mean numbers taken per skua pair and percentage of the colony food taken.

	EF Block	H Block
Foraging effort (pairs × hours)	7 pairs × 74 h	13 pairs × 36 h
Average number of eggs/skua pair		
Deserted eggs	25 (35%)	3 (16%)
Defended eggs	2 (3%)	0 (0%)
Average number of chicks/skua pair		
Dead chicks	17 (24%)	8 (42%)
Living chicks	27 (38%)	8 (42%)

and the amount of time spent in fossicking for this food around the breeding groups through January indicated its importance to them. Spilled food could not be utilised by skuas during the guard stage. It became important during the post-guard stage because the skuas could venture among the chicks and the few adult penguins on the breeding area to retrieve it. Food spilled by parents during chick feeding outside the breeding area, lost during the jostling between the chicks during the feeding chases, was immediately available. Some skuas actively helped to produce this food by flying so aggressively about the feeding parent and chick that food was spilled as the parent broke off feeding to chase the skua away. These flights were not aimed at the chick, they were clearly harassing the parent.

A measure of the value skuas placed on this food can be gained from the large amount of time they spent in this feeding activity. This is available in the observation logs of skua behaviour in January. Those of 1967–68 and 1968–69 provide the most complete records (Table 10.10).

In each year there is the expected trend for the foraging activity to switch from attacks on defended nest to attacks on unguarded chicks as the proportion of these two foods change through January. Skuas show a disproportionate interest in the last defended nests in a breeding group as these become exposed to attack from all sides as other nests are abandoned. It is these attacks that hold up the proportion of attacks on defended nests so far into January, even though few defended nests occur. Of greater interest is the amount of effort taken up with scavenging in each year, at first on eggs and dead chicks lying about on the breeding area or

Table 10.10. *Skua interaction with penguins as percentage of total activity to show the importance of scavenging late in the breeding season*

The categories interest in the penguins (2) and searching (3) are excluded.

Date	Attacks on defended nests[1]	Attacks on crèche chicks[2]	Scavenging[3]	Flights at feeding parents to spill food[4]
1967–68				
2 Jan.	42	0	58	0
8 Jan.	32	10	55	4
13 Jan.	19	39	32	10
17 Jan.	19	48	19	14
1968–69				
3 Jan.	35	5	60	0
8 Jan.	21	21	58	0
15 Jan.	18	36	33	13
22 Jan.	0	13	25	62

[1] Categories 5–25. [2] Categories 28–38, [3] Categories 4, 16, 17, 18. [4] Category 27.

left in deserted nests and later in the year on dead chicks and on spilled krill and fish. In early January in both years the skuas spent more time scavenging than in active attack (c. 60% of their time). Moreover, they gave increasing effort through January in attempting to produce spilled food by flying at feeding penguins. For example, on 22 January, 1969 almost the whole of the skua activity on the colony was centred on this food and 60% of their attacks were directed at adults feeding chicks. At this time there was little other food available to them on the colony (virtually all chicks were too big to tackle) and most skua pairs were foraging at sea.

Kleptoparasitism

This is the scavenging of the remains of chicks killed by other skuas. Food taken from other skuas is a significant process on the penguin colony because pairs losing food to neighbours must forage again when hungry rather than feeding again on the carcass. The general term for this behaviour is kleptobiosis ('all interactions in which a resource is removed or stolen from an owner') (Volrath, 1984). Within this category pirates are aggressive thieves and take the food by force; pilferers are not aggressive but take the food by stealth, often without direct intervention of the owner. Skuas act in both ways.

This is clearly an important activity among the skuas on the colony and about its fringe. As described in Chapter 8.3 individual skua pairs are variably successful in keeping penguin carcasses once they have been initially satiated. Whether or not the carcass can be defended depends on where it lies in the territory and the ease with which it can be defended. In part it depends also on the dominance hierarchy among the local skua pairs. Some skuas were overwhelmingly producers of food, others were just as clearly scroungers. The male of 25 on H Block produced a large proportion of the penguin food eaten by pairs 23 to 27 on the west side of H4 breeding group. Pair 13 on the east side could invariably hold most of the few chicks they killed.

Because these skuas forage at sea when they are unable to obtain colony food kleptoparasitism must contribute directly to the number of chicks taken from the colony by requiring the capable predators to kill more food than they otherwise would need. By supplying other pairs these skuas increase the overall amount taken from the colony by the local skua group at the expense of sea foraging.

10.5.3 *Selection of live chicks*

Very few chicks were noted at this colony starving to death in any year – the older ones adopt a characteristically hunched posture with their disproportionally long flippers touching the ground and are immediately obvious when breeding groups are checked. These differences between a healthy and a starving chick are shown clearly in Fig. 8.5. Four chicks measured on 19 January demonstrate the discrepancy possible between normal and starving chicks. At this time when chicks weighed on average 2500 g these four, with foot lengths typical of the age of chicks on this date, averaged just 690 g. Such chicks attracted skua attack. This was especially evident in the direct observations made of skuas working among the chicks in the crèche and in the measurements made of prey killed by the skuas, which often seemed very light in relation to the average chicks on the colony (see Chapter 8).

There were many observations of skuas actively selecting younger or smaller chicks from among those in the crèche during January and February. In general, selection took two forms. In the first, skuas ignored larger, mature chicks in weak attack positions, with adults or close to the breeding group. In the second, the skuas targeted smaller or weaker chicks. For example, they commonly went past larger chicks on the crèche margin to attack smaller ones in a less favourable situation for attack. Late in the season skuas actively worked chicks in the crèche in order to isolate smaller, younger or starving chicks that could be attacked. In small crèches of fewer than a hundred or so chicks and almost no adults this grim ballet might go on for days. The skuas had clearly recognised suitable prey, the targeted chicks just as clearly recognised the threat from attack and kept moving away from the skuas to the protection of larger chicks.

The size, age and condition of chicks taken as prey

From late December the chick skeleton became increasingly hardened and the skuas left parts of the body after feeding, characteristically only the linked pelvic girdle and legs of half-grown chicks but most of the skeleton and skin of mature fledglings. The collection of these remains provided information on the *quality* of prey as well as its volume. As recorded earlier these prey could be aged and their mass estimated from regressions of mass on foot length. Unfortunately, relatively few chicks killed by the skuas were retrieved for weighing and measuring before feeding had begun and most were found with the guts already eaten. Attempts were made to estimate total mass from the carcass by adding in the likely gut-mass for a chick of that age but this technique could not adequately take into account differences in the amount of food, and was not persevered with. If comparisons were to be made with colony chicks with empty guts then the technique would be quite feasible. But this comparison would entail holding and starving colony chicks for two to three days before weighing. An alternative technique to assess the quality of the prey being taken was to weigh and measure the foot, in the expectation that starving chicks would have a lighter foot than healthy chicks of the same age. (The foot of eaten prey is invariably left.) However, in trials of this technique even conspicuously starving chicks could not be identified by this method. Two chicks weighing 980 g and 760 g taken on 19 January, 1969 should have weighed c. 3000 g each. Their foot mass were quite typical of chicks of this age.

The information on selection of prey by skuas assessed from prey remains is shown in Fig. 10.3. There are three phases to this assessment. In the first (a) the mean mass of chicks taken as prey is compared with the mean mass of chicks on the colony at the time. This plot shows that through January the skuas were taking increasingly lighter chicks on average of those occurring on the colony. By the end of the month the prey mass averaged little more than half the mass of penguin chicks on the colony. Clearly the skuas were selecting the smaller chicks from breeding groups, but were these chicks smaller because they were younger or because they were less well grown, or both? The second plot (b) compares

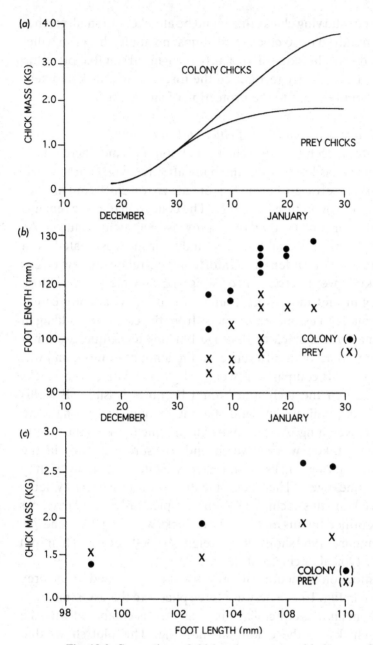

Fig. 10.3. Comparison of chicks taken as prey with those on the colony as a whole at the same date. (*a*) Comparison of mass of each class of chicks. (*b*) Comparison of foot length (as a measure of age) of colony and prey chicks. (*c*) Comparison of the condition of colony and prey chicks shown as mass/foot length, which is essentially, size for age.

foot length (as an index of age) of prey with foot length of chicks on the colony at the same date. On each occasion the prey had a significantly smaller foot on average, indicating that they were a younger sample from the colony population as a whole. But were they less well grown as well? Were they partly starved? The third plot (c) compares the mean mass of prey chicks for mean foot length with the estimated mass from the regressions of foot length and mass for chicks of the same foot length from the colony population at each date. There are again significant differences between the two samples.

In summary, it appears that the skuas were selecting smaller chicks than average during January and that these were both younger and less well grown than chicks in the population as a whole.

Selection during transitional stages of the penguin breeding season

There are three important transitional stages when the form of prey and availability to the skuas changes significantly. These are in order: from eggs to hatching chicks (during December); from guarded chicks on nests to post-guard chicks (from late December or early January to mid January); and at the end of the season as the chicks are becoming too strong to be overcome by skuas except in very favourable situations.

In each case differential predation on the prey available at the time could have an important impact on the survival of different age classes, and thus select both for laying dates and for breeding behaviour. For instance, differential predation of chicks rather than eggs at the start of the chick stage would select against early breeders while a preference for eggs at the end of the stage would select against late breeders. Similarly, the differential predation of early unguarded chicks would select against early breeders and/or parents leaving early whereas the differential predation of younger chicks through January would select against late breeders.

There is good evidence in the records of prey and from the observation of skuas attacking nests and chicks that skuas do exert pressure on early or late breeders at these critical points. Some of this is presented in the following sections and summarises information already given in the accounts of behaviour and food taken in Chapters 6–8.

Selection of chicks rather than eggs

Young chicks are an attractive prey for skuas, they are soft and flaccid, easily eaten, are more exposed to predation because of the need to be fed, and are much more readily hooked from beneath a parent than are

the eggs. Not surprisingly, therefore, chicks make up a significant proportion of the diet in the first week or so of the chick-hatching stage (Table
10.6). This selection would impact on early breeding penguins.

Selection of the first unguarded chicks

As described in section 10.4.3 the first unguarded chicks have an
increased exposure to skua attack as they are often chased out to the
breeding group perimeter by nesting penguins. On the perimeter they are
easy prey for skuas. Selective predation on these chicks also impacts on
early breeding penguins, or on those tending to leave their chicks early or
at a young age.

Selection of prey on defended nests at the end of the guard stage

The last guarded nests in a breeding group become exposed to skua
attack from all directions once a significant number of its neighbours have
been deserted. Post-guard chicks are not able to offer much protection,
retreating before skuas as they approach to begin attacks, so that sites once
well protected within the breeding group become dangerously exposed. In
these exposed situations skilled skua pairs invariably succeed in taking
chicks. Few if any penguins can protect eggs or chicks in the nest when the
attack is cooperatively mounted with one skua distracting from behind
(disorienting) and the other poised to attack from in front. The field
notebooks are remarkable for recording so often the loss of chicks from
the last defended nest on a breeding group. These were usually the nests of
late breeders, in which case the selection is against late breeding; but some
were also of broods being held longer than usual by delay in the parents
leaving for sea. Taylor (1962a) shows that there is a clear trend for later
breeders to abandon their chicks earlier. At this stage of the breeding
season the chicks are in fact much better protected among other chicks
than on the nest. Selection is also against late breeding by differential
predation of young chicks from the crèche. Skuas hunt for these younger
chicks with great perseverance in late summer.

10.6   Skuas: predator or scavenger – the exclosure experiments

So far the roles of the skuas in creating carrion by their activities
about the breeding groups have been highlighted in an attempt to redress
the balance after the inevitable emphasis in the earlier chapters of their
predatory skills. Provoking disturbance in breeding groups, whether
consciously or as a by-product of predation, inevitably threatens the
survival of eggs and young chicks – and any thrown out of nests or killed by

the penguins are likely to be gathered in later as carrion. But does all the skua activity witnessed at the colony significantly lessen penguin egg and chick survival?

The obvious test of skua impact is by the use of exclosure experiments, to compare the breeding performance of penguins protected from skuas with others fully exposed to their activity. In this way the combined effects of disturbance and predation should be readily demonstrated. However, as will be seen in the account that follows these experiments are not as easily arranged or interpreted as would seem probable at first sight.

10.6.1 *The design and analysis of the exclosure experiments*

In each of the five years of this study a proportion of the breeding groups on the colony were mapped and the breeding success of each pair recorded from regular observations made during the season. Experience with this work provided a good basis for the design of experiments and of the technique of observation and recording used for the experiments testing the significance of skua impact at this colony. This experience was further strengthened by an exploratory study carried out in the first year which was designed to show up the amounts of food becoming available to the skuas by protecting three breeding groups from scavenging. During this first year the technique of working from nest maps of the groups while remaining well outside the nest periphery was tested and perfected. At the same time the small amount of wire framework needed to keep skuas out of the breeding group was demonstrated.

Although exclosure experiments are the obvious way to test the impact of the skuas, in practice they were not all that easy to set up in this study. Even though this colony contains many small, isolated, breeding groups, there were still too few of these to design a faultless experiment each year. Ideally, pairs of breeding groups were needed representing the range of different circumstances of the penguin–skua association. In addition, study groups must not be too large or they cannot be mapped and the nest status recorded easily. Moreover, they need to be somewhat isolated from other groups otherwise the unguarded chicks will merge with those of other groups and be lost from the records, curtailing comparative records to the end of the guard stage. Finally, the groups being compared should be within the territory of a single skua pair to control for the different skua capabilities as predators. However, this skua pair must have many other penguins on which to prey otherwise fencing off a major part of their prey will force them to concentrate unnaturally on the control groups, and bias the impact recorded there. Not surprisingly, from this list of constraints there proved to be a limited number of possible experimental groups; a paucity exacerbated through not being able to place exclosures in either H or F blocks, which were kept for the behavioural observations of undisturbed skuas. In the event it was not possible to find comparable pairs of breeding groups each within a single skua territory and recourse was made to selecting different samples of breeding groups for each protected and control category.

The largest number of protected groups were in 1966–67 and 1968–69. In these years five protected breeding groups were compared with five unprotected ones in the colony centre and either four (1966–67) or five (1968–69) on the periphery. No protected groups were set up in the intervening year but the breeding success of all groups used earlier was again recorded. Colonies to be compared were roughly the same size and shape, with between 35 and 121 nests and circular or elliptical in outline. An important requirement was that they had to be able to have the nest contents recorded from outside the periphery to minimise experimental disturbance. In the first year the groups to be protected were assigned randomly from 15 groups selected for study and the remainder were assigned to central or peripheral control groups on the basis of location on the colony and skua access and likely intensity of impact.

The central groups were all located within large skua territories containing many other breeding penguins. In 1966–67 these ranged from 450 to 1764 nests (937 ± 225, mean and SE) and in 1968–69, from 276 to 573 nests (420 ± 58). In each case the control group fell entirely within a territory and thus could only be exploited by a single pair. It was more difficult to find suitable groups for the controls on the periphery of the colony and these were more variable in their likely intensity of skua impact. Ideally, all should have been exploited by several pairs in order to provide a contrasting maximum skua impact. In 1966–67 the groups selected were in territories where skuas shared from 32–270 nests (108 ± 54). In 1968–69 they were in territories with 13–162 nests (82 ± 29). In spite of the failure to find perfect groups for the periphery controls the two control sets stand nevertheless in good contrast: the central ones in territories with numerous other breeding penguins, the ones on the periphery in territories with considerably fewer.

At the beginning of summer all groups were mapped to scale using natural features to locate nest lines. Records of nest success and chick numbers later in the year were made at five-day intervals. At this spacing little difficulty was experienced in interpreting changes in nest siting and number from the maps revised on each visit. The study groups were visited for no more then 20 to 50 minutes on each recording occasion and were avoided at other times. Protected groups were fenced during early incubation using metal standards to support a mesh of widely spaced fencing wire. (See Figs 10.4 and 10.5 as examples.) The mesh over the penguins was c. 1.4 m above the ground with wires spaced c. 750 mm apart. Three wires were run about the perimeter stakes several metres from the penguins, with the lowest wire above standing penguin height. In spite of this apparently clear access for the skuas from around the edge of the groups no skuas were ever seen to go beneath the wire mesh. In some groups a tall pole was erected in the centre to keep flying skuas well away from the penguins, but from direct observation of skua and penguin behaviour of these and other protected groups this was considered unnecessary. The wire mesh above the protected groups prevented both hovering and flight attacks; the extension beyond the group perimeter prevented attacks from the ground.

Unfortunately, the uncertainty about the year-to-year continuity of this programme meant that the fencing had to be cleared away at the end of one season and erected again in the next. The penguins were clearly very seriously agitated during the hour or so needed for this work. Surprisingly none deserted its eggs.

In all analyses the numbers alive at set dates in the season have been compared rather than the more traditional numbers of eggs hatching and chicks surviving. These dates are 15 December, 1 January and 20 January; corresponding respectively to the dates of first egg hatch, first unguarded chicks and chicks first leaving the breeding areas (i.e., corresponding to fledging). Because the changes in the different stages are interdependent, so that a high egg mortality would bias the overall mortalities recorded at the end of later stages and mask differences that might occur during the chick period for example, comparisons need to be for each stage as well as at fledging. The shortcomings in the analyses are, however, inescapable even if they are carried out for each stage separately. For example, the loss of eggs of inexperienced parents would leave only the experienced birds in the colony, with the much higher expectation of their rearing chicks successfully, and heavy early nesting failure will also bring high numbers of parents back into the colony during the reoccupation period, resulting in a higher than normal disturbance at this time.

The breeding success figures given in this section are for living eggs and chicks at each date, with the non-hatching eggs being removed from the totals. Unless this is done a spurious apparent mortality occurs in late December and early January as these eggs show up through being left on

Fig. 10.4. Protected breeding group E5 at two dates in 1966–67. Upper photograph during early incubation, lower photograph during the post-guard stage.

Fig. 10.5. Examples of protected breeding groups, 1966–67. Upper photograph, group C2 (with the experimental test of a high pole to keep skuas further away than would a low screen). The control for this group, C1, shows in the immediate background. The lower photograph is of D4 after an exceptional summer snow fall. The penguins in the background are of B Block.

deserted nests or ejected by growing chicks. Two estimates of breeding success are thus possible: one calculated from the total numbers of eggs laid and the other from the estimated number of viable eggs laid.

10.6.2   *Results of the exclosure experiments to measure breeding success in the absence of skua activity*

These experiments were set up to answer two questions: is there significant egg and chick loss in a breeding group in the absence of skua disturbance and predation, and, is it possible to demonstrate an additional loss attributable to skuas? The first is readily demonstrated. Although the protected groups were overall more successful than either of the control groups in both years there was nevertheless substantial egg and chick loss from environmental and penguin effects in the absence of skua attack. In the first season c. 18% of viable eggs failed to give fledgling chicks in the protected groups, in the second, with its more difficult ice conditions for foraging, c. 30% failed.

The losses were, moreover, rather similar in each of the breeding groups making up the protected sample – between 10% and 24% in each group in the first year, and between 21% and 41% in the second.

Although some of the control groups had poor breeding success, notably, for example, C3 and H1 in 1966–67 which fledged only 21% and 24% of viable eggs respectively, others were much more successful, equalling protected groups in some cases, so that establishing an overall impact by skuas is problematical. Until this is determined it is clearly not possible to estimate the discrete skua impact on penguin breeding.

1.   *Comparison of the success of the same groups when protected or unprotected from skuas in different years*

The success of the same breeding groups protected from skuas by the fencing in 1966–67 and 1968–69 (with a single exception in each year) and unprotected in the other years demonstrates the intrinsic variability among the different groups and the degree that protection successfully enhances egg and chick survival. The same methods of monitoring breeding success throughout the season were employed in each year with nests being checked off against maps of location from outside the margin of the breeding group. Apart from the initial major disturbance of erecting the fencing at the beginning of each season for the protected groups the level of disturbance was similar in each year so that comparing across the years demonstrates the benefit of protection and at the same time provides reassurance that the disturbance caused during the erection of the fencing

Table 10.11. *The breeding success (percentages) of the experimental breeding groups in years with and without protection from the skuas*

Year	A17	E4	E5	D4	C2	C3
Success of total eggs laid to end of the guard stage (to 1 Jan.)						
1965–66	—	—	—	68	56	—
1966–67	**93**	**84**	**84**	**78**	**79**	25
1967–68	57	81	71	77	76	41
1968–69	**71**	**79**	**63**	74	**60**	**60**
1969–70	76	76	62	—	75	47
Success of total eggs laid to fledging (to 20 Jan.)						
1965–66	—	—	—	58	46	—
1966–67	**85**	**81**	**81**	**69**	**69**	19
1967–68	54	80	67	74	70	29
1968–69	**64**	**74**	**59**	62	**60**	**53**
1969–70	74	71	59	—	73	41

Breeding groups shown bold were protected from skua disturbance and predation.

was not grossly detrimental to breeding success. Comparisons of success to the end of the guard stage are the most germane to discussions of the impact of skuas, and both this record and the success to fledging are shown in Table 10.11. (Until the chicks are unguarded and begin to follow parents outside the wire cage during the feeding chases they are well protected from skuas. Later in the season chicks from protected groups were taken by skuas when outside the cage.)

The first year the groups were protected (1966–67) recorded very high success, and this was repeated for some groups in the following year without the protection. However, even when protected these penguin groups universally achieved only moderate breeding success in 1968–69 compared with the previous two years. Of these groups A17 and C3 were clearly the most vulnerable to skua impact, with the latter recording poor success in all three years in which it lacked protection. This group was the most closely invested by skua territories of all those compared. Overall, however, it is apparent that these breeding groups achieved much higher success when protected than when they were accessible to the skuas, in spite of the obvious disturbance caused in erecting the cages.

The conclusion from this part of the analysis is that there is appreciable egg and chick mortality within penguin breeding groups that is independent of skua disturbance or predation. In a favourable season penguins

protected from skuas were able to raise c. 80% of eggs to chick independence, in a poorer year c. 60%.

There is a variable effect of the protection on breeding success in these
groups, attributable both to environmental differences among the years
(for example, in comparing 1966–67 with 1968–69 environmental effects
were the most significant factor in the different success) and to differences
in location (i.e., largely to skua differences) within each year. This effect
was confirmed again in the results from 1969–70 of breeding success of
single breeding groups divided into protected and open areas. For A28,
survival recorded to the end of the guard stage of the total eggs laid was
82% and 87% for protected and open areas respectively compared with
86% and 47% for the same treatments in A32. In the first group skua
impact was insignificant, indeed the open area had higher survival, in the
second the impact was severe and fewer than half the eggs laid survived to
the end of the guard stage.

2.   *Comparing breeding success of protected and control groups within
      the same year*

The previous section examined the performance of the same
breeding groups in different years, showing that groups performed better
on average when protected than when open to skua attack. Because of the
very great differences in environmental conditions in each year it was not
possible from that analysis to measure the extent of skua impact. This may,
however, be done through comparing protected and unprotected groups
within the same year. The data for this analysis are summarised in
Appendix 5 and Table 10.12 with the statistical analysis in [4].

The result of these experiments (summarised graphically in Fig. 10.6),
in which largely the same breeding groups were used in both years, was
that although the protected groups recorded significantly higher survival
than either control group in 1966–67, and this was evident both at the end
of the guard stage and at fledging, their record was not significantly better
than either control group in 1968–69. In this year both entry to the colony
of breeding birds at the start of the season and breeding success through
the egg stage was much lower than usual, because broken ice hampered
migration and foraging; this major environmental impact on breeding
apparently swamped any differential impact of skua predation. The lesser
skua impact might have resulted from the greater overall breeding experience of the penguins in the second year. Only the most mature and
experienced birds would be able to reach the colony at the start of the
season, and only the fittest of these would be able to sustain incubation in

Table 10.12. *The breeding statistics of protected and control groups of penguins*

	Group type		
	Protected	Central control	Edge control
1966–67			
Total number of eggs	462	361	291
Number of viable eggs	434	339	268
Survival: number (% of total; % of viable)			
15 Dec. (first chick hatch)	423 (91; 97)	322 (89; 94)	240 (82; 89)
1 Jan. (first unguarded)	382 (83; 88)	273 (76; 81)	144 (49; 54)
20 Jan. (chick fledging)	350 (76; 81)	250 (69;74)	120 (41; 45)
1968–69			
Total number of eggs	220	267	297
Number of viable eggs	205	247	280
Survival: number (% of total; % of viable)			
15 Dec. (first chick hatch)	160 (73; 78)	217 (81; 88)	233 (78; 83)
1 Jan. (first unguarded)	150 (68; 73)	167 (62; 68)	187 (63; 67)
20 Jan. (chick fledging)	141 (64; 69)	153 (57; 62)	157 (53; 56)

Protected groups: Protected from skua disturbance and predation by a wire mesh cage. Central control groups: Breeding groups located in the central area of the colony. Edge control groups: Breeding groups located about the colony periphery and close to or among concentrations of skua territories. Small breeding groups in these situations had generally a lower breeding success than comparable groups within the body of the colony.

the harsh conditions. These birds also would be the least susceptible to skua disturbance and attack.

3. *Estimation of the differences among the experimental breeding groups as a measure of skua impact on the colony*

The losses within the breeding groups can be used as a basis for estimating overall skua impact on the colony, as in all other respects the experimental groups were identical. However, as statistically significant differences between the protected and unprotected breeding groups could only be established for 1966–67, estimates must be confined to this year. (In the second year (1968–69) the small impact of skua predation and disturbance was apparently masked by high egg losses, caused by unfavourable pack-ice conditions.)

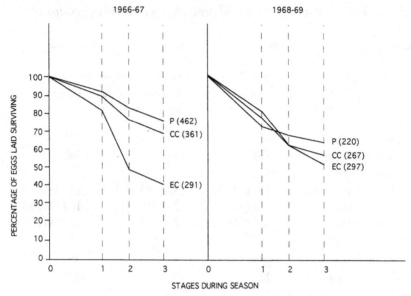

Fig. 10.6. Survival of eggs and chicks to three dates during the breeding season. Breeding groups protected (P) or unprotected (CC, controls in colony; EC, controls on colony edge) from skua disturbance and predation. 1966–67 and 1968–69. 0. 15 November; 1. 15 December (c. date of first eggs hatching); 2. 1 January (c. end of chick guard stage); 3. 20 January (c. date of chick fledging). Total numbers of eggs at start of season in each class shown in brackets.

In 1966–67 differences between the protected and central colony control groups, representing the skua effect, was about 3% by 15 December, when the first penguin eggs begin hatching, rising to between 7% and 8% for the records in early and mid January (for measurements from viable and total eggs laid). The differences between the protected groups and those on the colony edge were much greater in this year, of total egg numbers, 10.1% by 15 December, 34.9% by 1 January and 37.9% by 20 January, with similar figures for viable eggs (Table 10.13). The central groups, however, represent the overwhelming majority of penguins on the colony so that in this year it is reasonable to infer that skuas were responsible for about 8% of all egg and chick losses. The estimates of difference between protected and unprotected groups, however, have large standard errors, which when translated to 95% confidence intervals indicate that the 'true' mean difference in survival, attributable to skuas, may lie anywhere between 0% and 22%. Clearly there is not much point in further speculation on the precise impact of these skuas, measured in numbers of eggs and chicks.

Table 10.13. *Mean percentage differences between the protected breeding groups and control breeding groups, 1966–67*

Data are difference between means and standard error of difference.

	15 Dec.	1 Jan.	20 Jan.
Of total eggs laid			
Protected versus central colony	3.3 ± 2.8	8.5 ± 5.9	7.9 ± 6.0
Protected versus edge of colony	10.2 ± 3.9	34.9 ± 14.1	37.9 ± 14.8
Of viable eggs laid			
Protected versus central colony	3.2 ± 1.8	7.2 ± 6.6	8.0 ± 6.5
Protected versus edge of colony	9.7 ± 1.9	37.4 ± 14.3	40.9 ± 14.6

10.7 *Predator or scavenger: conclusions on the effectiveness of predation: evidence from comparing penguin breeding success in areas with high and low predator to prey ratios*

It is to be expected that where there are significantly variable skua/penguin ratios on the colony the breeding success of the prey would correlate with the density of the predator. Where skuas are attempting to gain as much food as possible from small numbers of penguins their impact should be greater than in areas where their effort is spread over larger prey numbers. In contrast, if skuas are merely scavengers on the colony, taking up food as it becomes available yet contributing little to its genesis, then the apparent impacts on the prey in different areas should be equal. This test of whether the skua is principally a predator or scavenger can be carried out by comparing the survival of penguin eggs and chicks on EF and H blocks. These study areas were selected with this comparison in mind – they contrast low and high predator/prey ratios. This comparison appears in Table 10.14.

For this test the important comparisons are of the numbers of eggs and chicks lost from each set of breeding groups in relation to the numbers of skua pairs, and of numbers lost in relation to those laid. Although the numbers lost to (or taken by) the six pairs on EF was much higher than on H (about ×10 in the first year and ×6 in the second), suggesting that the skuas on EF were the more successful predators with a greater impact on the penguins, the proportions lost were almost identical. This latter comparison suggests very strongly that overall these skuas were essentially scavenging, and that where they had more penguin nests in the territory

Table 10.14. *The survival of penguin eggs and chicks in relation to different predator/prey ratios. Comparing EF and H blocks*

Data are estimated means and standard errors. Breeding success and egg and chick survival estimated from scaling of study breeding group records in each place and overall counts of breeding numbers on 8 December and chicks alive on 20 January.

	1968–69		1969–70	
	H Block	EF Block	H Block	EF Block
Number of penguin nests (on 8 Dec.)	396	1547	584	2068
Number of skua pairs	14	6	13	6
Mean number of penguin nests/skua pair	28.3 ± 23.6	258 ± 129	44.6 ± 8.8	345 ± 52.7
Numbers of eggs laid	715 ± 17.1	2792 ± 67	1102 ± 23	3902 ± 81
Numbers of eggs and chicks lost	289 ± 17.1	1177 ± 78.5	403 ± 23	1167 ± 101
Numbers lost per skua pair	20.6 ± 1.2	196 ± 13.1	31 ± 1.7	194 ± 16.9
Numbers lost of each 100 eggs laid	40.0 ± 2.4	42.0 ± 2.8	36.6 ± 2.1	29.9 ± 2.6

they had access to greater volumes of scavengeable food. There is certainly no evidence in these egg and chick survival figures that the H skuas had become relatively more severe predators than those on EF, causing a proportionally more severe impact on their penguins.

This result is further evidence of the primarily scavenging role of skuas and the small role of predation overall on this colony.

STATISTICAL TESTS

[1] The success of breeding groups on the periphery and in the centre of the colony

1. 1968–69. Overall breeding success 0.98 ± 0.16 (mean and SD) chicks nest^{-1}. Comparison of central and peripheral breeding groups of all sizes showed that the central groups were significantly more successful. ($t = 3.25$, df $= 90$, one-tailed test, $p < 0.002$.) Significant difference occurred in tests of groups with more than 40 nests (moderate to large groups) ($t = 1.98$, df $= 71$, one-tailed test, $p < 0.02$) or restricted to groups with less than 50 nests (small to moderate groups) ($t = 1.947$, df $= 44$, one-tailed test, $p < 0.05$). These tests excluded the outside groups in block A along the sea edge, which are really part of the inner colony area in lacking a skua nesting margin, and those in G which sprawl and merge over a series of ridges and cannot be readily assigned to these two categories.

2. 1969–70. Although the overall breeding success was higher in the following year (1.28 ± 0.24 chicks per nest) the distinction between central and peripheral breeding groups was maintained in comparisons including both small and moderate groups ($t = 2.356$, df $= 74$, one-tailed test, $p < 0.02$) but not for moderate groups (with more than 40 nests) alone ($t = 1.158$, df $= 54$, one-tailed test, $p < 0.25$).

[2] Ratio of numbers of adults to unguarded chicks at the start of the post-guard stage. From counts on selected breeding groups 3 and 5 January, 1969–70.
Spearman rank correlation, $r = 0.12$ and 0.074 respectively.

[3] Tests of numbers of eggs and chicks taken as prey in relation to numbers in the colony population at the same time. Chi-square test of differences.
16–20 December, $\chi^2 = 4.50$, $p = 0.034$.
21–25 December, $\chi^2 = 0.643$, $p = 0.42$.
26–30 December, $\chi^2 = 3.857$, $p = 0.05$.

[4] Statistical analysis of the exclosure experiments comparing breeding groups of penguins protected from skuas with unprotected groups in the centre and edge of the colony.
Analysis by general linear models procedure; repeated measures analysis of variance comparing the survival statistics for numbers of eggs laid at three different points of the breeding cycle. Significant differences ($p < 0.05$) shown in bold.

Wilke's lambda statistic is presented together with its significance based on an
F approximation.

1966–67

Survival of total eggs laid

Protected versus central colony 0.892, $F_{3,9} = 0.363$, $p = 0.781$, ns.

Protected versus edge colony 0.385, $F_{3,9} = 4.78$, $p = \mathbf{0.029}$, s.

Central versus edge colony 0.455, $F_{3,9} = 3.58$, $p = 0.059$, ns.

Survival of viable eggs laid

Protected versus central colony 0.759, $F_{3,9} = 0.949$, $p = 0.45$, ns.

Protected versus edge colony 0.323, $F_{3,9} = 6.272$, $p = \mathbf{0.014}$, s.

Central versus edge colony 0.223, $F_{3,9} = 10.404$, $p = \mathbf{0.003}$, s.

1968–69

Survival of total eggs laid

Protected versus central colony 0.571, $F_{3,11} = 2.756$, $p = 0.092$, ns.

Protected versus edge colony 0.533, $F_{3,11} = 3.206$, $p = 0.066$, ns.

Central versus edge colony 0.871, $F_{3,11} = 0.544$, $p = 0.662$, ns.

Survival of viable eggs laid

Protected versus central colony 0.575, $F_{3,11} = 2.707$, $p = 0.096$, ns.

Protected versus edge colony 0.569, $F_{3,11} = 2.776$, $p = 0.093$, ns.

Central versus edge colony 0.844, $F_{3,11} = 0.673$, $p = 0.586$, ns.

11

Appreciating the penguins: is it worth living with the penguins and do skuas appreciate the advantage?

11.1 *Introduction*

Skuas prey on each others' eggs and chicks and are just as opportunistic in this predation as they are when scavenging and preying on penguins. The closer the nests are together, the more compact the territories, the greater the opportunities for picking up neighbours' eggs and chicks. For the same reason any situation that attracts outside skuas into the breeding area also heightens the risk to breeding birds. Penguins also pose a threat to incubating skuas, forcing them from the nest, and exposing the eggs to damage by pecking or trampling. Each of these threats is exaggerated for skuas nesting within the penguin colony or within its immediate surroundings. But at the Northern Colony the area about the penguins is more sheltered from the wind and blowing snow drift than the more exposed slopes and basins behind the colony.

There seems, therefore, to be a trade-off for the skuas between breeding at the colony, with its greater shelter and the prospect of penguin food, but heightened predation risk; and breeding away from the colony, with less disturbance, less risk of egg predation but greater risk from wind and snow drift.

This section considers two simple propositions: are skuas which have penguins within the territory at an advantage (because of the availability of food) and do skuas appear to appreciate this advantage by the selection of territories and their subsequent behaviour during the season?

A complete assessment of these questions is not possible here because ultimately it must be measured in the long term, in survival, in the length of occupancy and in life-term productivity. These factors need long-term study of a known-identity population so that comparisons can be made of the ages skuas are able to obtain territories in penguin and non-penguin areas, of age-related breeding success and life span in different habitats. The present study ran for just five seasons, with the breeding experience of some pairs known from earlier banding.

What can be attempted is an evaluation based on breeding success and breeding behaviour in each season of the study and on the way the two species interact. In the study of interaction special attention is given to ways that skuas are both advantaged and disadvantaged by living in close proximity to the penguins. The longer term study carried out at Cape Crozier (Ainley *et al.*, 1990) provides information on the demography of a comparable breeding population in this area that can be used in the interpretation and evaluation of the Cape Bird data. That study was not completed either – it ran for only 10 seasons – so that a life-term account is not yet available for this species.

11.2   *The breeding success of skuas in different locations within and near the penguin colony*

It is far more difficult to obtain secure data on the breeding success and causes of egg and chick mortality for skuas than it is for penguins. There are good reasons for this. First, skuas are both predatory and cannibalistic so that eggs and chicks can be lost without leaving much trace. Parents will certainly eat broken eggs and dead chicks. Second, eggs can be laid and lost within a few hours, and their presence unrecorded. Third, skuas, in contrast to penguins, may re-lay one or more times in a season following the loss of a clutch so that not only is it necessary to maintain a close watch on breeding pairs to record egglaying, but this watch must also be maintained for all the unsuccessful pairs for much of the season in case a new clutch is produced. Without close surveillance the unrecorded loss of one clutch can be incorrectly covered up by the appearance of another. Fourth, breeding skuas respond far more actively to disturbance by observers, most abandon the nest temporarily during nest checking, and during this upheaval in a crowded area with so many birds in the air at one time, eggs and chicks can be taken by opportunistic predators. This high nest desertion rate is in contrast to the strong nest holding of penguins, which only rarely desert their nests during routine observations of the breeding group. Finally, a large group of breeding penguins can be observed from a single observation point and the nests checked off readily against a map. Skua nests are more widely spaced, it would be fortunate to be able to see more than ten or so clearly from any vantage point, so that most have to be visited on the ground. Although they need to be visited regularly for accuracy of record keeping, each visit measurably impacts on the breeding success itself. Without extreme care a detailed skua study will only achieve a precise documentation of breeding failure.

In most cases it is simply not possible to assign a mortality factor to egg loss in skuas. On one day the eggs are present, on the next the nest is empty. Rarely, egg remains may be sprayed across the ground, suggesting a muffed predation attempt, or even more rarely an attack on a nest may be directly observed. But such attacks are over in a few seconds; a bird flying over drops into the nest and is off again flying hard back to its territory, sometimes pursued by the parents, often not.

Although there are a fair number of records of cracked, crushed or punctured eggs on nests, or of eggs frozen into the ground or submerged under water, it is not known what proportion of all damaged eggs these few records represent, as parents often eat their own damaged eggs.

The problem of obtaining good records of numbers of eggs laid, eggs hatching and chicks surviving and the problem of determining causes of egg and chick loss are the two major hurdles to producing informative records for interpreting skua breeding success and failure.

At Cape Bird the problem of gross disturbance of the breeding birds by the observers was tackled in two ways. First, attempts were made to habituate the skuas there to observers through a rigidly maintained policy of minimal handling and cautious entry into the territories (Young, 1970, 1990a) so that within two years from the start of the research most skuas remained within the territory, close to the nest or on the nest, when the territory was entered. During incubation the eggs are held on the feet within the brood pouch. It is important, therefore, that incubating skuas are not flushed from the nest in sudden alarm as the eggs can then be thrown out and broken. Nest approaches need to be cautious and un-alarming, but not secretive. It is possible that some of the egg mortality attributed to muffed skua predation from egg remains sprayed on the ground away from the nest is in fact from sudden flight from the nest by the parent itself.

The second way disturbance was minimised was by spacing out the number of nest and brood checks. At best, nests were checked every second day, in some years and for different areas every third day. The loss of accuracy was considered to be more than compensated for by the lessened impact on the breeding birds, and the consequent enhanced validity of the breeding records. Because of this cautious approach it would be nice to think that the breeding data recorded here is of a natural population, not greatly influenced by human interference or disturbance. It is not possible, however, to test this. What is certain is that very few eggs or chicks were lost during nest checks (there is only one certain record of the loss of an egg) and that the foraging behaviour of all skuas appeared to be normally maintained. An earlier study at Cape Royds (Young, 1963a) and a study of skua pairs watched from too close at hand on the upper basin at Cape Bird provided good experience in recognising when nesting skuas were being affected by observers. An indication of gross impact is evident in the way incubating birds appear unsettled on the nest, with far too many comfort movements to settle the eggs and with incubation spells that are far too short, and in the way the male is reluctant to leave the territory to forage at sea. Once these behaviours become manifest it is all too certain that the birds have been seriously affected by observer presence. It is worth noting as well that none of these skuas was fed from the field station and the rubbish pit was well screened. The field station did not attract loafing skuas that could also prey on the local breeding birds.

Although the breeding success of most of the skuas in the area was followed to some degree the critical comparisons were among selected groups occupying different locations about the colony. There were three important comparisons: 1. of pairs close to the colony and distant from the colony; 2. of pairs with penguins and lacking penguins in the same part of the colony; and 3. of pairs with territories on the colony with nests located among the penguins and those with nests outside the colony perimeter. The first measures the overall effect on skuas of being near the penguins, with all its advantages and disadvantages; the second measures the effect of having access to penguin food; and the third attempts to measure the affect of nest location among skuas having similar advantages of access to the penguins. It is worth stressing again that skua territories covered the entire colony area and only those skuas with territories containing breeding groups of penguins could obtain significant amounts of penguin food. There were two exceptions to this general rule. The first relates to the small 'club' area on A Block which allowed general access by skuas to the surrounding breeding penguins, but these were little exploited. The second refers to the fact that any skuas could attempt to forage anywhere in the colony from flight, but were at risk from defending skuas at all times and could only attempt to take eggs and young chicks, that is, food small enough to be lifted by flight easily and immediately from the breeding group. Only a very small number of local skuas attempted to forage widely over the colony in this way. The most notorious was the female of pair 11, who took almost as many skua eggs in these forays as penguin eggs. Fierce, uncompromising, territorial defence precluded any skua from foraging from the ground in another bird's territory. Only the most local territorial intrusions of this sort were ever recorded. The record of the way the male of H25 worked on H4 breeding group is a good example of how an aggressive skua may encroach on to other territories (see section 7.7).

It is important to be clear on what should be measured in these comparisons of skua breeding. One measure is the proportion of pairs producing eggs, and their survival to fledging. Skuas in this region have an unusually low breeding success (Ainley *et al.*, 1990; and this study) and this measure is probably too crude to be of great use in the sort of comparison contemplated here. In a poor breeding year no chicks might fledge at all. Most interest should focus, therefore, on four intermediate stages of the breeding cycle: the proportion of pairs (or territories?) producing eggs; the numbers of eggs produced and their laying date (in part at least reflecting the nutrition of the female and the skill of the male in providing food); the survival of the vulnerable eggs, prey to skuas and to the interest and clumsiness of penguins, as well as the ability of the pair to forage; and, finally, the growth and survival of the chicks, reflecting the availability of food and the experience of the parents. In these skuas, as far as one can be certain, all of the food needed by the female during pre-egglaying and incubation in pairs without penguins is given in courtship feeding by the male. To some degree both egglaying date and egg quality depend on food amount. Pairs with abundant food should, on this premise, lay earlier, and be less likely to desert the nest during incubation. Skuas nesting near the penguins are at risk from both penguins and skuas attracted to the colony. Their eggs are especially at risk, they are immobile, thin-shelled and poorly camouflaged. One obvious comparison is that of number of days survival in the different areas, a finer measure than simply whether they hatch or not. Finally, the amount, frequency and quality of food might affect both the survival of the chicks and their rate of growth. It might also determine whether one or both of the chicks hatching survive through the critical early period when chick sibling aggression is most pronounced (Young, 1963a; Williams,1969; Procter, 1975). Of great importance to this analysis, however, is not just the proportion surviving but the causes of egg and chick loss. It would be wrong, for example, to conclude on the basis of survival statistics that pairs away from the penguins had a low success because they lacked penguin food if the reason for the low survival of eggs was because of wind or snow! Unfortunately, this sort of information is much more difficult to obtain than the basic statistics of egg and chick survival. Many of these difficulties were referred to in Young (1963a) and little seems to have occurred in the intervening years to make their resolution more certain. Skua breeding remains a labour-intensive study.

In all the statistics that follow it should be remembered that nests were not checked regularly enough to guarantee that all eggs laid were in fact recorded. Such accuracy would really need nest checks several times each day. All that can be guaranteed is that all eggs surviving for two days or more were recorded. For this reason not much emphasis has been given to differences in the proportions of one- and two-egg clutches in each area. It is quite likely that some of the one-egg clutches in these records are really the surviving eggs of pair clutches. Finding nests and recording the date of egglaying is facilitated by recognising the distinctive calling of skuas with broods. This 'brood defence' calling (quavering call, Andersson, 1973; alarm call, Young, 1963a; Spellerberg, 1971a) only occurs after the eggs have been laid and persists for no more than a day or two after the eggs or chicks are lost. Incubating skuas can also be distinguished from resting birds by the way the wings are partly opened to accommodate the eggs in the brood pouch so that the wing tips are distinctively raised above the tail. This distinctive posture allows sitting birds to be checked from a distance using field glasses, but of course this method does not discriminate between one or two eggs or eggs and chicks. Such distance-checking does, however, alert observers to the need to visit the nest if changes are noted.

There were about 250 pairs of skuas north of the Waterfall Stream in the triangular area of bare ground between the Cape Bird ice cap and the sea. Of these, 185 were in the immediate environs of the penguin colony, with 95 or so with territories holding penguins. The remainder were in three quite separate areas on the high moraines along the very edge of the ice cap immediately to the east of the colony (27 pairs), in a high basin about the head of the Lab Stream (20 pairs), and on South Beach, the narrow coastal strip between the cliffs and sea between the Lab and Waterfall streams (18 pairs). Skuas in the first two locations were more exposed to the southerly gales than any other skuas in the area. Those on the South Beach were similarly exposed to those on H Block. Those nesting in the high basin were furthest from the colony and their extensive territories and widely spaced nests attracted few skua overflights (Williams, 1969).

11.2.1 *Breeding success over five seasons of skuas at the penguin colony*

The breeding success of the 185 pairs in the immediate environs of the penguin colony were recorded in each year (Table 11.1). The two certainties in the table are the number of occupied territories and the numbers of chicks surviving into the last weeks of January. When interpreting the table it should be remembered that second chicks were invariably lost soon after hatching; the proportions of first chicks surviving is thus considerably better than appears in the table. (In common with virtually all recent programmes on skuas the field station was closed well before the end of the skua breeding cycle. Brian Reid at Cape Hallett in 1959–61 was probably the last ornithologist in this region to record the flight of fledgling skuas from a breeding area at the end of the season. Ian Spellerberg at Cape Royds and Bob Wood at Cape Crozier have also seen the early stages of this phase in the 1960s but since then only members of Greenpeace at Cape Evans, also on Ross Island, would have been in residence on a skua breeding colony in this region this late in the year.) A small proportion of eggs laid (perhaps 5% of first eggs of clutches and of re-lays) would not have been recorded. The survival figures given, even though seemingly very low, are nevertheless slight over-estimates; the true survival rates would have been even lower.

The numbers of occupied territories remained remarkably constant from year to year (the table records the numbers checked, not the numbers present) but the numbers of eggs surviving to hatching and the numbers of chicks surviving into late January varied appreciably with mean numbers per territory ranging from zero in 1966–67 to 0.38 in 1968–69. The number surviving was only weakly related to either numbers hatching or numbers laid, and the differences among the years were statistically different even when 1966–67 was deleted from the analyses (for numbers fledged of those laid, $\chi^2 = 30.52$, df 4, $p < 0.001$; for numbers fledged of those hatched, $\chi^2 = 27.97$, df 4, $p < 0.001$). More chicks survived than expected in 1968–69 when compared for numbers of eggs laid or numbers hatched.

As is now becoming increasingly well documented, skua breeding success in high Antarctic latitudes is remarkably low. In part this is because of the almost universal (and still largely inexplicable) loss of the second-hatching chick of broods with two chicks, but at Cape Bird there was an even greater mortality during the egg stage, a steady attrition right through

Table 11.1. *The breeding success of skuas in the immediate environs of the Northern Colony, 1965–66 to 1969–70*

Year	No. of territories checked	Eggs laid[1]	Chicks hatched		Chicks surviving	Percentage eggs surviving[2]	Chicks surviving per territory	Chicks per breeding pair with eggs
			Number	%				
1965–66	183	289	121	42	37	12.8	0.20	0.23
1966–67	150	199	14	7	0	0	0	0
1967–68	135	310	57	18	28	9.0	0.21	0.25
1968–69	204	329	122	37	77	23.4	0.38	0.45
1969–70	102	140	32	23	23	16.4	0.23	0.30

[1] Includes replacement clutches.
[2] Chicks surviving to mid to late January in each year.

incubation. Even so, the breeding success for these five years is extraordinarily low, even in comparison with the usual poor success of Ross Island skuas. In one year there was total failure, in three only one in five pairs on average raised a chick, in the remaining year (1968–69) 40% of pairs raised a chick. To the best of our knowledge this low, but variable, success could not be attributed to differences in observer impact. Indeed, the best year was the one in which the most detailed recording was made using several observers. This low success was not, however, confined to the skuas about the Northern Colony. In 1967–68 Williams (1969) carried out a study of breeding success of all skuas north of the Middle Colony. In this year only 42 chicks were alive at the end of summer from 293 pairs that laid eggs; a nesting success rate of 0.14 chicks per pair with eggs.

Nor were these years as a whole exceptional. Ensor (1979) recorded a similarly low breeding success for these skuas of 0.30 chicks per pair laying eggs in 1977–78.

The breeding success recorded here is comparable with that at Cape Crozier. At this site Ainley *et al.* (1990) recorded a success of 0.12–0.25 fledglings per pair during the three years 1966–68 of pairs of unknown age and 0–0.15 fledglings per pair for birds aged from four to seven years old. The success rates at Cape Bird and Cape Crozier are both lower than has been recorded for skuas at Cape Royds, the only other population on Ross Island which has been intensively studied. In 1959–60 Young (1963a) recorded a breeding success of 0.46 chicks per pair (representing 22% of eggs laid, 28.7% of chicks hatching) and this record was matched later by Spellerberg (1971b) for the same area with breeding successes ranging from 0.3 to 0.78 chicks per pair over three years. The consistently lower (and similar) rates at Capes Bird and Crozier compared with those at Cape Royds point to real differences in the breeding conditions at these places. The poor success at Cape Crozier was attributed to the mass mortality of chicks during persistent strong winds (Ainley *et al.*, 1990), which overturn any advantage this population has through being closest to the polynya (and open water for sea foraging) that forms each year to the east of Ross Island.

A similarly clear cut mortality factor has not been identified at Cape Bird, where a steady attrition of eggs and chicks occurred through the season irrespective of weather conditions. Most of the mortality in fact occurred during incubation, and the survival of chicks was high after their first few days. In 1966–67 the records for seven pairs nesting close together along a narrow rocky slope on the beach beneath B Block penguins is illuminating in showing how severe egg mortality was in this year and for indicating where most of this could be attributed. Each of these pairs laid two eggs but within a few days all had been lost. Five eggs were found broken on the nest – these had certainly been pecked by penguins – and two others were found with their contents sprayed out from the nest, broken during a skua predation attempt or thrown out by the parent when leaving the nest precipitously. There was no sign of the remaining seven. These nests were no more than a few metres apart because the skuas had packed onto this rocky slope for shelter from blowing grit and sand on the beach and in an attempt to avoid penguins moving to and from the beach. There were few chicks still alive anywhere in the study area in mid January of this year when the base was closed for the season, and only one of these was judged to be mature enough to be able to fledge before winter. The special feature of this season was that as so many eggs were lost at the start of breeding most pairs re-laid at least once and chick hatching overall became very late. Only one chick had hatched in this year by 25 December. Skuas in January give up incubation and the care of chicks too young to fledge so that few of these late re-layings had much chance of surviving at all.

There are two obvious differences between Cape Royds and the other two areas. First, the topography there is more broken, with the skuas distributed in small numbers among basins and valleys, well protected from the full force of gales that can run freely across the smooth contours at Cape Crozier, and across the upper slopes at Cape Bird. Second, the breeding densities at Cape Royds are generally lower. There are fewer breeding pairs and many are quite isolated, so that opportunities for intra-specific predation are much less.

There is less value in comparisons with other studies at more northerly sites (and in more benign climates) where success is usually much greater, and where the second hatching chick usually also survives. For example, Le Morvan *et al.* (1967) at Terre Adélie reported 30 eggs hatching of 41 laid (71% survival of eggs), of which all except one chick fledged. At King George Island (South Shetland Islands) Peter *et al.* (1990) recorded 1.92 eggs laid per pair of which 54% fledged, a fledging rate of 1.03 chicks per pair. Even in these more favourable conditions, however, skua breeding is exceedingly variable (Trivelpiece *et al.*, 1990) and breeding may fail completely in some years.

With such universal very low breeding success on Ross Island it is clear that differences and causes will only be established safely with longer-term studies and that comparisons based on single-year data will more often mislead than illuminate. Moreover, studies at different sites also need to be based on agreed guidelines and methods, with similar levels of observer disturbance, to ensure that exact comparisons can be made.

Overall, the very low success recorded for these skuas was both a disappointment and a matter of great interest. It is also a matter of some personal embarrassment as Young (1970) set out a guide for how best to do this sort of study and how to ensure low observer impact. These years and this study have now the poorest breeding success on record for skuas over a number of years. All groups of skuas performed badly every year irrespective of where they lived and nested. The similarly poor record for Cape Crozier skuas over these same years offers little consolation.

What is most interesting in the Cape Bird records, however, is not so much their evidence of variability of chick production over these years but that most pairs failed here in most years irrespective of whether their territories contained penguins or not.

11.2.2 *Breeding success of skuas in different geographic areas at Cape Bird*

More detailed study was made of skua breeding in a number of places about the colony to compare the breeding performance of the skuas when closely associated with or well isolated from the penguins. In Table 11.2 the series shown runs from those most closely invested by the penguins (A Beach skua territories are entirely within the colony and the nests are among the penguin groups) to the most isolated for the region on the high moraine and basin. Relative breeding success of skuas in these locations should reflect the different advantages and disadvantages of these locations; for those at the colony of access to penguin food but greater disturbance from people, penguins and other skuas, for those away from the colony no access to food but less disturbance. Unfortunately for the symmetry of this comparison, those farthest away from the colony are also the most exposed to strong winds.

Comparing different seasons

The very wide differences in breeding success apparent among the seasons was independent of the numbers of eggs laid, which was fairly constant (1.37–2.29 eggs per breeding pair, including re-layings). The anomalous year was 1966–67 with complete breeding failure, irrespective of location or of access to penguin food.

Comparing pairs with and without penguins

This comparison may only be validly pursued with the sure knowledge of where each of the pairs was foraging. This understanding

Table 11.2. *The breeding records of skuas in different locations of the Cape Bird area: percentage eggs laid that survived as chicks to late January*

Data are percentage survival with number of eggs laid in brackets.

	1965–66	1966–67	1967–68	1968–69	1969–70
A Beach (15[1])	4.5 (22)	0 (17)	3.6 (28)	18.2 (22)	10.7 (28)
EF Block					
With penguins (9)	31.6 (19)	0 (14)	18.7 (16)	35.7 (14)	16.7 (18)
Without penguins (9)	—	—	—	14.3 (14)	10.5 (19)
EF total (18)	—	—	—	25.0 (28)	13.5 (37)
G ridge (18)	19.2 (26)	0 (21))	37.9 (29)	23.5 (34)	—
H Block					
With penguins (14)	28.6 (21)	0 (25)	13.0 (23)	47.6 (21)	10.0 (30)
Without penguins (12)	11.1 (18)	0 (13)	21.4 (14)	28.6 (21)	0 (23)
H total (26)	20.5 (39)	0 (38)	16.2 (37)	38.1 (42)	5.6 (53)
South Beach (18)	—	—	38.9 (18)	31.0 (29)	3.8 (26)
High moraine (28)	—	—	—	28.6 (35)	—
High basin (20)	—	—	—	31.3 (32)	—

[1] Number of occupied territories in each area.

and certainty about these pairs came from the long hours of direct observation of banded birds.

> *Skuas with penguins nesting among and away from the penguin breeding groups*

The skuas nesting among the penguins would seem to have the most favourable territories of all the skuas on the colony – they were completely surrounded by breeding penguins. In fact these skuas performed poorly in each year and many of their eggs were laid to be lost almost immediately. Clearly, although living on the colony may be excellent for adult living, it is less suitable for producing young. If the difficulty with these sites is because of interference from penguins, and from the attractiveness of the area to foraging skuas which also prey on skua eggs, then pairs that hold territories on the colony, but can nest outside its perimeter away from most of the penguins, would seem to have the best of both worlds: available food with little heightened risk. The territories on EF Block, and some of those on H Block, are of this sort.

Pairs with these territories performed consistently better than those with nests among the penguins, with about 25% of eggs surviving on average compared with 10.5% surviving in nests on the colony, but the differences were not statistically significant [1]. These pairs can be compared as well with those breeding in the same places that lacked access to penguin food. There are five years' data for H Block skuas (14 pairs with penguins, 12 without). In one year skuas without penguins were more successful, in three those with penguins did better and in the remaining year no skuas in either group raised chicks. However, the mean numbers for the four years when chicks were raised were not significantly different [2]. Nor could differences be shown for the nine pairs with and without penguins on EF during 1968–69 and 1969–70 [2].

From these comparisons it does not seem that close access to the penguins as a food source conferred a great advantage to skuas breeding in these years. Those that nested right down on the colony among penguins were generally unsuccessful, and those that nested about the periphery of the colony, although more successful than if nesting among the penguins, were scarcely more successful than neighbours which lacked penguins. Even though close proximity to the colony was not advantageous for skua breeding, perhaps it was still better to breed there in the shelter than well away from the penguins in the hinterland?

This was examined by including in the comparisons breeding groups that were so far away from the colony that they were generally untroubled by the penguins or by skuas attracted there. The pairs on South Beach were included to provide an exact comparison with the H Block pairs. These skuas nested along the base of the same cliff slope and their territories faced out similarly towards McMurdo Sound, but they were well away from the penguins. These skuas were equally as successful as the H pairs in two years, and failed equally with them in the third [3]. Their record, therefore, fails to demonstrate an advantage or a disadvantage in living away from the colony. Finally, to test whether the benefit of wind shelter about the colony (and the greater opportunity for gaining food) was eroded through nest predation and disturbance two groups of skuas on the exposed heights behind the colony were included in the comparisons in 1968–69. Seven locations were tested for chick production. These included 25 territories in the midst of the colony penguins, similar numbers at each of three places on the colony perimeter, and three sites away from the colony altogether. Differences in chick production among them were not significant [4].

This has been a very general comparison of breeding success among groups of skuas breeding in different locations about the penguin colony. The low success in these years made the outcome of each breeding attempt somewhat of a lottery, and there were too few pairs in each place to overcome the statistical variability inherent in such low success. Overall, however, it is possible to conclude that access to penguins did not appear to confer great advantage. In these five years skuas with penguins did not produce significantly more chicks than the other skuas in the local area. These comparisons run across the full breeding cycle, from egg production through incubation and hatching to chick growth. But the egg stage is the most vulnerable part of the cycle both for penguin damage, either through pecking or through being trodden on in the nest, and for skua predation. Comparisons based on egg survival, either as days surviving or as proportion hatching, may well provide a more useful test of the benefits and risks of breeding near the penguin colony (Table 11.3)

Hatching success from eggs laid, a measure of egg survival, was markedly lower for pairs on the colony in this sample than for those in the other areas, which were equally successful [5]. In only one year (1968–69) did many of the pairs nesting among the penguins manage to hatch chicks successfully, and this was the year with the lowest penguin numbers. These records indicate, contrary to expectation from considering food availability, that there was little benefit from being near the colony (comparing peripheral and away pairs) during incubation and that nesting among the penguins was decidedly disadvantageous (comparing pairs among penguins with all other pairs).

The implication from the bald records of egg and chick survival is that the colony area is hazardous for nesting skuas. Do the records of the *causes* of egg loss, the most vulnerable stage, support this view?

As noted above most skua eggs simply disappear from the nest leaving little evidence of the cause of loss. Recovery of eggs on territories and the interpretation of damaged shells and remains of eggs on the nest site does, however, implicate both penguins and skuas. There are also direct observations of skua and penguin attacks on nesting birds and on deserted eggs.

Furness (1984) considered in his study of great skua on Foula, Shetland Islands, that eggs had been taken by conspecific predators 'where an egg disappeared between 2 visits'. Ultimately, this is the formally correct diagnosis, as all eggs lost from the nest for any reason will be eaten by skuas, either by the parents or by others. In skua habitats there are few if any other predators. But this attribution subsumes a variety of contributing factors, and the distinction between parental and other skua predation

Table 11.3. *Percentage of eggs laid that hatched within specified nesting locations about the penguin colony*

Data are percentage hatched of number laid (given in brackets).

Nest location	1965–66	1966–67	1967–68	1968–69	1969–70
Among penguins[1]	18.9 (37)	—	6.9 (43)	35.1 (37)	8.6 (35)
About the colony periphery[2]	53.4 (88)	—	48.9 (88)	51.0 (104)	28.9 (51)
Away from the colony[3]	—	—	66.7 (18)	61.5 (96)	47.1 (26)

[1] Pairs with nests on the colony area (inside a line joining peripheral breeding groups).
[2] All pairs on E, F, H and G blocks with nests outside the colony area.
[3] Pairs on South Beach, and on the high moraine and basin.

is important. In many instances predation should more correctly be considered scavenging, as the eggs are already dead, have been deserted by the parents, or have been damaged. The only way to distinguish between the two causes is by inspecting the eggs regularly at the nest while at the same time maintaining continuous observation of the behaviour of the sitting birds. Under these, admittedly impractical, circumstances the causes of egg loss would be perfectly clear. To some degree both these approaches were attempted at Cape Bird. Eggs were inspected when the nesting records were made and groups of nesting skuas were observed. The lack of any account of the loss of most of the eggs missing from this population indicates the inadequacy of this attention.

Table 11.4 gives data on egg loss during 1966–67 and 1968–69 for three areas on the colony where skua surveillance was most intense and where the most complete accounts of loss could be expected. Of the 118 eggs lost from nests in these years, 33 were cracked, punctured or crushed on the nest or found broken or as splattered content near the nest and nine were seen being taken by predatory skuas. In total the loss of 42 (36% of eggs lost) could therefore be attributed to penguin and skua depredation. In contrast only five (4%) were lost through failure to hatch (were addled), were deserted or were seen being eaten by parents. Seventy-one eggs (60%) simply disappeared from the nests. If the only records available were of *missing* eggs predation by skuas as a factor would have been given even greater significance. But with the evidence of high levels of breakage and puncture of eggs on the nest the significance of scavenging by parents assumes greater importance. Most of these damaged eggs would have been eaten by the parents, not lost to other skuas. This scavenging

Table 11.4. *Information on the causes of egg loss in skuas*

Data pooled for 119 nests in 1966–67 and 1968–69 for three areas of the colony.

	Pairs on A beach	EF Block pairs	H Block pairs	Total
Numbers of eggs lost	30	35	53	118
Broken on nest				
Penguin damage	8	1	2	11
Skua or penguin damage	3	4	2	9
Cause uncertain	2	4	7	13
Total found broken				33 (28%)
Deserted eggs	0	0	1	1
Addled eggs or fail to hatch	0	0	2	2
Parent eating own egg	0	1	1	2
Total deserted				5 (4%)
Skua predation observed	3	2	4	9 (8%)
No explanation for loss	14	23	34	71 (60%)

Numbers in brackets are percentages of eggs lost.

behaviour is clearly quite common. Although only a small proportion of time was spent in direct observation of these pairs, on four occasions parents were seen eating their own eggs.

Clearly the incubation stage for these skuas is a difficult period, marked in this population with high levels of conspecific egg predation. At the same time many sitting birds were attacked by penguins. Most extricated themselves safely but in a small proportion of attacks the eggs were broken or punctured.

Attacks on sitting birds by penguins were commonly seen during the routine work on the colony and during the observation logs. The most serious and sustained attacks were by groups of wanderers, young penguins exploring the colony (Sladen, 1958). Single young penguins lacked the confidence to attack a skua and breeding penguins appeared generally to be so intent on their own activities that they seldom bothered. Sitting skuas responded to penguins standing over them by gaping (Fig. 11.1), the same response as given in the club to another skua in the same circumstances. Most simply stayed on the nest unless the penguins charged them. They seldom deserted to penguins standing over them, even when several penguins were involved and the confrontation lasted for several minutes.

Fig. 11.1. Wandering penguin and incubating skua on the nest.

If forced from the nest most skuas retreated in front of the penguins. This retreat left the nest exposed between the penguins and the skua so that if the penguins then charged the skua to chase it away they inevitably ran across the nest and eggs. Some eggs were broken in these attacks. Once the skua was forced into flight then one or both birds of the pair would then fly about the penguins attempting to chase them from the nest area. If these attacks on the penguins occurred near the nest there was an even greater likelihood of egg breakage as the penguins rushed to and fro in their attempts to peck or catch the skuas.

In other attacks, seemingly motivated more by curiosity than aggression, the penguins having displaced the skua from the nest then poked at the eggs, pushing them about the nest bowl. Characteristic holes up to 6 mm in diameter resulted from these 'attacks'. Penguin egg shells are much heavier than those of skuas and penguins blundering across them, or even pecking at them, seldom cause damage. Most skuas attacked by penguins at the nest gave the 'alarm call'. This call, more usually given by nesting skuas towards intruders, serves to alert and call the mate. Apart from these circumstances the only other times it was heard from skuas at Cape Bird was when sibling chicks were fighting and, more rarely, when one of a pair on a territory was being attacked by other skuas.

In widely dispersed nesting groups eggs and young chicks are generally safe from skua predation. There are no records from this study of direct attacks on sitting birds to get at their eggs and chicks. Even single birds alone on the territory seem immune from this sort of attack. Where the two birds are present the bird off the nest can intercept and deflect skuas flying near or into the territory, and deflect neighbours flying near the nest area.

In the crowded nesting areas about the colony the ideal situation of having a large space about the nest no longer obtains. In the most densely settled areas the nests are so close together that even a momentary lapse in attention can allow a neighbour to take the eggs. Conversely, overzealous territorial defence by the pair against one neighbour can provide an opportunity for another at the deserted nest. In crowded areas there are also far more birds in flight about the area or alertly watchful close at hand.

The risk of nest predation is further exacerbated on the colony by the numbers of skuas attracted there to the penguins. Most of these are local birds and although concentrating on the penguin groups they are also alert to the chance of picking up skua eggs. Skua eggs are smaller than penguin eggs, and although more fragile, are much easier to pick up by a skua in a swift attack from flight. Several skilled and aggressive predatory skuas with penguin territories commonly flew widely over the colony and its near surroundings. Two stood out. The female of pair 11 on H Block took at least seven skua eggs in one season from skuas on or near the colony during its regular sorties there for penguin eggs. Most of these records came from observation logs of H Block skuas and as the eggs were taken in the main colony human disturbance was not implicated. In a second year this pair accumulated four skua eggs during 8/9 December, and at the same time lost one of their own, found broken near the nest. The second of these predatory skuas was the male of pair 153. During some logs as much time was spent in flight about the colony outside its territory as within. On 26 November, for instance, during three hours study this bird undertook 11 flights over the colony outside its territory totalling 44 minutes flight time. No skua eggs were seen taken by this bird during any of its flights around the colony, but skua egg shells were found on the nest area. Skua egg shells are so light and so easily broken up and blown away that their numbers do not provide a reliable index of skua egg predation.

The enhanced possibility of nest predation by skuas at the colony comes about because of four special circumstances. First, many of the most favoured sites were crowded and nests were closely packed. The nest spacings for pairs in different locations shown in Table 11.5 indicate the extent of preference for nesting around the fringe of the colony. In some places there the nests were only a metre or two apart. Second, most skuas

Table 11.5. *Nest spacing of pairs in different locations about the Northern Colony. Distance to nearest neighbour (mean and SE in metres)*

	Mean	SE	Number of nests
Pairs with nests among penguins			
A Beach	64.6	6.9	10
H Beach	21.0	1.9	22
Pairs nesting on the colony fringe			
G Block ridge	7.9	0.65	23
Valley between B and D blocks[1]	9.2	1.1	20
Pairs nesting away from the penguin colony			
South Beach	33.0	3.2	9
Slopes behind H Block	28.2	3.4	12
High Basin[1]	28.1	3.2	15

[1] From Williams (1969).

nested on the slopes along the landward margin of the colony so that all their flights to the sea for foraging or down to McDonald Beach for bathing flew across the colony and the skuas there. Third, the club of non-breeding birds was located in the middle of the colony on A beach. This group of birds provided a natural focus for any stray skuas flying about the coast. Flight into the club took them across a string of territories. Fourth, a group of local skuas foraged widely across the colony and its peripheral areas. For example, in 21 hours in December, 1967 there were 39 of these wide-ranging flights by pairs 146 and 153. In late December, 1968 there were 43 foraging flights over the colony by pairs 145, 152 and 153 during 36 hours of observation. Many of these flights were for more than five minutes each, involving several circuits over and around the penguin colony.

Under some circumstances skua eggs are easier to take than penguin ones. A skua's own foraging away from the nest at the penguins often provides opportunity for predators. Temporary desertion of the nest to feed from a penguin egg taken by the mate, or to help kill and feed from a penguin chick, is a common reason for the nest being deserted. The different skua pairs responded very differently, however, to the attraction of penguin food or to the protection of the penguins within the territory. For example, on 21 December, 1966, pair 144 lost their own egg to an outside skua while both birds were 40 m away from the nest defending their penguins. In contrast to this casual defence of the nest, the male of

Table 11.6. *Nest position in relation to the penguin groups within the territory (1969–70)*

Nests on flat ground among penguin groups	35
Nests on higher, broken ground above the penguin groups	23
Nest below the penguin groups	14
Nests on the beach away from the penguins	14
Nests beyond a gully from the penguins	10

pair 152, although alone on the territory, did not leave the nest at any time in spite of numerous incursions by outside birds into the territory.

From the foregoing analysis skuas breeding at the colony seem to be at greater risk from egg loss during incubation than are skuas nesting further away. Not only are there more skuas around to take advantage of any lapse in nest defence, but these lapses are more likely than on the barren slopes away from the colony because of the attraction of the penguins. Preying on penguins and feeding on any food captured takes the skuas away from the immediate nest area. The more isolated the nest is from the part of the territory with the penguins the more hazardous is any temporary desertion. On the other hand, locating the nest among the penguins increases the risk from attack by penguins.

In an ideal skua breeding area the nest is located near the centre of a large territory and is overlooked by a roost from which the resting bird of the pair can fly to deflect any intruders before they reach the nest area. The roost should preferably be the highest local point. There are few such territories at Cape Bird. The ones that conform most to this ideal are those strung out along the beach cliffs where the nest and the higher roost can be located on the slopes of the cliff. On the colony itself the highest undulations in the local area will have been taken by the nesting penguins so that no roosts are available for the skuas. In other places the nests are safely tucked away from the penguins in rough ground among the moraine boulders backing the colony but are so crowded together that the roosts of one pair also overlook the nests of neighbours lower down the slope.

The most difficult situation for these skuas is where both the nest area and the penguin food within the territory are to be defended. In many sites these different priorities split the focus of territorial defence and weaken the defence of the nest and brood. Unless these skuas have established an absolute dominance over their immediate neighbours any temporary nest desertion to chase skuas from the penguins exposes the nest to predation.

Table 11.7. *Protection of nest area and penguins*

The position of the non-incubating parent in relation to nest area and penguins on H Block. Data are from behaviour logs of nine pairs over 5.5 h on 29 December, 1967 and 4 January, 1968.

Non-incubating parent	Number of 5 min intervals	Percentage of time
At the penguins	168	23
Closer to the penguins than the nest	175	24
At the nest	19	2
On the far side of the nest from the penguins	65	9
Absent from the territory	307	42

The most awkward of these 'split-focus' territories are those which have the nest in the lower slopes of the stream valleys with the penguins to be defended largely out of direct view on the plateau beyond the valley lip. Of the 96 pairs with territories containing penguins in 1969–70, 35 had nests among the penguins (of which 23 were on the flats among the breeding groups and 12 were on rocky slopes) and 61 had nests outside the periphery of the colony of which only 23 were on broken ground well above the penguins. The rest of these were on the beach below the penguins, so that penguins needed to walk past them on their way to and from their breeding group, or were safely out of the way from the penguins beyond a major gully (Table 11.6). Only one pair, however, defended geographically separate feeding and breeding territories as described at Cape Hallett by Trillmich (1978). There was also one of these territories at the Middle Colony at Cape Bird (Williams, 1969).

Records made of the position of non-incubating birds of pairs on H Block in relation to the nest and the penguins within the territory illustrate how these birds have weakened their nest defence by needing to watch and defend their penguins (Table 11.7). These birds were on flat territories running out on either side of the penguin groups on the shallow beach ridges. The nests were between 15 m and 35 m from the penguins. For much of the time these skuas were closer to the penguins than to their own nest area.

Increased costs of territorial defence by skuas with penguin territories

From the earlier accounts of the numbers of intruders on territories near the penguin colony there is clearly an additional direct cost in time and effort in defence of territories with penguins, in addition to the costs of breeding failure.

Breeding skuas maintain a rigorous defence of the territory at all times. This is done through display, through flight chases and through fighting. Territorial skuas display at birds in flight over the territory, especially by the bent neck and long call displays and evict birds on the ground through upright displays, long calls and fighting. Intruders on the ground within the territory are also attacked from flight by 'stoop attacks' in which the intruder is hit on the ground by the trailing legs. The significance of skua displays has long been recognised and they are now well described and their roles well understood (Spellerberg, 1971a; Andersson, 1976; Furness, 1987). These do not need to be considered further here. More costly for the skua, and imposing some risk from injury, are the overt attacks on intruders, by stoops at birds on the ground, by chases and attacks on flying birds. Most of the defensive attacks recorded for skuas with penguin territories are from flight. There are two reasons for this. First, most of these territories are large and extend well beyond the immediate nest area. Moreover, many have the nest in a narrow salient on the slopes above the penguins with the bulk of the territory running down on to the penguin colony. Flights are the only quick way of reaching or intercepting intruders. Second, territorial boundaries are commonly within penguin breeding groups so that their defence at these points is possible only from flight.

Because of the differences in territory size and shape there is a clear difference between the defence of the small breeding territories around the fringe of the colony, defended from the ground by displays and ground fighting, with only the occasional stoop by a bird returning from foraging or bathing, and the defence of the large penguin territories defended predominantly from flight – by flights, moreover, away from the nest area and above the penguins. These flights, always carried out at great speed and fierce intent, may extend out to several hundred metres from the roost where they began and clearly add measurably to the cost of breeding in these territories, offsetting some of the advantage gained from the food there. They are very distinct from the leisurely wheeling that the birds use when searching for food among the penguins.

The records made over one season of all defence flights undertaken by the EF Block skuas during the logs indicate the extent of this additional effort. The territories of these skuas ran out to as much as 120 m from the nest on to the colony and most chases went well beyond this distance. For the eight pairs on this area there were a total of 3.5 territorial flights h^{-1} on 11 December rising to 7.8 h^{-1} by the end of December to fall to 2.0 h^{-1} by mid January. The highest number coincided with the highest numbers of deserted eggs appearing on the penguin breeding groups at the onset of the post-guard stage. Although these may seem modest totals each one imposed an additional energy cost on these skuas that was not incurred by neighbours on smaller territories. As noted earlier, they also took the defending skua well away from its own nest and brood.

The response of these skuas to intruders depended very much on intruder behaviour and demeanor. Birds in direct flight high above the territory were not challenged whereas intruders obviously searching for food among the penguins, in flight circuits or hovering, would immediately be attacked, as were any skuas attempting to land anywhere in the territory. The attacks varied both in frequency and intensity among the skua pairs. Although all would attack birds landing in the territory, or showing any intention to do so, they may or may not attack others flying over or near the territory even if flying slowly and searching the ground. The different responses arose in two ways. First, some skuas were recognised by defending birds and were chased any time they came near the territory. These chases could be pursued for great distances, the two birds flying hard out of sight along the coast from the colony. Second, some birds chased at skuas in other territories as well. This behaviour constituted group or area defence. There were many examples of this, too many to be discounted as aberrations. H3, H11 and H12 males were very prone to this behaviour and all at times would chase at birds landing elsewhere in the local area or flying about other territories. On EF Block there are observations of 145 male flying all the way down to A beach to attack the non-breeders in the club, of 153 male overflying 152 to attack a bird landing in 150 territory, of 146 chasing a skua out of 149 that had been harassing the sitting bird there.

Williams (1969) compared the numbers of times intruders flew across one of the close-packed colony areas (the valley between D and C blocks) with numbers flying over the widely spaced territories of the high basin. His figures illustrate very large differences in disturbance to territorial skuas: on average there were 104 flights h^{-1} across the colony area compared with 23 h^{-1} in the basin during egglaying, 102 compared with 15 during incubation and 150 compared with 22 during late chick rearing. Moreover, on average six times as many intruding skuas landed in the colony area than in the basin – and each of these had to be chased off.

11.3   *Date of egglaying*
     The overall timing of the breeding season for the skuas in this region is no doubt determined proximately in response to photoperiod or from an endogenous circannual rhythmicity through environmental zeit-gebers (as will be seen in Chapter 12) but there is considerable variability within the population. Skuas arrive later on the breeding areas than the penguins and penguin eggs are already present on the colony when they arrive. Does the availability of the food on the colony confer a breeding advantage to pairs with penguins within the territory by allowing them to breed earlier? Pairs without penguins depend on courtship feeding by the male foraging at sea for the production of eggs. The records of ease of foraging for fish under the best conditions of wind and ice suggest that greatest advantage may be disclosed during seasons in which sea foraging is difficult.
     The first skua egg was found at Cape Bird during the five years of the study on 17 November and the last eggs of first clutches occurred on 5 January; 52 days later. The last eggs of replacement clutches were found on 12 January.
     Mean laying date for the first or single egg of clutches was in early December, varying little from year to year (Table 11.8). Laying dates were not significantly different in any year (single factor analysis of variance of frequencies in six-day intervals, $F_{4,30} = 0.54$, $p = 0.70$). The close and broken pack-ice characterising the early part of the 1968–69 season, causing the penguins such difficulty in their movement to and from the colony, clearly did not much affect the skuas.
     In each year the laying dates of pairs with >100 pairs of penguins in the territory were compared with those of pairs having few or no penguins to

Table 11.8. *Mean laying date of the first egg of clutches for all pairs at the Northern Penguin Colony*

| Year | Date | | |
	Mean	SE (days)	*n*
1965–66	7 Dec.	0.6	173
1966–67	4 Dec.	1.0	99
1967–68	8 Dec.	0.6	119
1968–69	4 Dec.	0.4	197
1969–70	8 Dec.	0.9	82

Table 11.9. *Laying dates of skua pairs with and without access to penguin food*

The same 31 pairs were used for each place in each year. Laying dates shown are for number of days from 21 November in each year. (Mean and SE.)

Year	Pairs with access to penguin food	Pairs without access to penguin food	Difference between mean dates
1965–66	12.9 ± 1.36	18.8 ± 1.24	5.9
1966–67	16.2 ± 1.72	20.1 ± 1.91	3.9
1967–68	13.8 ± 1.16	18.2 ± 1.14	4.4
1968–69	12.9 ± 1.37	15.4 ± 0.98	2.5
1969–70	13.2 ± 1.85	19.3 ± 1.41	6.1

determine whether access to penguin food was associated with earlier egglaying. The same 31 pairs on the colony were compared each year with an equal number of pairs on H and B areas lacking penguins. In each year the mean laying date of the colony pairs significantly preceded the others by between 2.5 and 6.1 days ($F_{9,300} = 3.921$, $p < 0.001$) (Table 11.9).

Ideally, this comparison, as with that of breeding success, should be carried out with selected pairs of birds of the same age and breeding experience, as both markedly affect breeding (Ainley *et al.*, 1990). Although the ages of these skuas was not known, some pairs had been banded at the nest as breeding birds by Bob Wood in 1964 and were therefore known to have had at least one years' breeding experience before this study began (Wood, 1971). Most of these skuas were still present during this work suggesting that overall this was a very stable, and experienced, breeding population.

There are a number of possible explanations for the consistent differences in laying dates shown here between pairs with and without penguins in addition to that of better nutrition. Colony skuas might take up their territories earlier in the year, advancing their entire breeding season, or they might be older or more experienced birds on average, with more stable pairs. Wood's banding records indicate that at the very least the H Block pairs were not newly established; a higher proportion of these birds carried his bands than anywhere else at Cape Bird. There were, moreover, only a small number of changes in these pairs each season during the five years. For the 21 pairs on H, 16 were unchanged, two had changed one bird of the pair, and three had changed both by the end of the study. This stability compares well with that of pairs with penguin territories on EF in which three of 11 pairs were changed. Thus, the differences shown in the breeding statistics are unlikely to be because these two groups of skuas have significantly different breeding backgrounds. Even if they had this factor does not appear to affect breeding date all that much. It is in fact quite difficult to identify pair changes from inspection of the laying date records. Failure of a mate to return at the beginning of the season may well cause a break in the breeding record or delay egglaying but changes in which a new pair forms promptly at the start of the season are scarcely recognisable. This is not surprising when judged against the variability in laying date of stable pairs across these five years. In 11 of these pairs on H Block (which has the most accurate records as first eggs of clutches were unlikely to have been missed in this area) the mean difference in laying dates from one year to the next was a surprisingly high 7.27 ± 6.12 days (standard deviation), range 0–26 days ($n = 41$). Nor was there any consistent trend for earlier laying during these years, or of the variability in laying date to decline.

In summary, there was a consistent difference in the mean laying dates of groups of skuas with and without penguins that could not easily be accounted for by differences in breeding age or experience, or in the stability of pairs. A further possibility is that the laying dates directly correlate with the dates the pairs were re-established on the territory at the start of the season, with egglaying occurring at a fixed interval after pair formation. This relationship could arise either because territories with food *allowed* skuas to breed early in the year, or because the penguins merely attracted the birds to settle earlier. Information on pair establishment and the lag between occupation and egglaying is needed to distinguish between these possibilities. Some of this information is shown in Table 11.10. Specifically with regard to the dates of egglaying, the consistent early occupation of the territory by EF skuas is particularly revealing. EF pairs, mostly with large territories and many penguins, arrived earlier in the year and had higher occupation levels subsequently than those in either H Block or the South Beach. The table, however, does little more than formalise the subjective appreciation gained from working with these skuas: one expects to find both birds of colony pairs on the territory for most of the day at this time of the year.

Is there in fact a more or less constant interval between pair formation and the date of the first egg? Data from two years seem to indicate this. In spite of the real uncertainty about recording the first time the pair occur together on the territory, there seems good agreement for two years that most pairs lay their first egg about 20 days after coming together on the

Table 11.10. *The occupation of H and EF territories at the start of the breeding season*

Percentage of birds on the territory at the time checked.

	13 Nov.	17 Nov.	20 Nov.	22 Nov.	24 Nov.	26 Nov.
1967–68						
H (24 pairs)	37	37	58	67	79	—
EF (10 pairs)	78	80	80	—	100	—
1968–69						
South Beach (17 pairs)	—	35	41	—	—	71
H (22 pairs)	32	68	52	—	—	77
EF (11 pairs)	—	68	91	—	—	—
1969–70						
South Beach (17 pairs)	—	—	—	47	41	42
H (24 pairs)	—	—	—	80	80	65
High Basin (20 pairs)	—	—	—	67	53	67

territory (for H pairs 23.5 ± 1.0 days (mean and SE, $n = 23$) in 1967–68 and 22.3 ± 0.8 days ($n = 15$) in 1968–69; for eight pairs on EF Block in 1967–68 the lag was 23.5 ± 0.9 days).

These records suggest, therefore, that pairs that can settle on the territory earlier in the season will also lay earlier in the season. Thus the advantage gained by skuas with penguins over those foraging at sea at this stage may arise not so much from their better nutrition but that through attraction to the penguins they settle on their territories earlier in the season.

11.4 *Do skuas appear to appreciate the advantage of nesting close to the penguin colony*

11.4.1 *Taking up territories at the beginning of the breeding season*

The penguin breeding groups are obviously attractive to skuas and deserted and displaced eggs would be very apparent to any skuas flying over the colony on their arrival at the breeding area. Is there any evidence that the penguin food attracts skuas to settle nearby at the start of the season or are all the territories in the general area settled at about the same time, irrespective of proximity to the breeding groups? This is a different question to the one examined in the previous section which discussed pair establishment and egglaying dates.

This question can be tested in several ways: by examining the order in which territories in different places are occupied each year; by seeing the way skuas on territories with penguins first settle and re-establish the territorial boundaries at the beginning of the year; and by observing their interest in foraging for food. If skuas settle at the same time in territories with and without food, and if those with penguins settle first at the nest area rather than near the penguins and, moreover, seem indifferent to the penguins, then it may be safely concluded that skuas in fact occupy their territories for the universally common requirements of breeding rather than for foraging.

Comparing the occupation rates of areas near and away from the penguin colony

In each year surveys of birds on territories were made two or three times a day from the beginning of field work about mid November until the breeding group was fully established. In the last three years this routine survey was augmented by a more detailed record in which colony areas were also compared with outlying sites at South Beach and the High Basin in an attempt to measure the relative attractiveness of breeding areas to skuas on the fringes of the colony. Skuas are about the region well before these surveys could begin, however. Records of these were made for the 1968–69 season by Morgan Williams and Eric Spurr. The first birds were seen by them to land on H Block on 1 November with the first pair present on 8 November and the first copulation attempt on 9 November. From this date there was a rapid build-up in numbers so that by 13 November when the routine checks of occupation and pair identity began most territories had at least one bird present.

The most telling comparison was thought to be of the order in which skuas settled along the beach to the south of the colony. Part of this beach was occupied by H Block penguins while a comparable section, with similar topography and aspect, further south lacked penguins. It was hypothesised that if skua distribution in this area was strongly influenced by the penguins, or if penguin territories allowed earlier settling, then the area immediately about the penguins would be occupied first. This comparison was made over two years.

In Table 11.10 the records are of proportions of resident skuas present on territories on each date. Although indicating an apparent tendency for the skuas to settle on H Block earlier than on South Beach the cumulative records of sightings of individual returning skuas in the two areas show little difference, with a rapid increase in appearance through the last ten days in November to reach the full pair complement in each place by the end of the month. In 1968–69, for example, by 21 November 85% and 95% of skuas were back on the territories of South Beach and H Block respectively, rising to 91% and 95% respectively by 26 November and all had returned in both areas by 1 December. These similar return rates were however, for the close fringe of the colony. Identical rates occurred also in the High Basin, far removed from any likely penguin influence.

There is little support, therefore, in these data for the suggestion that skuas first occupy territories containing penguins and only then take up territories away from the colony. The higher recorded occupation rates shown by the EF skuas suggest nevertheless that skuas on these territories spend a higher proportion of the early season on the territory.

Evidence for the attraction to penguins in the way the territories on H Block were first occupied at the start of the breeding season

Movements of birds and pairs at the start of the season before the pattern of territories and occupancy stabilised for the year provided numerous examples of the flexibility of this skua's fidelity to breeding areas and mates. At the very start of the season their propensity to form temporary liaisons and to shift about the local area belied the very stable breeding order that is soon established for these breeding groups. At the same time gaps in the defended pattern of territories are exploited temporarily by neighbours or occupied by outside birds or pairs. Any tendency of the skuas to gain access to penguins or contrarily to gain more secure nesting sites should be evident in these early manoeuvrings for breeding partners and breeding space. The records of the way H Block was taken up by the breeding skuas at the beginning of the 1967–68 breeding season illustrate some of these tendencies. Figure 11.2 shows skua occupancy at four dates from mid November. There were already 20 birds on the area when the first records were made. At the start there was little territorial behaviour and at this time most birds were simply resting on or near the preferred nest area. Three of the 20 birds present, all apparently in pairs, were not, however, established skuas of the area. In a three hour log of the behaviour of these skuas only pair 11 took any interest in the penguins, and gained two eggs. No other skua approached the penguin breeding groups. By 18 November, territory boundaries were being defended and pairs 11, 14, and 26 were foraging at the penguin groups. Pairs 11 and 12 were already defending their sections of H3 and H4, while 14 had greatly extended its possession of H4 into the areas to be reclaimed subsequently by pairs 25, 26 and 27. In the absence of pair 9, H2 attracted

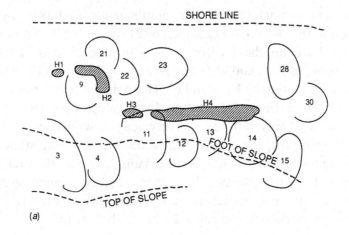

(a)

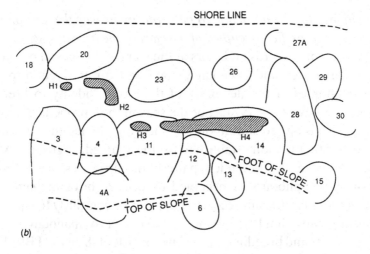

(b)

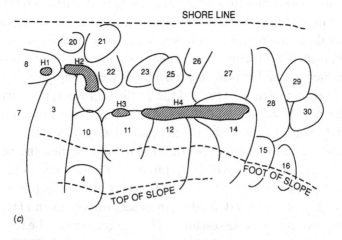

(c)

Fig. 11.2.

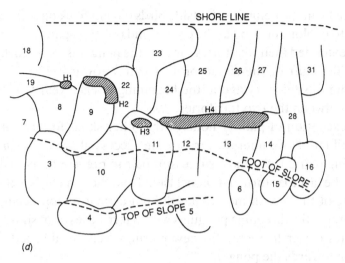

(d)

Fig. 11.2. Skua occupancy of territories on H Block early in the 1967–68 breeding season. (*a*) 13 November; (*b*) 16 November; (*c*) 18 November; (*d*) 28 November.

pair 3 to shift 40 m down the slope towards penguins and the single bird present of pair 7 joined with an outside bird to form a temporary pair on the northern corner of this group. At this date 10 of 31 skuas present were outside birds, six of these were in temporary pairings – defending territories and carrying out courtship feeding with some nest building and attempted copulations. By 28 November the historical pattern of territories and pairs was fairly well established. Even so there were still 10 outside birds present at this late date and pairs 20 and 21 were still away and their territories empty. In subsequent changes during the early season pairs 19–22 moved to occupy the western margins of H1 and H2 groups and 26 pushed towards H4, separating pairs 25 and 27, the identical shift occurring also in the following year. In contrast to this evidence of movement towards the penguins, pairs 12 and 13 both pushed their territories up the steep slope to give roosts above the nests. Movement away from the penguins for nesting was also recorded in the replacement experiment on EF Block by the new pair coming onto territories 146/148 after these pairs had been removed (Young, 1972). This new pair settled first among the penguins in the centre of the original area and then after establishing these boundaries with neighbours began pushing back the skua pairs on the periphery of the penguin colony to gain a nest site.

On H Block in this season there was clearly ambivalence by skuas towards the penguins. Birds tended to settle first and occupy nesting areas some distance from the penguins. Outside birds did not flock onto the area and attempt to exploit the food at the colony and only two pairs (3 and 14) temporarily extended their territories towards the penguins. Not much can be made of the shift of the single bird of pair 7 toward H2 as it was originally from this territory and at the beginning of each year tended to move freely between the two territories.

With the exception of pair 11 none of the other skuas foraged at the penguins until late November. Several of the successful predators in this area (including pair 25, the most successful pair in preying on penguins) were among the last to return. The conclusion from these records is that at the beginning of the season most skuas did not appear to appreciate the value of incorporating breeding penguins within territories or of shifting to new territories closer to them. No new pairings were established from skuas shifting towards the penguins.

This account is for a single year and was of particular interest because of the exceptionally large movement towards the penguins by pair 3. Similar records were prepared in each year to examine longer term changes but they confirmed that in each year skuas did not begin to occupy areas close to the penguins or dramatically shift towards them either in the absence of neighbours or through conflict with them. Indeed, there was surprisingly little territorial conflict during the settlement period, certainly in comparison with the very severe fighting, with birds bloodied and exhausted, that occurred when skuas arriving later in the season fought to re-gain mates. In this same year prolonged fighting occurred in three pairs (10, 12, and 13), with the victor standing ground on the territory in such exhaustion that the plumage remained bloodied and in disarray for hours afterwards.

11.4.2 *Evidence for attraction to the penguins later within the season in changes to territories*

Boundary changes of territories on the penguin colony were made for two reasons: to provide a more secure nest site and to reach breeding penguins. Usually these two changes were readily interpreted and distinguished. Attempts by skuas to extend the territory up into broken ground behind the penguin colony or to push a boundary back from the usual resting place of a chick, or to gain a roost above the nest were manifestly all concerned with nest and brood protection. Alternatively, boundary changes that carried the territory onto the penguin colony, especially changes away from the nest, were equally clearly to gain access to penguin food.

Only small changes were recorded during the breeding season among the pairs with territories on the colony or about its periphery and the maps of colony territories needed scarcely to be altered from the start of the season to its end. The few changes that did occur were by the slow attrition of one territory and the expansion of another. No dramatic changes were recorded during the five years of observations. On H Block the only significant change each year was the movement of pair 26 onto the margin of H4 between pairs 25 and 27. Also, in each year pair 9 was able to move from the eastern side of H2 to forage also along the western side during the post-guard stage. Pairs 20–22 took little interest in the penguins at any time and allowed this expansion of pair 9's range – provided it was confined to the penguin breeding area. As noted earlier, pairs 12 and 13 both pushed their territories up the steep slope each year to give roosts above the nests.

In EF the largest change each year was the movement of pair 142 along the margin of E9 (at the expense of 146 new pair) and the incorporation of F16 and F17 into pair 154's area. These changes were only possible on EF because of the very large areas being held by the skuas there and because 146 new pair showed little interest in the penguins as a food resource. In contrast to these changes among the penguins there was almost no change in the boundaries close to the nests. For most skuas, and especially those crowded together along the moraine slopes of the inland edge of the colony, this area had to be constantly defended and maintained. Loss of ground here could lead to the loss of nests and broods.

11.4.3 *Changes across seasons: the longer term changes*

The maps made in each year on H and EF blocks chart trends in the movement of pairs about the colony (Fig. 11.3). The simple plan of territories on H Block was of two rows of territories running across the beach flat on either side of the penguin groups spread out along the low ridge in the middle of the beach. Such a pattern, uncluttered by complex topography, was an excellent one with which to consider long-term changes. There were few significant changes and the plan recorded in

---

Fig. 11.3. (*overleaf*). Changes in the pattern of territories on H and EF Blocks during the study period and to 1987. (*a*) Changes on H Block. Note especially the maintenance of pattern among the skua territories even though the penguin breeding groups have been markedly reduced. (*b*) Changes on EF Block. Major changes during the study period were the experimental replacement of pairs 146/148 by 146 new pair and the movements of a number of pairs nesting along the foot of the moraine on to the colony area (145a, 146a, 149a, 150 and 154). Subsequent changes have been the splitting of 146 into two territories again and the division of area 152.

(*a*)

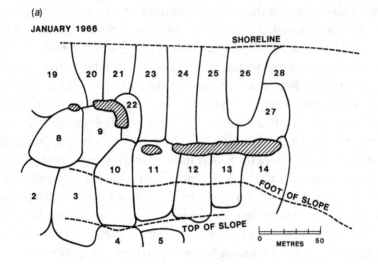

JANUARY 1966

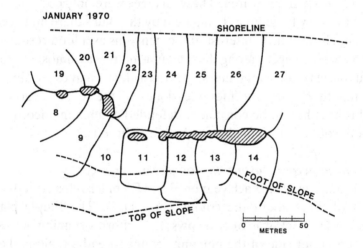

JANUARY 1970

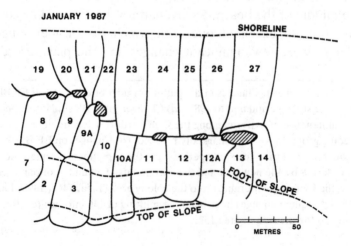

JANUARY 1987

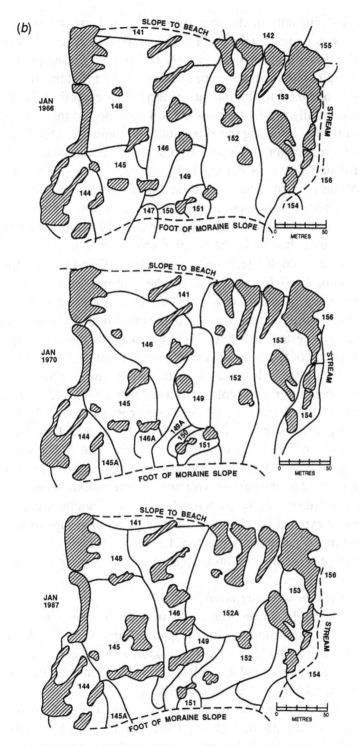

Fig. 11.3.

1965–66 remained virtually unchanged through to at least the 1974–75 season, when it was copied again by Paul Sagar (P. Sagar, unpublished records). Twelve years later in January 1987 the area was re-mapped to record a surprisingly similar pattern, one so similar in fact that the original numbers were re-assigned to the individual territories even though only two of the birds initially resident here were still present (Young, 1990b). The junction of territories along the ridge line in the middle of the beach was maintained again even though the penguin groups occupied a smaller length and provided less of a barrier between pairs than before. The maintenance of this pattern of territories over this length of time indicates its primary dependence on topography rather than the influence of penguin breeding groups. This overriding factor ensured a rather constant pattern irrespective of changes in the penguin groups or skua identities.

There was a more complex territorial pattern on EF and interpreting changes was complicated further there by the removal experiment in December 1968 when two pairs (146 and 148) were removed and were replaced by a single pair (146 new pair). With the exception of 146 new pair all skuas on this area nested among the broken rock of the moraine slopes along the inland edge of the colony. A major trend from 1965 to 1970 for pairs to encroach onto the colony was nevertheless apparent. Three pairs (145a, 146a and 149a) entered the colony area and pair 150 shifted over the years to incorporate the inner edge of F9. The largest gains were by pair 145, which enlarged its territory to reach E9, and by pair 154, which succeeded in pushing 153 back off three colonies. The latter was a surprising change as pair 153 were the most aggressive predatory skuas in this area and depended on the penguins for much of their food. In contrast, 146 were weak predators so that the encroachment onto their large area on all sides was more explicable. They lost parts of penguin groups to pairs 145 and 149 but more surprisingly much of E9 and F1 to pair 141, which already held a large territory with numerous penguins on A Beach.

11.4.4 *The club of non-breeding skuas*

Finally, it should be recorded that in spite of being in a very favourable area for expansion across penguin groups on A Block this location has not been exploited by the local skuas and the club area has remained much the same size and in much the same location over the more than 20 years since it became established here.

When the study began in 1965–66 the non-breeding club was located above the penguin colony on the high point between H and B blocks well away from the penguins. In the following year, probably because of the

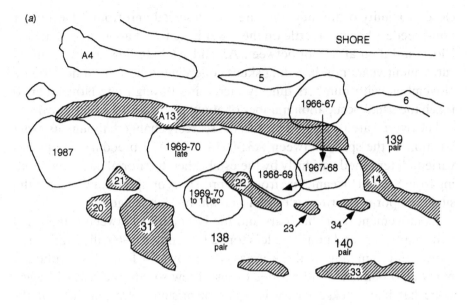

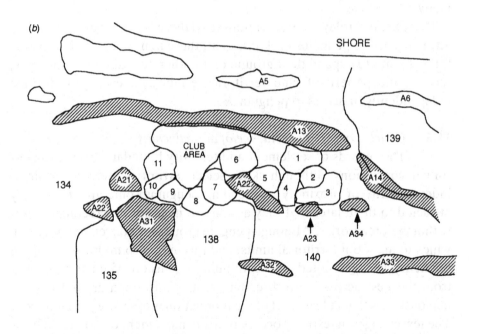

Fig. 11.4. (*a*) The movement of the non-breeding club area among penguin breeding groups on A Block, 1966–70 and 1987. (*b*) The relationship between club area, club pairing territories and established breeding territories in December 1969.

close proximity of the hut, this site was deserted and from 3 December non-breeders began to settle on the lower beach, first about a pool near B Block and then at a pool between A5 and A6 breeding groups. During subsequent years non-breeding skuas congregated here from mid to late November, with numbers quickly increasing through late November to reach the usual occupation number of about 80 birds present at any time.

The overall area occupied by the club scarcely changed, though its exact location in the space between A31–A13–A32–A33 breeding groups has varied, determined variously by the established territorial pairs, by flooding from Second Stream, and from the pressures on space coming from the small temporary territories established on the club fringe.

Its movement over the years and the boundaries between established pairs, club pairs and non-breeders for 12 December, 1969, illustrating the typical intermingling of these areas, are shown in Fig. 11.4. Although many club pairs, with territories of just a few square metres, laid eggs, none hatched chicks. This was a not surprising failure in view of the immense traffic of penguins through this area and its close proximity to so many other skuas.

There are two relevant observations from these records in relation to the attractiveness of penguins to skuas. First, these non-breeding skuas made little attempt to expand their ground on to other penguin breeding areas; second, the pattern of club territories established each year formed independently of access to penguins.

11.5   *Conclusions on the evidence of attractiveness*

The records of settlement and territory boundary change suggest similar conclusions: although skuas select territories and nesting areas independently of proximity to breeding penguins some are so strongly attracted to them later that they attempt to incorporate breeding groups within their territories. Although penguin groups are convenient points on which to mark out territorial limits, and this may account for some of the boundary changes noted, the continuing extension of territories away from the nest across entire breeding groups into open ground beyond demonstrates that at least part of the motivation is to secure penguin food. The few changes in territory occupation and movement of individual birds and pairs recorded here is in accord with the overall conclusions reached by Ainley *et al.* (1990) on breeding area stability at Cape Crozier. They concluded that skuas almost always breed themselves in their natal locality, that most never change their breeding areas once established and that those few birds changing breeding sites were all relatively young birds.

STATISTICAL TESTS

[1] Comparing the breeding success of pairs nesting among the penguins with pairs which although having penguins within the territory nested outside the colony perimeter. Survival within colony 10.5%, on H Block 24.8%, on EF study area 24.7%. Single factor analysis of variance $F_{2,9} = 1.77, p = 0.217$, ns.

[2] Comparing percent survival to fledging of pairs on H and EF areas with and without penguins within the territory.
H Block, over four years, 14 pairs with penguins, 12 without. $t = 0.89$, 6 df, $p = 0.405$, ns.
EF Block, over two years, nine pairs in each category. $t = 1.42$, 2 df, $p = 0.284$, ns.

[3] Comparing breeding success of 26 pairs on H Block with that of 18 pairs on South Beach for three years. $t = 0.32$, 4 df, $p = 0.76$, ns.

[4] Comparing numbers of chicks surviving of eggs laid for all areas in 1968–69. $\chi^2 = 3.88$, 6 df, $p = 0.67$, ns.

[5] Comparing hatching success of skua pairs nesting among the penguins with that of pairs nesting on the periphery or well away from the penguin colony. Single factor analysis of variance $F_{2,8} = 11.6, p = 0.004$.

12

Associating together: the longer-term implications

12.1 *Introduction*

The association between penguin and skua in Antarctica could have existed from the earliest re-colonisation of continental Antarctica following the amelioration of life conditions as the last glacial maxima gave way to warmer temperatures. Colonisation was possible for the southern Ross Sea from c. 7000 years before present (BP), and on the Antarctic Peninsula from perhaps 10 000 years BP. Of course the association may have been for much longer than this with the two birds shifting ranges in concert with the ebb and flow of the glacial and interglacial intervals that have characterised the Pleistocene.

Even the shorter interval, however, represents a substantial evolutionary period for these two species. The level of selection on the penguin populations would no doubt be determined by the predation pressure exerted by the skuas and on the skua populations by the level of specialisation towards this prey (Vermeu, 1982). The conclusions from this study are that skua predation is an important component of penguin egg and chick mortality in some circumstances, that its impact would be much greater but for the antipredator behaviour of the penguin adults, and that although all skuas potentially (through their strength, agility and weapons) could be strong predators on the penguins only a small proportion of them act as such. Nevertheless, the relation between the two species in terms of natural selection is clearly that of predator and prey. This conclusion does not, however, imply an ecological dependence on penguins by skuas. Many skuas breed perfectly well without penguins and the occurrence of the two species together on the same geographic place may be largely fortuitous: both need sheltered snow- and ice-free breeding grounds within range of open sea for access and foraging.

Although it is not possible to do more than speculate on the formative changes of this association as it became established, the current level of predation, and more importantly, the current potential for predation on eggs and chicks, significantly constrains penguin breeding.

In this environment neither species is able to hide from the other, there is no physical refuge, so that the prey must protect itself by other strategies, by colonial breeding, by antipredator defence, by breeding season timing.

Obvious constraints imposed on the penguins by skuas are the necessity for colonial breeding, the difficulty of establishing new breeding groups, both within a colony or on new sites, the necessity for continuous defence of the brood, and also, perhaps, the need for close breeding synchrony.

12.2 *Breeding site selection and disposition of penguins and skuas within the breeding areas*

Both species nest on bare ground and nests fail on ice and snow.

The principal distinction apparent when considering the association between the two species in this region is that whereas the skuas are loosely colonial, occupying territories ranging from a few square metres to many hundred metres spread widely across the area, the penguins are tightly aggregated in just a few colonies. That is, skua breeding is not confined to the penguin colony areas. Nor can skuas generally forage at the penguin colonies unless holding territory there. At Cape Royds and the three Cape Bird colonies the territorial skuas entirely prevented others from foraging among the penguins. At Cape Crozier, although parts of the colony are accessible to outside skuas, in practice this colony can serve only the local population as long distances isolate it from other skua breeding areas. The association of skuas with penguins at different sites will be considered in more detail in the following chapter.

As has been demonstrated in the previous chapter (see Table 11.5) the density of skua territories about the penguin colony is determined more by shelter and land form than proximity to penguins. The upper slopes have few breeding skuas. At Cape Royds the concentration of skua territories into the sheltered basins along the shore line was a very obvious feature of this population, with the immense slopes rising up from the peninsula towards Mt Erebus containing few pairs.

For this skua, therefore, selection has focussed on shelter, on the need to nest on bare ground, and on the universal need in skuas for space around the nest and roosting points.

The characteristics of penguin breeding are aggregation within a small number of colonies and aggregation within these into the tightly-packed territories of breeding groups. Although there may be subsidary functions for the close packing of nesting penguins (in social stimulation or to communicate foraging information, for example), the prime function is for nest defence; providing a good example of Hamilton's (1971) 'selfish herd' principle, with each penguin protected by neighbours. Direct observation of skuas attacking nesting penguins demonstrates the overwhelming significance of colonial defence. Once skuas are given the freedom to attack from the sides or from the front and back of the nest simultaneously the odds on success are enormously enhanced. Isolated nests are easy game for even moderately aggressive skuas. Conclusions on the importance of the colonial breeding habits reached from direct observation are strongly supported as well by the records of nest survival in the breeding groups: nests with the widest access and exposure to skuas are the least likely to survive. During incubation and chick guard stages the breeding groups tend to 'round off' as the most isolated nests are predated.

These conclusions would be supported as well by evident stability in the pattern of penguin breeding groups in the colony, because unless a fair number of pairs settled together simultaneously into a new site there is little likelihood of their breeding successfully and of the new group persisting and enlarging. Small groups have such a high proportion of nests on the margin that their relative exposure to skua attack is much greater than for larger compact groups. For the same reason groups that have shrunk over time to small numbers will also have a limited survival. The survival of both depends, however, on the interest and aptitude for predation of the local skuas. Small groups may well survive and grow in the territories of weak predators, which are either unskilled or indifferent to the penguins, or, less certainly, in territories which already have high penguin numbers. Information on the distribution of penguins and on the size and stability of breeding groups is available for this colony over a 25-year period. The penguin breeding groups were mapped in the first year of the study and have been counted and re-mapped many times sub-sequently. During the first few years the original maps could be used with few obvious modifications but in the later years with the rapid build up in penguin numbers more changes have occurred.

During the first years routine observations were made on the survival of small groups and a lookout was kept for attempts to establish new groups. In early December 1965 when the first maps and counts were made of the colony there were 28 groups with fewer than 40 nests. Only three of these were abandoned over the following five seasons. These were the smallest groups counted during the first survey: A29 with two pairs, A19 with eight pairs and A7 with 13 pairs. All others survived these first years, and all were still present when re-surveyed in 1986–87 (after 21 years). A single new group of just two surviving nests in December was established during these first five years. It was located on the beach 4 m outside A5 and did not survive the season. It did not re-appear in the following year.

The few changes found during these five seasons were indicative of the longer-term stability of this colony evident in its later history. Since 1965 the numbers of nests counted annually in early December first fell from c. 25 000 to a minimum of c. 15 000 in 1968 to be followed by an unbroken sequence of annual increases to reach 41 000 in 1987 (Taylor *et al.*, 1990). In spite of these large population changes there have been few changes in the status of individual breeding groups. Eight small groups have been abandoned; three of them (A7, A19 and A29) by 1969, the remainder by 1979. Eleven new groups were established, but none was large and none was outside the original area of the colony. Of the 32 287 nests counted in 1988 (K.-J. Wilson, 1990), a doubling in number from 1968, only 660 (1.8%) were in new breeding groups. The overall pattern of breeding groups on the colony has, therefore, been remarkably maintained and a naive observer would have little difficulty identifying most of the groups at the present time (1991) from the original plan drawn in 1965. This stability comes, to a large degree, from the underlying topography of the area and the specific requirements of raised, snow-free ground for nesting. It comes also, nevertheless, from the propensity of the penguins to join existing breeding groups rather than attempt to establish new ones. The small changes in the pattern of breeding groups on H and EF areas across 21 years (from 1965–66 to 1986–87) demonstrate the contribution of these two components of pattern stability (Fig. 11.3).

A sharp decline in numbers of breeding penguins on H Block has, nevertheless, left the overall pattern of breeding groups along the beach ridges undisturbed. H1, which was one of the smallest groups in 1969–70, had survived but H3 had been abandoned and H4 had broken into three separate groups. The decline in numbers overall in this area, both in absolute numbers and relative to the colony as a whole, has been attributed to disturbance from the field station (Young, 1990b). Not surprisingly, no new groups were established here. The changes on EF are more illuminating. Although the pattern of groups in 1986–87 was immediately recognisable from the maps of 1969–70, there had clearly been a substantial increase in numbers of nests in the interim. Many groups had enlarged; two of them, E2 and E11, merging with smaller neighbours. The greatest increase was in E4 and E5 (groups used several times during the study for experiments comparing protected and control groups of penguins) which had merged and swollen over the intervening years from about 150 nests to well over a 1000. In spite of the clear evidence of a great increase in numbers on this area only two new groups had become established. A small group existed a few metres to the south of F9, and a substantial new group was present on the flat between E18 and F4. The first had c. 10 nests, the second 80–100 nests. Both could have arisen by the earlier extension of the neighbouring groups. Nine of these groups were identified as being at risk in 1969–70 (with fewer than 40 nests in December) and all of these had survived. F8, a much larger group from this period with about 75 nests, was unaccountably missing.

These changes to the colony over time seem to support the contention that it is difficult for penguins to establish completely new breeding groups. Penguin numbers more than doubled since 1968 yet the pattern of breeding groups was little altered and only a tiny proportion of the increased numbers has been accomodated in new groups.

A similar pattern of early abandonment without later recolonisation and of increasing density among established groups has occurred also at the West Colony at Cape Crozier. Although the numbers there have also doubled over the past 20 years (Ainley *et al.*, 1983; Taylor *et al.*, 1990), the breeding groups on the western margin of the colony that had been abandoned earlier have not yet been recolonised. Their identity markers from the earlier research period stand as sentinels on the mounds of leaching guano (Young, 1989, personal observation). It was this sort of observation that led Oelke (1975) to implicate the changed ratios of penguins to skuas as critical for penguin occupation in areas that had suffered from human disturbance. He postulated that when the numbers of penguins decline significantly the survivors could no longer defend themselves against skua predation.

In summary, the longer-term response by Adélie penguins to skua predation seems to have been selection for return and breeding within the natal area and for the close packing of breeding territories within defined breeding groups. Ainley *et al.* (1983) provide impressive statistics of how accurately this selection has focussed to return naive breeders to their birthplace.

Even so, none of the breeding groups at Cape Bird developed the optimal circular outline that provides the minimum number of vulnerable nests on the periphery. To a great extent the shape of breeding groups reflected the underlying topography suggesting that the requirement to locate the nest on higher ground to ensure that it would not be flooded (or to select snow-free higher ridges in spring) outweighs selection for predator impact.

12.3 *The breeding seasons: dependent synchrony or fortuitous coincidence?*

12.3.1 *Relationship between the breeding seasons of the two species: seasonal timing and breeding season synchrony*

There is no possibility of the prey hiding in space from the predator in its breeding area: there is no shelter, no camouflage is likely to deceive, and the flight range of skuas would always allow discovery on the coastal fringe of the continent and its islands. It is possible, therefore, that the prey might respond to predation by hiding in time, by breeding when the skuas are absent or by adjusting their season to minimise predation on the most vulnerable stages. They are constrained, however, by other environmental factors, by their own demands for food and for favourable breeding conditions, and predation is but one of a variety of factors acting to select for highest reproductive output. Individual penguins might also be able to reduce predation impact by diluting its effect, by establishing a temporary refuge among other nesting conspecifics. This would be achieved by a closely synchronised breeding season, in which the dual effects of being hidden among others and of 'swamping' the predator during the most vulnerable life-cycle stage would confer individual advantage. These two aspects of the timing of the breeding cycles are considered in the following sections. Although some of these aspects can be examined at Cape Bird there is value also in comparing the relationship between the two species through their entire breeding ranges. In the following sections the timing of the seasons and their synchrony is examined in relation to the habitabi-lity of the breeding area, the time each species needs for breeding, and the timing of each in regard to the apparent dependence of the skua for penguin food.

Since Lack (1968) there has been almost universal acceptance that 'each species has presumably evolved its timing of breeding so that it raises more offspring'. A similar statement, for example, is at the core of the review by Gwinner (1986) of circannual rhythms. This timing is achieved through a complex interaction of endogenous and environmental components which Baker (1938) first recognised as being of two sorts: ultimate factors, such as food, cover and breeding site condition, which characterised the 'right' time of the year for breeding, and proximate factors, which provided the timing mechanism. As with other aspects of breeding the timing of the cycle is sensitive to natural selection, adapting the species to local con-ditions across the species' distribution.

In skuas and penguins proximate factors must act to ensure these two species independently and in association breed when conditions are most favourable. In high Antarctic latitudes this must mean in practice as soon as reproduction is likely to be successful: specifically, when the breeding areas are largely free of ice and snow after winter, when food is readily and consistently available at sea, and when temperatures are at their most benign. Murton and Westwood (1977) consider that food availability is the most important ultimate factor – 'that most birds breed around the time when food supplies for themselves and their young are most readily available'. As with virtually all temperate and high latitude species the proximate control of the timing of the breeding season is likely to be a response to photoperiod, rather than temperature, interacting with endo-genous rhythms. Proximate factors also control the timing of the end of the season, determining how late into the summer egglaying may persist. Overall, these factors demarcate a 'window' of breeding opportunity, of egglaying, for each year. Pairs that are too early or too late will be equally disadvantaged.

12.3.2 *The Cape Bird summer: the duration of suitable breeding con-ditions*

The conditions on the breeding area: snow and ice on the colony and surrounding areas

Cape Bird has extremely low annual snow precipitation levels and is protected by the slopes of Mts Bird and Erebus from snow blown off the

Ross Ice Shelf. In five summers' records between mid November and late January the colony was lightly powdered with snow on only a few occasions and was covered by a heavy fall (c. 100 mm cover) only once. Few skua and no penguins were recorded as abandoning nests because of snow. This record contrasts with that from Cape Royds, 25 km to the south, where blown snow may regularly cover nests.

There is apparently not much snow through the winter either. In early November the penguin colony is largely snow-free, there are few drifts on breeding areas and in only one year for one breeding group did the penguins occupy a snow-covered area. This was in 1969 for B14, an area subsequently abandoned.

The skua breeding areas about the colony were more regularly affected by snow. The nesting areas of pairs 18 to 31 along H beach were invariably covered in 100–200 mm of frozen snow at the start of each season so that the birds were at first claiming snow surface rather than ground. The river valleys dissecting the colony were also invariably filled with ice and snow after winter and some skuas there regularly laid eggs on the snow and just as regularly lost them, as they and the eggs sank below the snow surface during the first days of incubation. This was the usual fate of the first clutches each year of pairs 154, 155 and 156, for example. Territories in the high basin above the laboratory and on the high moraine above the colony against the ice cap were much more prone to snow and ice accumulation than further down on the colony. Even so few nests were lost because birds laid on the snow. Nor was their laying date significantly delayed.

Thus, the amount of snow and ice cover on the breeding areas, either carrying over from winter or during the breeding season itself, does not seem to be an important ultimate factor in reproductive success for this area and consequently in setting the time of breeding.

Access to the breeding ground and the availability of marine food

There is no information on the seasonal abundance and accessibility of the krill and small fish food of penguins and of Antarctic silverfish for skuas. It is assumed that these species are present throughout the year in the upper water column. The early arrival of skuas into the Ross Sea indicates that at least for this species some food is available. The same conclusion cannot be reached for the penguins so easily because at first these are fasting.

For the Cape Bird skuas and penguins snow and ice cover is likely to be more significant at sea than on land. Its extent and form governs access to the area for the penguins and access to food for both species. The two species are, however, differently constrained by sea-ice cover in early summer with the advantages of flight in skuas offset by fasting in penguins. Skuas can fly more easily to open water, but need to forage two or three times each day. Penguins travel much more slowly, but through being able to fast and undertake long incubation and foraging bouts they can advance their breeding season ahead of the time that the open sea would normally reach the breeding area. Once the chicks hatch there must be much closer foraging access.

There are two ways to judge the earliest possible onset of breeding in relation to sea-ice. Both depend on historical records: on the dates of the formation of the southern Ross Sea polynya, the ice-free pond that sustains all the sea birds in the Ross Sea, and on the history of breeding success and failure over time.

The importance of polynya (areas of open water surrounded by ice, Smith and Rigby (1981)) has been stressed by Stonehouse (1967a) for high latitude Antarctic species and subsequently by Stirling (1980) for the Canadian Arctic. The analysis by Sturman and Anderson (1986) of sea-ice decay in summer and development in autumn and winter gives a broad picture of the seasonal change in ice cover for the area. The small amount of ice in summer in the Ross Sea has long been recognised from the earliest voyages of discovery in the region. To be significant for breeding birds the Ross Sea polynya must be within the normal foraging range throughout the breeding cycle. This seems generally to be the case. Recent research (Davis *et al.*, 1988; Sadleir and Lay, 1990) using radio telemetry has for the first time established that routinely during incubation the range for the penguins in this region extends out to 30–40 km to the northwest of Cape Bird. Longer trips were also recorded carrying the penguins far to the north and east of Beaufort Island. There are no records of this sort yet for skuas but direct observation showed birds foraging throughout McMurdo Sound to the north and west of Cape Bird. The flight paths of birds leaving to go foraging when the Sound was packed with ice were tightly about the ice cap to the north of the colony suggesting that these birds were foraging about the north and east of Ross Island. Flights of up to several hours could project these birds out to 50 km, even 100 km, distant.

Local ice conditions in McMurdo Sound and between Ross Island and Beaufort Island are more more variable and depend much more critically on the local weather. As described in the account of foraging (section 9.2), pack-ice is carried into McMurdo Sound around the northern face of Ross Island by the easterly flowing current. For most of the summer whether the Sound is ice-packed or has open water depends on the frequency of southerly winds. Strong winds push the ice northwards out of the Sound. This pattern of steady filling of the Sound and dramatic clearing continues until mid January. From this date the southern part of the Ross Sea is essentially ice-free and the flow into McMurdo ceases. For a few weeks until freezing begins the Sound is largely ice-free in all winds.

The first breaks in the uniform ice cover in the southern Ross Sea are recorded each year in mid to late October. In 1968–69, stigmatised as a difficult year for penguin breeding because of severe ice, the first breaks were not evident until a month later, in mid to late November. Open water to the west and the immediate north of the colony did not occur until much later in most years. The route taken by the penguins through the Ross Sea to Ross Island is not known. It is apparent, however, from the ice records showing continuous pack-ice cover at this date that they are able to move freely towards the colony through dense pack.

The experience of the 1968–69 season demonstrates that it is not so much the *extent* of pack-ice cover which is significant for penguins but its topography. Penguins walk and toboggan easily and quickly across smooth ice and snow; Taylor (1962b) recorded speeds of 3–3.8 km h^{-1} on ice. They have great difficulty negotiating tumbled and ridged ice. It was apparently the broken, jumbled and heaped ice surface that was the real problem in this year, not its greater local extent.

The influence of the extent of fast-ice about the breeding colony is a quite separate problem. Continuous ice cover would prohibit breeding by either species. Its role in breeding behaviour and success at Cape Royds has been examined in general by Stonehouse (1963) and for specific impact on incubation spells and egg and chick mortality by Yeates (1968). The extent of fast-ice is, however, certainly less significant at Cape Bird because even at the very start of the season its edge usually lies no further north than the colonies there and in some years lies well to the south. In only one year of the study was it possible to walk out on to fast-ice from the Northern Colony to the ice edge a kilometre or so from the shoreline. In the other years the ice edge was already several kilometres to the south in November.

The end of the summer conditions at sea comes very suddenly. From mid January to the end of February the Ross Sea is essentially ice-free. But from this date the sea begins to freeze over very quickly to reach 80–100% cover except for a narrow strip of clear water along the Victoria Land coast. By 1 April even this last vestige of open water has gone. (Except for narrow leads and breaks that sustain Weddell seals and the colonies of Emperor penguins.)

The timing of the season for penguins at Cape Bird would appear to be related both to the conditions en route to the colony and to conditions offshore during incubation and chick rearing. The pack-ice at these stages is, however, merely a quantitative variable: until the ice dissolves entirely in this region in mid January the degree of cover and the topography depend on wind conditions. But as Yeates (1968) has demonstrated, access to sea close at hand is crucial for successful breeding, especially during chick rearing. Open sea is not routinely achieved in McMurdo Sound until late December through to mid January. The earliest that egglaying should occur is, therefore, the end of November. Laying earlier than this will invariably occasion risk. The second important parameter for life-cycle timing is the sea and ice condition at the end of summer. McMurdo Sound begins to freeze over in mid to late March. The difficulty of passage for penguin chicks through this ice could also influence the timing of egglaying. On this basis the latest that egglaying could take place (to ensure a three-month breeding cycle), is in mid December.

Living conditions on the breeding areas

The important factors under this heading are sunlight hours, temperature and storm frequency. In conjunction they determine living conditions and comfort on the breeding areas, contributing to the survival of eggs and chicks. At this latitude the sun first appears above the horizon in mid August, there is 12 h sunlight by 20 September and continuous sunlight from late October to mid February. Changes in the autumn are similarly rapid. By 10 March sunlight hours have again fallen to 12 h and by 25 April the sun has dropped below the northern horizon, leaving a shortening twilight into winter.

Air temperatures mirror the sun's passage. Temperatures begin to rise in September, reach their highest levels in December and January and fall again to winter levels by May. Mean daily air temperatures when the penguins come on to the colony in October are a bracing $-20\,°C$, the chicks hatch into about zero degrees and leave to sea in February when temperatures average c. $-8\,°C$. Skuas take up their territories later in slightly warmer conditions but many are still on the territories into March when mean temperatures of c. $-17\,°C$ at the time have already begun to fall quickly towards the winter minimum. This asymmetry in the breeding of each species in relation to the temperature cycle presumably reflects essential differences in their breeding and migration abilities.

The final environmental factor to be considered is the frequency and duration of southerly storm conditions. Although the immediate Cape

Bird colony area misses most of these they nevertheless can have a devastating impact on the breeding birds. The damage sustained to the timber on the south wall of the laboratory at Cape Bird from wind-blown sand and gravel over winter indicates that fierce winds do occur here at times. Strong south winds are more common in winter than summer. At Scott Base winds gusting to over 61 km h^{-1} occur on average 11 days each month from March to October compared with only three days each month during summer. More significantly, severe winds, gusting over 95 km h^{-1}, which are more likely to impact on Cape Bird than do lighter winds, occur on three days a month in the winter but on average only 0.2 days a month in summer (Sinclair, 1982).

In summary, living conditions are much more settled and warmer during the few summer months than at any other time. The colony is bathed in sunshine, the air temperatures may even exceed zero at times and storms are uncommon. But none of these environmental factors places a qualitative barrier on the timing of the cycle or places an absolute limit on its start and end.

Conclusions on habitability

The important shortcoming in this assessment of the constraints and determinants of the breeding cycles is the lack of information on the amount and availability of food in the sea, during the migration to the breeding areas, while the birds are breeding, and for the adults and chicks during their winter migration away from Ross Island. For both species, however, the breakup of the winter ice cover and the formation of polynya near to Cape Bird is clearly a critical determinant for foraging. The dates of ice breakup in the spring and ice formation in autumn seem therefore to set the limits for the breeding season of both species – roughly from mid October to the end of March. The ground conditions at the colony for nesting are probably favourable throughout the year, while living conditions, defined by sunlight hours, temperature and storm frequency are less rigidly defined and constraining than the breakup and onset of pack-ice. These different factors are not, however, all symmetrically located about the date of the sun's zenith on 20 December (corresponding to the longest day in temperate regions). Although the changing photoperiod is symmetrically placed, and the season's warmest air temperatures lag only marginally, the timing of open water is grossly skewed (Fig. 12.1). Substantial open water is not achieved until mid December but then extends through to the end of February.

From this review it appears that the maximum time span available for

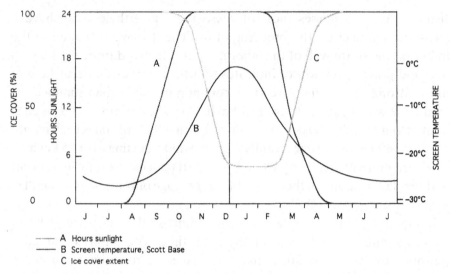

- - - - - A Hours sunlight
———— B Screen temperature, Scott Base
           C Ice cover extent

Fig. 12.1. The seasonality of environmental factors of photoperiod, temperature and sea-ice cover at Cape Bird.

breeding at this site is about 160 days, and that the critical determinant for these species is sea-ice condition.

12.3.3 *The minimum time needed for breeding*

What freedom is there for the penguins and skuas to adjust their breeding cycle to the most favourable time within the habitable range? From comparisons of the times of breeding with latitude Young (1977) suggested for skuas and Lishman (1985b) for penguins that these species were unable to breed any later in the season at high latitudes because the summer there was so short. Can this assertion be tested? What is first needed is a measure of the minimum time needed by each species to complete successful breeding. This calculation is shown in Table 12.1.

Traditionally the penguin cycle has been calculated to the date of chick departure in late January and early February, and on this basis the penguin cycle (mean duration 96 days) is appreciably shorter than that of the skuas (mean duration 139 days). A significant proportion of penguins remain after chick departure at the colony in order to moult. As this analysis is concerned with fitting the breeding cycle within the habitable summer season the moulting period should rightly be added to the breeding time. When this is done the periods penguins and skuas occupy breeding areas are closely similar.

Although skua chicks fledge at between 49 and 60 days old there is a further long span on the colony before departure. Departure is so late in

Table 12.1. *The durations of breeding season stages and the overall estimated length of the breeding cycles in Adélie penguins and skuas (days)*

	Penguins			Skuas		
	Minimum	Mean	Maximum	Minimum	Mean	Maximum
Pre-egglaying	5[1]	10	20	13[4]	22	30
Incubation	36.5[2]	36.5	36.5	31[5]	31	31
Chick rearing						
to fledging	—	—	—	49[5]	54	60
to departure	40	50	60	25[6]	32	43
total				74	86	103
Breeding total	81	96	116	118	139	164
End of breeding to moulting	1[1]	9	15			
Moulting	15[3]	20	25			
Breeding and moulting total	97	125	156			

[1] Taylor (1962a), Spurr (1975c).
[2] Astheimer and Grau (1985), Lishman (1985b).
[3] Penny (1967).
[4] Young, E. C. (unpublished).
[5] Young (1963a).
[6] Neilson (1983).

the season that it has rarely been observed. There are no modern records for Ross Island where wintering bases (except for that of Greenpeace on Cape Evans) are distant from skua breeding areas. The time spans cited for the post-fledging parental care period are from Neilson (1983) who overwintered at Palmer Station on Anvers Island on the Antarctic Peninsula. Brian Reid (pers commun.) recorded skua chicks leaving Cape Hallett in 1959. In that year many chicks could fly by mid February and virtually all could fly well by the end of the month. Adults and chicks left the colony during March with the last two chicks still present on 20 March. The last of the local breeding birds departed between 22 March and 2 April but others were seen flying past Cape Hallett on 6 April; these were the last skuas for the season.

The estimated minimum times for breeding are 81 days for penguins and 118 days for skuas. If moulting is included in the penguin cycle their minimum occupation is then 105 days. These times might be shortened further in both species by reducing the pre-egglaying period, for skuas by beginning egg formation even before reaching the colony as occurs in the penguins (Astheimer and Grau, 1985); in both species by raising chick growth rates; and in skuas by shortening the length of time chicks are supported by their parents on the breeding area after fledging. Raising chick growth rates would demand higher foraging rates by adults. Shortening the post-fledging period in skuas would no doubt lower chick survival at independence. The penguin cycle could also be significantly shortened if all adults moulted at sea.

The length of time south polar skuas are resident on the breeding grounds is, however, inexplicably constant throughout its breeding range. They are present in McMurdo Sound from early October to early April, c. 190 days in total – for the same duration as at Pointe Géologie, Terre Adélie (mean 183 days, range 178–188) (Le Morvan *et al.*, 1967), at the Windmill Islands (190 days) (Eklund, 1961), at Anvers Island where they are on the territories for 173 days and in the area for about 200 days (Neilson, 1983; Parmelee *et al.*, 1977) and at Signy Island, far to the north (190 days) (Hemmings, 1984). Few authors distinguish between the times skuas are first and last seen in the area, when they will be generally in flight passage, and when they are on territories or on the ground. This distinction is clear in Ainley *et al.* (1978) for Cape Crozier who record for one year the first bird seen on 19 October and the first seen on the ground on 27 October. But the time these skuas are about the breeding areas seems remarkably uniform even when some records are of birds on the territory and others are merely of when they were seen in the general area. At all sites, in all latitudes, they appear first in October and leave finally in April.

The breeding cycles at high latitudes seem, nevertheless, very tightly constrained by the time available for breeding and this urgency is recognisable in the breeding behaviour of both species. One significant manifestation of this urgency is the readiness to accept partners onto the territory at the beginning of the year if established mates are late returning. In the penguins this tendency leads to quite high levels of mate changing over seasons (Ainley *et al.*, 1983) – and to fighting when the mate of preceding seasons returns. For skuas there is proportionally much lower rates of mate change between seasons; at Cape Bird, such changes were uncommon, with a greater tendency to delay breeding at least for a week or more in the absence of an established mate. There was also less fighting within the pairs because of mate changes. One reason for the difference may lie in the timing of egg formation in these two species: penguins will have well developed eggs on reaching the colony so that pairing and copulation cannot be delayed. Skuas form eggs during courtship when nourished by the mate. However, skuas also show a strong propensity to take up new mates in the absence of the established one. For example, between 9–13 November, 1967 on H Block, male 9 was found with four different mates, male 10 and female 13 were each found with three while in territory 12 two males and two females were variously present. In 1968 the female of pair 10 accepted three males in turn between 16 and 20 November. In the following year different males were found with the female on territory 12 nearly every day for a fortnight before the true mate returned. These commonplace records suggest that it is also advantageous for resident skuas to begin breeding as soon as possible.

Practising breeding techniques and practice pair bonding occur in both species. In penguins it takes place during the reoccupation period when new and failed breeders occupy and defend territories and develop nests. At least in some northern sites there is also reoccupation of the colonies by some penguins again very late in the autumn. At Anvers Island (64°46' S, 64°03' W) from 18 April Adélie penguins were observed in 'a remarkable 3-week period of sexual activity that was characterised by courtship displays and nest building but no egg-laying' (Parmelee *et al.*, 1977).

In skuas similar behaviour occurs at the end of summer when both successful and failed breeders again exhibit nest building. This resurgence of pairing and nesting behaviour was first described by Young (1963b) at Cape Royds but apart from Spellerberg (1966) has scarcely been recorded subsequently in the literature. It is no doubt widespread and the lack of records indicates the difficulty Antarctic scientists have in being able to carry out field work late into the summer.

Skuas but not penguins may replace clutches. In part this difference reflects differences in the behaviour of the two species. The penguins are precluded from re-nesting because the female leaves the colony for a long spell immediately the eggs are laid. Re-laying would only be possible for these birds following the female's return. Such replacements c. 30 days later at best (10 days foraging, 21 days for egg formation) would place the new clutch in mid December on average – delaying fledging until well into February. Penguins do, however, produce a third egg to replace the first of a clutch lost during egglaying. This inflexibility contrasts markedly with the success of skuas to produce replacement clutches and to fledge chicks from them. Over 5% of chicks in each year at Cape Bird hatched from replacement clutches. Spellerberg (1966) also records significant numbers of replacement clutches at Cape Royds, but his figures for the time needed to produce a new clutch of between four and 11 days seem too short. In nine instances at Cape Bird where there were precise dates for the loss of one clutch of eggs and the appearance of the first egg of the next, and where there did not seem to be undue delay through mate loss or change, the replacement clutch appeared between 11 and 18 days later (mean 13.4, SE ±0.8 days). In these pairs the first clutch was lost between three and 11 days after the second egg had been laid. Skuas appear, therefore, to have rather more time to complete their breeding than do penguins. The demands to fledge chicks and to time departure from the breeding area are less stringent apparently. It must be appreciated, however, that the facility to produce replacement clutches in skuas is governed by more than time available for breeding. Brown skuas on Chatham Islands do not usually produce replacement clutches even though living in a benign environment with an extremely long breeding season (Young, unpublished).

There are other observations, however, which suggest that the skua breeding season will not be extended indefinitely. The 1969–70 season was characterised by an especially spread-out initial egglaying and with many pairs re-laying the cycle overall was later than average. In mid January although some chicks were almost flying there was still a large number of nests with eggs and recently hatched chicks. By 17 January it was apparent that some young chicks were being starved by parents and the forlorn whistling call of hungry chicks could be heard everywhere. Several young chicks simply disappeared. Two others were weighed over two days; both lost weight rapidly and starved on the nest. Other nests with eggs were abandoned, with the parents sitting off the eggs. In others, both parents were away from the territory together, an exceptional event for breeding skuas at this colony. The immediate research response to these observations was to confirm that the skuas could forage at sea normally. An observation log on 19 January recorded foraging times of 14, 39, 54, 70 and less than 130 minutes for skua pairs with well grown chicks. These are good foraging times. It is hard, therefore, to avoid the conclusion that some late nesting pairs were abandoning breeding because of the low probability of raising the chicks to fledging before winter. This exceptional incident suggests that egglaying, and especially re-laying can go on for too long in the summer; that the onset of the refractory period is too long delayed. The disproportionately high mortality of chicks recorded by Wood (1971) and Neilson (1983) through late summer storms indicates that any delay in breeding will heighten chick mortality. From his experience of late summer mortality of chicks Neilson concluded that early breeding 'is the most optimal strategy in adapting to the polar environment'.

Comparison of times needed for breeding with the time available suggests that there is in fact some room for adjustment of the timing of breeding cycles of both species within the habitable time span set by the

Table 12.2. *Comparative seasonal breeding cycles of penguins and skuas at Cape Bird*

	Adélie penguin	Skua
First arrive on colony	10–15 Oct.	Late Oct.
First occupy territories	10–15 Oct.	1 Nov.
All territories occupied	24 Nov.	Early Dec.
First eggs in season	4 Nov.	16 Nov.
Mean laying date (of 1st eggs)	14–16 Nov.	4–9 Dec.
Last eggs in season	5 Dec.	10–12 Jan.
First chick of season	6 Dec.	20 Dec.
Mean hatching date	15–18 Dec.	4–12 Jan.
Last chicks hatched	5 Jan.	10 Feb.
First chicks fledged	28 Jan.	15 Feb.
First chicks leave area	28 Jan.	Early Mar.
Last chicks leave area	20 Feb.	Late Mar.
Last adults recorded in region	?	3–6 Apr.

seasonal availability of open water. This conclusion remains valid even if breeding is not as efficiently prosecuted as the minimum times require – the mean times also leave space.

12.3.4 *The timing of the breeding cycles at Cape Bird within the habitable period*

The key points of the two cycles are now well established and are briefly summarised in Table 12.2. At all stages of the year the penguins are ahead of the skuas: they take up territories earlier, mean egglaying and hatching dates are three weeks earlier and chicks fledge and go to sea a month earlier. The times needed for incubation and chick fledging are similar, however, so that the essential differences are in the timing of the start of the season and in the lengthy post-fledging period of the skuas.

What evidence is there that the delay in skua breeding is consequential on the penguin cycle; that breeding is delayed to ensure the greatest benefit from the penguins? In theory how should this predator cycle be related to this prey (presumably so that egg formation, incubation and chick rearing can be best supported)? This linking can be indicated by bringing the two cycles together to ensure that the time the skuas need most food for their chicks (from about 20 to 40 days old), corresponds to the stage of the penguin cycle when most food is available, during the late guard and early post-guard stages. These times correspond to days 66–86 of the breeding cycle of Adélie penguin and to days 73–93 of the skua cycle, points based on the mean cycle durations shown in Table 12.1. For the best fit of the two cycles to give most food the skuas should begin breeding at the same time as the penguins each year. Delaying breeding would certainly be favoured for both egg formation and incubation in skuas, as penguin eggs and young chicks would then be abundant, but the more the delay the more difficult chick rearing becomes. Penguin chicks are mostly too strong to be killed by skuas after mid January, yet the natural skua cycle has still 6 to 8 weeks to run. On this colony, as found at Cape Royds during the earlier research, there was no carrion left at the end of summer for the skuas.

These same two species are associated together throughout an enormous geographical and latitudinal range (17 degrees of latitude, equivalent to 1700 km). Comparison of the way the cycles of the two species vary across this latitudinal range might usefully define the constraints on each species and illuminate further the way the two are associated. If it is assumed that the habitable period is longer towards the north than for Ross Sea populations then this would offer a wider span for each species' requirements and offer greater flexibility for the selection of different seasonalities. In the circumstances of predator dependence the synchrony between the cycles of the two should be constant for all habitats: in the absence of strong dependence the two could vary separately. Table 12.3 brings together the available information on how the two cycles interrelate in different localities.

This exercise proved to be much more difficult to carry out and much less useful than expected because in spite of the numerous studies that have been carried out on these birds, with long-term programmes on several colonies, the basic data on breeding dates are often lacking. Data do not seem to be available at all in the literature of the Adélie penguin for the Palmer Peninsula or for Cape Hallett, or for skuas at Cape Crozier. In other places only sparse information exists: comments on peak laying, on the start or end of laying, for example. Few places provide consistently uniform data on mean laying dates and variance. Some general and specific observations are nevertheless possible. First, the skua cycle everywhere follows that of the penguins. On Signy it lags six weeks behind; on Ross Island three to four weeks behind. On Signy, however, the brown skuas are more closely associated with the penguins and these lag behind by about three weeks. Where these two species breed sympatrically brown skuas breed ahead of south polar skuas.

Table 12.3. *The timing of breeding seasons, recorded as egglaying dates, of sympatric Adélie penguin and south polar and brown skua in different localities*

Adélie penguin	South polar skua	Brown skua
Signy Island South Orkney Islands (60°43′ S, 45°38′ W)		
mean 3 Nov. (1st eggs)[1]	mean 19 Dec.[2] range 5–13 Dec.[2]	first eggs 27/28 Nov.[2]
King George Island South Shetland Islands		
1. Fildes Peninsula (62°12′ S, 58°58′ W)		
	median 9 Dec./30 Nov.[3]	4/6 Dec.[3]
2. Point Thomas (62°18′ S, 58°30′ W)		
peak 3 Nov.[4] range 29 Oct.–19 Nov.[4]	mean 21 Dec.[15]	mean 2 Dec.[15]
Anvers Island, Antarctic Peninsula (64°46′ S, 64°03′ W)		
mean 3/14 Dec.[5]	30 Nov./5 Dec.[5]	
Windmill Islands, Vincennes Bay (66°15′ S, 110°32′ E)		
mean 21 Nov.[6]	peak 27/29 Nov.[7] range 21/26 Nov.–19 Dec.	
Pointe Géologie, Terre Adélie (66°40′ S, 140°01′ E)		
?	range 15 Nov.–10 Dec.[8]	
Ross Island		
1. Cape Bird (77°14′ S, 166°28′ E)		
mean 11/13 Nov. (1st eggs)[9] mean 17/20 Nov. (singles)[9]	mean 4/9 Dec.[10]	
2. Cape Crozier (77°27′ S, 169°14′ E)		
mean 18 Nov.[11]		
3. Cape Royds (77°36′ S, 166°10′ E)		
range 4 Nov.–4 Dec.[12]	median 4 Dec.[13] range 27 Nov.–27 Dec.[13] mean 27 Nov./7 Dec.[14] range 18 Nov.–22 Dec.[14]	

Because of the variability in the ways the cycles have been described by these authors it has not been possible to give the laying dates in a standard form. In the table means and medians, peak laying times and ranges are given at different places. An entry such as 27/29 means dates for more than one season.

The dates shown are from: [1] Lishman (1985b); [2] Hemmings (1984); [3] Peter *et al.*, 1990; [4] Trivelpiece and Volkman (1979); [5] Neilson (1983); [6] Penny (1968); [7] Eklund (1961); [8] Le Morvan *et al.* (1967); [9] Spurr (1975c); [10] Young (this report); [11] Ainley *et al.* (1983); [12] Taylor (1962a); [13] Young (1963a); [14] Spellerberg (1966); [15] Trivelpiece and Volkman (1979).

There is, therefore, little evidence in any region in this comparison for selection bringing the skua cycle into a more favourable relationship to that of the penguin in order to ensure greater food resources for chick rearing.

12.4   *Degree of synchronicity of breeding seasons*

An alternative for prey unable to avoid predation through hiding from the predator is to 'hide' within groups of conspecifics; to breed synchronously to protect vulnerable stages. On the other side a predator 'keyed' to the prey would also be expected to show synchronicity, to be able to target and exploit prey during its own most nutritionally demanding life-cycle stages. Comparisons of degree of synchronicity of breeding cycles for the two species across the breeding range might well be able to indicate the relative role of predation compared with other environmental factors. There are, however, as noted in the previous section, surprisingly few data for this comparison. Lishman (1985a) gives 3 November ±0.4 days, $n = 60$ (mean and SE) for 1980–81 and 3 November ±0.3 days, $n = 74$ for 1981–82 for the Adélie penguin at Signy, but these figures are confounded for considering their relation to skuas through competition there with chinstrap penguin.

On Ross Island Ainley *et al.* (1983) report a standard error for mean laying date of 4.9 days ($n = 1996$) and Spurr (1975c) for Cape Bird one of four days in each of two seasons for first eggs of clutches.

There are more data for skuas. Hemmings (1984) gives 3.3 days, $n = 13$ for Signy Island skuas. Neilson (1983) for skuas at the Palmer Peninsula records standard errors during four years of 1.83 days ($n = 23$); 1.42 days ($n = 22$); 1.55 days ($n = 42$); and 2.88 days ($n = 6$), respectively and Spellerberg (1966) for skuas at Cape Royds gives mean laying dates of 27 November ±5.5 days ($n = 62$), 30 November ±5.9 days ($n = 57$) and 7 December ±8.7 days ($n = 57$) for three years respectively. (In Spellerberg's text these errors are stated to be sample standard deviations but from inspection of the data it is more likely that they are in fact SEs.)

With the small amounts of information available it is prudent to conclude only that both species appear to breed synchronously throughout their ranges and that they show little if any clinal trend in their degree of synchronisation.

12.5   *Timing of the cycles in relation to universal proximate factors*

An alternative approach to assessing the independence of the two cycles is to relate breeding season directly to latitude. Clear trends in the timing of the cycles in relation to latitude would suggest a predominate role for a universal proximate factor, which is likely to be photoperiod (Gwinner, 1986). The absence of any clear trend, or of gross distortions in any trend, would, alternatively, suggest that local factors are more significant. Local factors would include possible interactions between predator and prey.

Young (1977) for skuas and Lishman (1985b) for the Adélie penguin have both demonstrated a general trend relating season and latitude, with populations at higher latitudes breeding later. In each species, however, breeding appeared to occur at high latitudes earlier than predicted from the trend lines and both authors suggested that this was because the summer there was too short to allow later breeding to occur. Later breeding would jeopardise the survival of young birds leaving the breeding area in autumn. From these analyses the two cycles do indeed appear independent. They offer little understanding, however, of the underlying causation. A better appreciation of this for skuas can be gained by re-plotting the dates of breeding against photoperiod rather than latitude itself (Fig. 12.2). It is usual to record photoperiod in these analyses at the

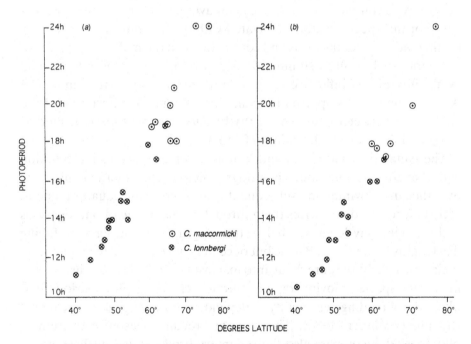

Fig. 12.2. Relationship of egglaying date to latitude and photoperiod in different populations of brown and south polar skuas. (*a*) Photoperiod at laying date (clutch initiation); (*b*) photoperiod at the start of egg formation.

date of clutch initiation (laying of first egg) but photoperiod would act as a cue well before this in promoting gonadal development. Thus, a second plot has been produced showing the photoperiod 15 days before clutch initiation. (There does not seem to be any published information on the time needed for skuas to produce eggs. The time given here is the same as quoted for large gulls by Roudybush *et al.*, 1979.) These plots reinforce the view that photoperiod, as a universal proximate factor, has a predominant role in the control of seasonality of both species. This is not unexpected: it is scarcely probable that an abstract phenomenon such as latitude could directly influence anything! Such plots using photoperiod are especially illuminating for the skuas for indicating a single trend line extending across the entire latitudinal range of skua populations in the Southern Hemisphere.

The finding of Astheimer and Grau (1985) that Adélie penguins begin developing eggs *before* coming ashore to breed markedly limits the range of local proximate factors influencing gonadal development for this species – and also strengthens the perception of the significance of photoperiod. Their Cape Crozier sample (from the same latitude on Ross Island as the colony at Cape Bird) came ashore on average 1–2 days before the yolk was complete, only 9.9 days on average before the first egg was laid. For this species factors associated with the breeding area itself (e.g. nesting, other breeding birds and consorting with the male of the pair) are not implicated at all in stimulating egg development. Nor is local food availability of any influence as these birds are fasting at this time. The Adélie penguin thus appears to exhibit an extreme manifestation of the interaction between endogenous rhythmicity and photoperiod, moderated possibly by age and condition factors.

The Antarctic skuas have a much more complex seasonal interrelation with changing photoperiod. Their migration to the Northern Hemisphere over the austral winter and subsequent return across the equator immediately before breeding carries them through a wide range of photoperiods and changing day lengths. Their return migration coincides with falling day lengths in the North Pacific but once they have crossed the equator on their return flight they are flying into increasingly long day lengths. In such migratory species moving rapidly through changing photoperiods it is likely that breeding seasonality is dependent on endogenous circannual rhythms (Gwinner, 1986). The plots of breeding season in relation to photoperiod encompass also the sedentary brown skua but these populations nevertheless demonstrate the same close link between photoperiod and breeding date. In contrast to the migratory south polar skuas these are more likely to be responsive to local photoperiodicity.

Local factors at the breeding areas moderating the proximate controls response are likely to be much more significant for skuas than for penguins as skua eggs are apparently produced after arrival on the territory. Nutrition, pair bonding and social stimulation of other conspecifics are possibly all implicated. Nutrition is clearly a significant factor. Breeding females are fed by males from their arrival on the territory until the eggs hatch. Pietz (1984) recorded that when foraging was disrupted by ice the skuas laid later on average but that pairs fed artificially at the research station bred even earlier than usual. A controlled experiment of artificially supplementing the diet of females has not yet apparently been attempted but all the evidence of the success and failure of breeding in this species indicates that the response to the overriding proximate environmental factors is conditioned by the strength of the pair bond, mediated through courtship feeding, and by local food availability.

The evidence adduced for the control of breeding season through endogenous rhythms and photoresponse adds further weight to the belief that the cycles of these species are not directly associated. The broad trend of breeding season with latitude for populations in strikingly different local conditions, and for different skua species, argues against such an association. It would seem from the trends disclosed that more general ultimate factors are being favoured; factors that apply apparently in all populations of both species.

The key to understanding the timing of these two cycles is that of the feeding and migration ecologies: when on average food is most readily and abundantly available and when needed by each species, how the differences in forms of movement, of flight compared with swimming and walking, affect both the migration on to the breeding grounds in spring and the departure of adults and especially of chicks in autumn, and how fasting provides the opportunity for penguins to begin breeding earlier. On these considerations the timing of the skua cycle seems likely to be independent of the penguin one: it looks much more certainly to be related to the seasonal asymmetry of open water for feeding.

12.6   *Conclusions on association*

At the conclusion of the earlier study of these two birds at Cape Royds Young (1963b) concluded that the skua cycle was not at all linked to the penguin cycle and that for much of chick rearing little food was in fact available to the skuas from the penguin colony. This conclusion was admittedly based on a single season's study of the two birds at a single small penguin colony. Moreover, it arose from the first demonstration that most skuas were dependent on fish rather than penguins and in the first flush of this new appreciation may have been overly dogmatic. How does this initial conclusion on the relationship between the breeding cycles now stand – in the light of this more detailed study, and in the light of a much greater knowledge of the two species throughout Antarctica?

In general, the initial conclusion seems still remarkably secure. The timing of the life-cycles of the two species has been confirmed, especially the later breeding of the skuas which carries their cycle well beyond that of the penguins into late summer, and the seasonal pattern of penguin food availability and need is now generally well understood.

13

Synthesis

'but under ordinary conditions and when penguins' eggs and chickens are not in season, they catch fish for themselves at sea or chase the petrels that have fed themselves till they disgorge.'

E. A. Wilson (1907).

'it is with the beak and habits developed during the winter at sea that the bird must manage at the rookery, and for this role it is poorly adapted.'

E. C. Young (1963b).

This book has provided an account of an association between two species sharing the same hostile Antarctic environment and nesting together in one of the small number of suitable breeding locations in the southern Ross Sea.

Both are marine species – the penguin feeding on zooplankton and fish, predominantly *Euphausia crystallorophias*, the skua feeding largely on the Antarctic silverfish (*Pleuragramma antarcticum*). The place of each species is thus securely located within the food web of the Ross Sea; it is

when they breed together that their ecosystem roles are at all contro-
versial, specifically in the degree the two species are linked together in a
prey–predator relationship.

The skuas certainly take eggs from defended nests and predate chicks.
These behaviours are so conspicuous that for many years they misled
observers into thinking that skuas were entirely dependent on the penguins for their food during breeding. They also gave skuas a poor
reputation: their inadequacy as a predator of chicks being construed as
cruelty, of prolonging the killing unnecessarily.

In hindsight it is surprising that this belief of the dependency on
penguins was sustained for so long; sustained even beyond the discovery of
breeding groups of skuas distant from the penguin colonies, and appreciation of the intense territorialism of breeding skuas. That a reappraisal of
the skuas' independence of penguins was not made earlier at Cape Royds –
with its proximity to expedition sites – points rather more to the undevel-
oped nature of behavioural biology at the time than to poor observation.
The failure hints also of disruption of natural conditions. It is possible that
resident skuas were so harassed by the visitors that breeding was precluded
and that other skuas attracted to food waste and rubbish masked the small
size of the resident group and overwhelmed its attempts to defend the
penguin colony.

The focus of this present study was to describe the foraging and breeding
behaviour of the skuas at the Northern Colony at Cape Bird and to
evaluate their impact on the penguins. Equal emphasis had to be given as
well to the prey species – to measure the food resource on the colony
throughout the season and to judge to what extent breeding mortality
reflected skua attention. Nevertheless, by inclination this was primarily a
study of the feeding ecology of skuas.

The methods used in the research in this habitat were necessarily very
simple ones and almost all results of value came from direct observation of
skuas living and breeding among the penguins. Most of these were made
from a distance to give minimum disturbance, in an attempt to ensure
validity, and to allow observation of a large area of the colony at the one
time. Although the gross effects of disturbance on penguin and skua
behaviour had been recognised by this date the insidious effect of low-key
impact, as recently described for penguins (Culik *et al.*, 1990) was certainly
underestimated. There are also a number of reports of disruption to
colonies by research and other disturbance (Ainley *et al.*, 1983; Wilson *et
al.*, 1990; Young, 1990b).

It is not known to what degree skua breeding, skua foraging patterns or skua occupation was affected by the work. In any similar research it would clearly be prudent to take even more care; to use hides for all observations even from a distance and to keep well away from study breeding groups of penguins.

The single exceptions each year to a programme based entirely on direct observation were the measurements of prey size and the exclosure experiments to record breeding in the absence of skua predation and disturbance. These experiments gave important confirmation of the small impact that skuas seemed to be having on the overall breeding success at this colony.

There were five other Adélie penguin colonies on Ross Island: two others at Cape Bird, one at Cape Royds and two at Cape Crozier. The ones at Cape Crozier were also being researched at this time and the studies at Cape Bird depended on the work there for information on skua and penguin demography and age-dependent breeding, survival, and the amount of skua and penguin movement among these breeding areas. Because of this research at Cape Crozier no attempt was made to establish known-age research populations for either penguins or skuas at Cape Bird. Skua chicks were banded each year and enough adults were banded to allow their certain identification during the observation logs, and to be certain of breeding pair continuity in the study territories. Some skuas had been banded the previous year by Robert Wood as part of the Cape Crozier study, to check on breeding site fidelity, and these were an invaluable aid in interpreting breeding statistics. Penguin chicks were not banded.

Essentially, the study was of the performance of individual skuas, pairs of skuas and groups of skuas under the range of conditions occurring at this site. Interpretation depended a great deal on comparative results. From the outset comparisons targeted two groups of skuas. One of these was on H block, with about 30 pairs; 14 of these had territories with penguins, but these were small territories with few penguins. The other group was on EF Block. This comprised a smaller number of skua pairs in larger territories, each running from the nest down the slope into the colony. On average this group had much higher penguin numbers within the individual territories.

The Southern and Middle colonies at Cape Bird were little visited; the Southern Colony just once each year for a single day in early December to count the breeding pairs. The decision to retain this colony as a control against which to measure research disturbance at the Northern Colony

Table 13.1. *Numbers of nests in early December and proportional change from year to year in each of the colonies at Cape Bird during the term of this research (1965–66 to 1969–70)*

Year	Northern Colony	Middle Colony	Southern Colony	Total for Cape Bird
1965–66	25 056	—	—	—
1966–67	25 790[1] 102%	1419 —	10 577 —	37 786 —
1967–68	21 560 83.6%	1183 83.4%	8865 83.4%	31 608 83.6%
1968–69	15 067 69.9%	835 70.1%	5046 56.9%	20 948 66.3%
1969–70	19 768 131%	1260 151%	8237 165%	29 265 139%

[1] Estimated from comparative counts of study colonies in each year.

proved valuable. During the first four years of the study the numbers of penguins on the Northern Colony fell steeply, a fall exactly paralleled in the other two (Table 13.1). Without this evidence of a *regional* decline, it would have been difficult to reject criticism of the research methods and their impact on penguin breeding. In the final year of the study numbers of penguins breeding in all three colonies increased significantly, and this increase has continued until very recently (K.-J. Wilson, 1990, Young, 1990b). It is not known what drives the long-term fluctuations in penguin breeding numbers in this area. However, the reason for the low numbers of breeders on the colonies in 1968–69 was clearly evident. The broken ice cover of the early season, even though not quite forming the 'mountain range' described by Lishman (1985b) for the ice at Signy Island in 1980, still caused very great difficulties for penguin movement and access. Fewer penguins reached the breeding colony at the start of the season and many subsequently lost their eggs through failure of partners to return after foraging breaks. The significance of pack-ice density in relation to penguin breeding success was identified by Ainley and LeResche (1973) from their observations at Cape Crozier. The breeding failure at Cape Bird identified another significant feature: its surface topography is also important. It is worth noting that the skuas were little affected by these ice conditions and their breeding success at Cape Bird in this year was by far the best of the five years monitored.

The results for this programme fall into a number of areas. The first achievement was in establishing the site as a long-term research base. From the very first reconnoitre it was immediately apparent that the Northern Colony offered a uniquely favourable location for research on skua and penguin behaviour and ecology. In part this came from its isolation and protection from interference by visitors and Antarctic operations and in part from its topography providing numerous small breeding groups of penguins and skuas for experimental work and breeding group comparison. It also contained many excellent viewing sites for direct observation of skua and penguin activity. The maps of the area and the breeding groups within each penguin colony produced by R. H. Blezard during the first two years have supported all work subsequently. Without them it would have been much more difficult to maintain the continuity of observation and monitoring that has been one of the hallmarks of this research area.

The second achievement was that for five years the research team was able to estimate numbers of breeding penguins and the amount of food potentially available for the skuas through each season. This was the initial step in determining whether this was enough food for the resident skuas. The assessment of sufficiency came from estimating the amounts taken by the skuas and lost from the colony, and through direct observation of the behaviour of skuas for evidence of hunger and for foraging at sea, especially of pairs of skuas with known records of predation and scavenging. But two sorts of records had primacy in this evaluation. The first comprised observations of skuas on large territories containing many penguins having to forage at sea, implying that they were unable to gain food on the colony at the time. There were numerous such observations. The second was of the almost universal failure year after year of any pair to raise both chicks. At this colony, as at Cape Crozier and Cape Royds, the second chick is driven from the nest area by the first and is almost always lost. This sibling aggression appears to be an adaptive response to hunger. The most telling point for assessing food sufficiency here is that this response generally takes place during the period of peak food availability on the penguin colony, during early to mid January in the early post-guard stage. On this criterion it appears that food is never abundant.

The account of the biomass on the colony has drawn two clear distinctions: between the numbers of eggs and chicks and their food volume at the different stages of the season; and between this volume and the amount actually available to the skuas. The first distinction recognises the difference between numbers of prey and the changing volume of each prey type during the season. Although peak numbers of prey items occur during the penguin incubation stage, the food biomass then is relatively small compared with that of the chick stage later in the season. In the second distinction (which compares food present with food available to skuas) the discrepancy is most marked towards the end of the penguin cycle when the growing chicks are too strong and agile to be easily mastered by skuas, and are essentially immune to skua attack, although in some special circumstances skua attacks are favoured and may be successful, for example, at breeding groups on steep slopes where the chicks can be tumbled away down the hill or when chicks are caught in the open well away from adult penguins. Overall, however, there was little food for the skuas at this colony from mid January. Moreover, there was no food left over from earlier for scavenging either, as has often been proposed for these large colonies.

The third area of achievement has been in the description of the foraging behaviour of the skuas on the colony through the season. This research disclosed a wide variability in the interest of the skuas in the penguins, ranging from indifference to intense and sustained predation. In contrast to the early accounts of the ferocity of the skua as a predator on penguin eggs and chicks only two birds of the study groups, one each on H and EF blocks, could be categorised as effective and committed predators; and even these two gained much of their food from the colony by scavenging. At times both these birds could gain food at will. No others were anywhere near as effective.

The skuas at Cape Bird had several food sources so that decision rules about where to forage and how long foraging in each place should be sustained were required throughout the season. Attempts to understand how the decision rules could act for these birds at each point (summarised in Fig. 1.3) were frustrated by such obvious factors as being unable to follow the birds in foraging at sea, but more significantly through being unable to assess the nutritional state of foraging birds, in case a simple hunger 'switch' operated to terminate foraging bouts at one place and send the birds to another, or to see the signals or motivational indicators that birds might show at each decision point. It was clear nevertheless that for many of the birds on territories containing penguins foraging at the colony was preferred to foraging at sea. On the Antarctic Peninsula and Scotia Arc islands, however, this species is displaced from the penguin colonies by 'the larger and more aggressive' brown skua (Neilson, 1983) and is entirely dependent on marine foraging (Trivelpiece and Volkman, 1982; Neilson, 1983; Pietz, 1987).

The conclusion from this study is that these skuas are not especially adept at predation on penguins. They do not attempt to attack penguin adults and there are few successful attacks on mature chicks. The features that facilitate predation are agility and strength, single-mindedness and risk taking. These, however, produce success only because of the lack of attention by the penguins, so that eggs and chicks are exposed through disturbance among the breeding pairs, failure in incubation or chick-guarding routines, or through inexperience. Well nourished and experienced penguins within the shelter of a breeding group have little to fear from the skuas foraging among them: few skuas risk direct attacks on these penguins.

The difficulty of separating out the roles of skua disturbance and predation from other factors in penguin egg and chick mortality was tackled in this study primarily through the exclosure experiments. This is really the only way that the impacts of skua scavenging and attack behaviour could be isolated from the environmental factors of weather and sea-ice cover and from individual and group penguin behaviour. To date these appear to be the only experiments to have been done on this question. The results were of great interest in showing first, the levels of nesting failure that may occur in this penguin in the absence of any skua effect, and second, the small effect that skuas may have. In neither year was survival significantly enhanced in the protected groups compared with control groups in the centre of the colony. In the first year the protected groups had higher survival than those on the edge of the colony, but so did the central control groups. In the second year, following the initially poor breeding conditions with higher than usual loss of eggs, survival rates were similar in all three sets of breeding groups. The overall conclusion from these experiments is that over much of the colony skuas do not have a major impact on penguin breeding. For most breeding groups it is exceeded by the activities of the penguins themselves. That is, it is hard to demonstrate from these experiments that the skuas are in general strong predators on the penguins. The evidence suggests on the contrary that most of their food is taken by scavenging, and that most of this comes not from their activities disturbing the penguins (which was also excluded in the protected groups) but from the effect of penguin and environmental factors. In some places on the colony, however, there was undeniable predator impact with very high egg and chick losses. Invariably, these high impacts occurred in smaller breeding groups on the edge of the colony where the ratio of skua to penguin directs the predator on to just a few breeding penguins. The levels even here are quite variable and depend on the aptitude of the predators, on the interest taken in the penguins and on the skill and tenacity brought in tackling such large prey.

What is certainly clear from the study is that the colonial nesting habit and defensive activities protect the penguins from far greater impact. Without these interdependent activities even moderately aggressive skuas would have little difficulty taking sufficient eggs and chicks from the colony to support their own breeding for a major part of the season. Their prevalence ensures that all skuas must work assiduously to gain enough food even under the most favourable circumstances. The amount of food taken up by the skuas is nevertheless impressive and makes an important contribution to the diet of skuas nesting there.

To what extent are skuas with penguins within the territory advantaged? In years with extreme sea-ice conditions, as described by Parmelee *et al.* (1978) for the skuas at Anvers Island, or in periods of continuous fog or high winds, only those pairs with this alternative local food resource may be able to breed at all. Under less extreme conditions the advantages are less obvious. In this study it was not possible to demonstrate general advantage gained from nesting with penguins in early breeding date or in egg and chick survival, although at least one group with high average penguin numbers did breed significantly earlier. Other authors have been less equivocal. Trillmich (1978) reported that skuas at Cape Hallett with feeding territories 'began to breed about 8–10 days earlier than the other pairs'. It is difficult to evaluate this assertion in the absence of data and in the knowledge that all skuas nesting near this colony were able to gain penguin food there. Trivelpiece and Volkman (1982) also showed that brown skua pairs on 'optimal' territories at Point Thomas, King George Island bred earlier, but again the groups compared were not distinguished clearly by diet and both were able to feed on the penguins. Each study equated advantage with earlier breeding. As has been seen in the comparisons of food availability and requirement in this study, summarised in Fig. 13.1, earlier breeding should confer advantage by providing penguin food for a greater proportion of the chick stage.

The longer-term impact of each species on the other is not easily judged without a history of the association or of comparisons of penguins breeding in the absence of skuas. Even at Cape Bird such assessments are frustrated by the lack of detailed histories of each skua pair since its first encounters with penguins. It is not known, for example, whether the variation in abilities and interest shown by the skuas is from aptitude or experience, whether those ignoring them have been injured earlier, or are young birds, or whether the skills needed are learned and developed over a long

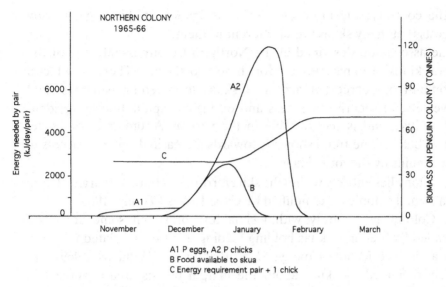

Fig. 13.1. The food available to skuas on the Northern Colony in 1965–66 and the relative amounts needed by a breeding skua pair through the season.

duration. What evidence there is suggests that the latter is most likely to be the true situation. The long-term impact on penguins is much clearer. For this species there has been selection 'for resistance to predation' (R. J. Taylor, 1984). Without the specific behaviours of colonial nesting and defence and anti-predator behaviour, egg and chick loss through predation would be much higher.

The interaction between the two species is not, however, entirely one way. Skua eggs and chicks are lost indirectly because of this interaction. Skuas at the colony lose eggs and chicks through negligence, and, as do pairs in any breeding colony, to conspecific predation, but the level is probably significantly higher than in areas away from the penguins, which attract fewer skuas into the area, and which are less often distracted from egg and chick care. Eggs are also lost directly through penguin attacks on nesting skuas. It is indeed arguable that a higher *proportion* of skua eggs are lost to penguins than penguin eggs are lost to skuas. But even so it would be a false conclusion to consider the penguins predators on skuas!

The broad conclusions of the study appear to confirm those of the earlier one at Cape Royds which concluded 'it therefore seems possible that proximity in nesting is to some extent fortuitous, and merely due to a shared habitat preference' (Young, 1963b). This conclusion can now be affirmed perhaps with a little more assurance. No further evidence has been brought forward to demonstrate a total dependence on the penguins, on the contrary, the more detailed investigation at this larger colony suggests that many skuas are indifferent to them.

The association described at the Northern Colony is only one of the several kinds recognisable now for these two species. There is in fact a continuum of degrees of access of skuas to breeding penguins, and conversely, of security of access and exclusive exploitation by resident birds. This trend is recognisable in the various Antarctic colonies described below. The trend shown is towards increasingly general access to the penguins by the local skuas.

1. Colony lies entirely within the skua territories about its margin. These are all small colonies, exemplified by Cape Royds (Young, 1963b).

2. Colony lies entirely within skua territories but some areas are defended by breeding skuas holding feeding territories isolated from the nest area. The Middle Colony, Cape Bird (J. M. Williams, 1969) was largely defended by skuas about the margin but its area contained in addition a single feeding territory.

3. Colony lies within the skua territories about the margin with additional breeding territories entirely located within the colony centre. These are also likely to have feeding territories defended by pairs breeding outside the colony limits. These are larger colonies with breeding groups sufficiently widely spaced that skuas may breed among the penguins of the colony centre. The Northern Colony at Cape Bird (this study) falls within this category, as does the Point Thomas colony (Trivelpiece *et al.*, 1980; Trivelpiece and Volkman, 1982). There were no separate feeding territories at the Northern Colony.

4. Colony lies partly within territories of skuas nesting about the margin or within the colony but parts are undefended (i.e. fall outside any skua breeding or feeding territory) and are thus accessible to any skuas of the area. Cape Hallett (Trillmich, 1978) and the Cape Crozier West Colony (Müller-Schwarze and Müller-Schwarze, 1977; Young, personal observation, 1989) fall within this category.

(It is possible that the open area described by Trillmich at Cape Hallett was artificial, resulting from the disruption caused by the adjacent research station at that time. The open area at Cape Crozier on the contrary appears natural: the penguin breeding groups in the colony centre are too closely packed to allow skua breeding among them and the colony area is too great for territories on the margin to extend very far into its interior. In the absence of breeding territories in the colony centre it appears inevitable that large tracts of this large colony will be generally accessible to local skuas.)

These different associations have significance for both penguins and skuas.

Colony areas defended by skuas, either in breeding–feeding territories or feeding territories, can generally be exploited only by the defending pair. Although eggs and small chicks might be taken from these areas by outside birds this will comprise only a small proportion of losses. Intruders cannot forage on the ground in defended areas so that larger chicks are immune to attacks from these birds. The overall effect of this exclusive ownership means that breeding groups will be differently affected by skua predation depending on the interest and ability of the skua pair of their area and on the numbers of breeding pairs within its territory. Penguins within the territory of an aggressive pair working hard to obtain food from a few breeding penguins will inevitably be severely affected by skua predation and disturbance. At the opposite extreme penguins within a large territory of a pair that had little interest in foraging for penguin food would suffer little from this pair, while at the same time being protected from intruders.

It is not yet clear whether penguins breed more successfully on average in defended or undefended colony situations. The extreme contrast is between hordes of skuas flocking about the breeding group, harassing the penguins in turn from every side, compared with sheltered protection within the territory of a benign pair. Preliminary data from Cape Crozier (Müller-Schwarze and Müller-Schwarze, 1977) suggest the contrast is less extreme than this, with the breeding groups in defended and open areas recording closely similar success. The large numbers of penguins moving about the central areas must surely hamper skuas attacking penguins there, while the high density of groups reduces the chances of skuas being able to overcome chicks before their escape back into a crèche.

These conclusions from Cape Bird on the exclusive utilisation of breeding groups by territorial skuas are supported to a degree by research at other colonies. Trillmich (1978), for example, specifically notes that intruders were unable to kill chicks in any of the defended areas at Cape Hallett. Similarly, Müller-Schwarze and Müller-Schwarze (1977) drew a clear distinction between the behaviour of skuas and the amounts of food taken in the defended areas about the margin at Cape Crozier colony and in the colony centre. The position at Point Thomas (Trivelpiece and Volkman, 1982) is less certain. Although the breeding groups are defended by skua pairs, intruders seem nevertheless to be able to gain food there and did not appear to forage at sea, the alternative food source.

The form of the association between the penguins and skuas at a colony has significance also for non-breeding skuas and skuas breeding on territories away from the colony.

1. Where the colony lies entirely within breeding territories these skuas are denied access. The exceptions to this general rule are foraging flights for scavenging and egg predation, access through establishing feeding territories, or access through a non-breeding club.

2. In larger colonies where there are undefended areas, occurring naturally because of breeding group concentration or where the natural association has been disrupted by disturbance or the persecution of skuas, all birds have access. These colonies can support a much higher proportion of breeding skuas than can defended ones and may serve to attract larger numbers of non-breeders.

It is not known how skuas behave when breeding in association with other bird species in Antarctica. Skuas have now been found with petrels on coastal and inland sites in many places, e.g. on the Theron Mts (Brook and Beck, 1972); Rockefeller Mts (Broady *et al.*, 1989); Muhlig-Hofmann Mts (Howard, 1991); Hop Island (Norman and Ward, 1990). Some are far inland so that unless the petrels can provide all the food needed by the breeding skuas long flights to the sea would need to be undertaken. These sites offer fascinating opportunities for research.

The Northern Colony at Cape Bird has proved to be an ideal natural laboratory for behavioural and ecological studies of this sort. Its advantages have now been well established. It does not seem prone to natural disasters; it is blessed with a benign climate; has a wide range of sizes of its penguin breeding groups, many of which are quite isolated; and it has good numbers of both skuas and penguins. Of high importance in this assessment is that it offers excellent conditions for observational research. There are numerous points from which to observe the penguins and skuas from a distance, contrasting markedly with the situation at say the Southern Colony or at Cape Royds where no such places exist. It was through the availability of such observation points that this study could be carried out under close to natural undisturbed conditions of the skuas breeding around and within the colony. It is important to ensure that such favourable research areas are protected so that observational studies of this kind can be continued.

Numbers of eggs and chicks on the Northern Colony and the estimated biomass at different stages of the penguin breeding season

1965–66

Date	Egg-chick/nest	SE	Numbers				Biomass(kg)			
			Egg	Chick	N (total)	SE$_N$[1]	Egg[2]	Chick[3]	Total	SE$_{NW}$[4]
5.11			4 500		4 500		468		468	
10.11			15 000		15 000		1 560		1 560	
15.11			35 000		35 000		3 640		3 640	
20.11	1.899	0.034	47 614		47 614	847	4 951		4 951	107
27.11	1.865	0.026	46 758		46 758	644	4 862		4 862	90
4.12	1.832	0.017	45 926		45 926	426	4 776		4 776	74
11.12[5]	1.778	0.020	40 543	5 022	45 565	501	4 212	723	4 935	117
18.12	1.652	0.028	13 161	28 246	41 407	707	1 368	6 524	7 892	611
24.12	1.505	0.037	3 802	33 916	37 718	925	395	18 855	19 250	1 571
1.1.	1.423	0.041	1 396	34 271	35 667	1 030	145	33 517	33 662	3 139
8.1	1.356	0.046		33 100	33 100	1 168		59 391	59 391	3 128
16.1	1.309	0.048		32 073	32 073	1 206		90 510	90 510	4 137
24.1				31 110	31 110			118 284	118 284	
1.2				27 262	27 262			103 650	103 650	
2.2				16 678	16 678			63 409	63 409	
5.2				10 170	10 170			38 666	38 666	
9.2				984	984			3 741	3 741	
13.2				250	250			950	950	

1966–67

Date	Egg–chick/nest	SE	Numbers				Biomass(kg)			
			Egg	Chick	N (total)	SE$_N$[1]	Egg[2]	Chick[3]	Total	SE$_{NW}$[4]
5.11			1 000		1 000		104		104	
10.11			7 500		7 500		780		780	
16.11	1.133	0.062	29 150		29 150	1 656	3 031		3 031	176
20.11	1.649	0.040	47 420		42 424	1 216	4 412		4 412	138
25.11	1.812	0.037	46 625		46 625	1 192	4 849		4 849	137
1.12	1.805	0.031	46 444		46 444	1 063	4 830		4 830	126
10.12[5]	1.765	0.025	45 400	208	45 608	949	4 699	18	4 717	115
15.12	1.720	0.029	36 656	8 597	45 253	1 008	3 708	1 238	4 946	204
20.12	1.627	0.039	13 566	28 284	41 850	1 187	1 410	6 633	8 043	665
25.12	1.447	0.060	3 389	33 848	37 237	1 654	352	18 819	19 171	1 463
30.12	1.363	0.059	418	34 650	35 068	1 614	43	33 887	33 930	2 528
4.1	1.292	0.049	53	33 189	33 242	1 368		53 305	52 305	3 507
14.1				32 300	32 300			86 079	86 079	
21.1	1.213	0.052		31 623	31 623	1 432		88 576	88 576	4 233
24.1				30 267	30 267			108 870	108 870	
30.1				26 523	26 523			92 538	92 538	
2.2				16 226	16 226			56 642	56 624	
5.2				6 240	6 240			21 666	21 666	
9.2				936	936			3 265	3 265	

1967–68

Date	Egg–chick/nest	SE	Numbers				Biomass(kg)			
			Egg	Chick	N (total)	SE$_N$[1]	Egg[2]	Chick[3]	Total	SE$_{NW}$[4]
5.11			6?		6					
10.11			6000		6000		624		624	
13.11	0.776	0.073	16726		16726	1582	1739		1739	166
17.11	1.280	0.751	27598		27598	1614	2870		2870	171
21.11	1.632	0.060	35293		35293	1296	3670		3670	142
26.11	1.799	0.054	38790		38790	1155	4034		4034	130
1.12	1.844	0.041	39748		39748	875	4134		4134	104
6.12[5]	1.836	0.035	39582		39582	754	4116		4116	94
11.12	1.789	0.031	37794	790	38584	670	3930	75	4005	132
16.12	1.753	0.033	27095	10695	37790	713	2817	1540	4357	150
21.12	1.647	0.038	12112	23390	35502	812	1259	4561	5820	435
26.12	1.540	0.044	2456	30757	33213	949	255	12087	12342	1072
31.12	1.436	0.047	418	30548	30966	1006	43	21322	21365	1512
5.1	1.396	0.049		30090	30090	1074		41014	41014	2689
10.1	1.374	0.052		29634	29634	1125		54935	54935	3041
20.1	1.332	0.065		28714	28714	1400		94555	94555	5257
25.1				28139	28139			97867	97867	
30.1				24407	24407			84887	84887	
2.2				17228	17228			59918	59918	
10.2				1148	1148			3992	3992	
15.2				0	0					

1968–69

Date	Egg–chick/nest	SE	Numbers				Biomass(kg)			
			Egg	Chick	N (total)	SE$_N$[1]	Egg[2]	Chick[3]	Total	SE$_{NW}$[4]
5.11			600		600		62		62	
10.11			3 500		3 500		364		364	
15.11	0.938	0.087	15 347		15 347	1 427	1 596		1 596	149
22.11	1.562	0.062	25 551		25 551	1 014	2 657		2 657	110
28.11	1.655	0.040	27 069		27 069	660	2 815		2 815	77
6.12[5]	1.536	0.042	25 130		25 130	697	2 613		2 613	79
13.12	1.472	0.047	22 813	1 262	24 075	772	2 372	181	2 553	86
18.12	1.412	0.048	13 822	9 284	23 106	786	1 437	1 847	3 284	258
24.12	1.276	0.039	3 654	17 216	20 870	646	380	5 336	5 716	
30.12	1.136	0.039	793	17 802	18 595	636	82	9 542	9 624	1 025
4.1	1.046	0.044	264	16 855	17 119	713	21	18 945	18 966	1 542
10.1	1.013	0.046		16 360	16 360	757		28 597	28 597	2 006
14.1	0.989	0.047		16 044	16 044	769		48 565	48 565	2 649
25.1				15 723	15 723			49 496	49 496	2 724
30.1				13 637	13 637			47 524	47 524	
3.2				9 626	9 626			33 546	33 546	
5.2				3 208	3 208			11 179	11 179	
10.2				641	641			2 234	2 234	

1969–70

Date	Egg–chick/nest	SE	Numbers			SE$_N$[1]	Biomass(kg)			SE$_{NW}$[4]
			Egg	Chick	N (total)		Egg[2]	Chick[3]	Total	
5.11			1 000		1 000		104		104	
10.11			11 000		11 000		1 144		1 144	
18.11	1.657	0.048	32 807		32 807	954	3 412		3 412	108
24.11	1.820	0.033	36 048		36 048	660	3 748		3 748	83
4.12[5]	1.790	0.024	35 455		35 455	483	3 687		3 687	68
15.12	1.704	0.038	22 056	11 699	33 755	759	2 294	1 684	3 978	180
19.12				19 516		612	75			
23.12	1.539	0.038	3 162	27 313	30 475	756				
27.12			1 490	28 000	29 490	656	145	18 714	18 859	1 119
30.12	1.458	0.032	553	28 301	28 854	639	57	27 678	27 735	
4.1				28 000	28 000			43 820	43 820	2 348
17.1	1.389	0.032		27 510	27 510	733		87 784	87 784	3 288
24.1				26 959	26 959			94 059	94 059	
30.1				23 383	23 383			81 583	81 583	
4.2				16 506	16 506			57 589	57 589	
10.2				1 100	1 100			3 838	3 838	

Notes

[1] Numbers of eggs and chicks. Standard errors could only be estimated for sampling dates, in each year from about mid November to mid January.

[2] The mean egg mass and standard error of the mean used in all the biomass estimates was 0.103 ± 0.00135 kg. This mass was calculated from the data given in Astheimer and Grau (1985) table 2 and is the mean of first and second eggs of clutches minus the egg shell mass.

[3] Mean mass of chicks on the colony on each date was estimated from the weights of samples of chicks on breeding groups (Table 4.1). Missing weights were obtained from chick growth curves.

[4] The standard errors for the biomass estimates were obtained from the expression:

$$\mathrm{SE}_{NW} = \sqrt{(N^2 \times \mathrm{SE}^2 + x^2 \times \mathrm{SE}_N^2 + \mathrm{SE}_N^2 \times \mathrm{SE}^2)}$$

where N and SE_N are the estimated numbers of chicks and their standard errors and x and SE are the mean mass and standard errors of the weighed chicks at the same date.

[5] The reference date tying in the colony-wide nest count with the productivity and survival data from the study colonies.

Mean numbers of eggs and chicks taken each day by the pairs on EF and H Blocks during the different stages of the penguin life-cycle

1. Incubation stage – determined by direct observation of foraging skuas

Pair	1967–68	1968–69	1969–70
EF Block			
144	0.89	2.53	0
145	0	3.78	2.08
146	3.56	—	—
148	2.67	—	—
146 new pair	—	0	2.08
149	0	1.26	0
152	1.78	6.32	4.17
153	3.56	7.58	10.4

2. Guard stage – determined by direct observation of foraging skuas

Pair	1966–67	1967–68	1968–69	1969–70
EF Block				
144	4.36	0.84	2.25	0.61
145	1.31	0	3.75	0.61
146	6.98	7.58	—	—
148	5.67	5.05	—	—
146 new pair	—	—	0	0
149	1.75	1.68	0.75	0.61
152	2.62	4.21	2.25	2.46
153	6.11	4.11	6.50	3.08

2. Guard stage (cont.)

Pair	1966–67	1967–68	1968–69	1969–70
H Block				
8	0	0	0	—
9	0	0	0	—
11	4.2	3.4	2.7	—
12	0	0	0	—
14	0	0	0	—
19	1.4	0	0	—
20	0	0	0	—
21	0	0	0	—
22	0	0	0	—
24	1.4	0	0	—
25	1.4	2.8	9.5	—
26	0	0	0	—
27	5.6	1.4	0	—

3. Early post-guard stage – determined from prey remains

Pair	1965–66	1966–67	1967–68	1968–69	1969–70
EF Block					
144	0.57	0.43	0.29	0.25	0.43
145	0.29	0.29	0.14	0	0.57
146	0.86	0.43	0.29	—	—
148	0.86	0.57	0.43	—	—
146 new pair	—	0.25	0.14	—	—
149	0.71	0.43	0.14	0	0.43
151	0.29	0.57	0.29	0.50	0.43
152	1.28	0.57	0.57	0.25	0.52
153	1.86	1.29	0.71	0.25	1.00
H Block					
8	0.57	0.28	0.29	0.25	0
9	1.29	0	0.14	0.25	0.33
11	0.43	0.28	0.57	0.50	0.33
12	0.86	0.71	0	0	0
13	0.71	0.71	0.28	0.25	0.33
14	0	0.14	0	0	0.33
19	1.00	0.71	0.29	0.25	0.67
20	0.14	0	0.29	0.25	0.33
21	0.14	0.43	0	0	0
22	0.14	0.43	0.43	0.25	0.67
24	0.43	0.28	0.14	0.25	0.33
25	0.29	0.86	0.43	0.25	0
26	0.57	0	0.14	0	0.33
27	0.29	0.28	0.14	0	0.33

4. Late post-guard stage – determined from prey remains

Pair	1965–66	1966–67	1967–68	1968–69	1969–70
EF Block					
144	0.6	0.3	0.2	0.0	0.1
145	0.3	0.3	0.0	0.0	0.1
146	1.2	0.9	0.3	—	—
148	0.5	0.4	0.2	—	—
146 new pair	—	—	0.2	0.0	—
149	0.2	0.5	0.0	0.0	0.0
151	0.5	0.3	0.2	0.1	0.1
152	0.6	1.5	0.3	0.5	0.1
153	1.2	0.7	0.6	0.7	0.3
H Block					
8	0.4	0.0	0.2	0.0	0.0
9	0.3	0.7	0.2	0.0	0.0
11	0.5	0.6	0.2	0.0	0.1
12	0.2	0.5	0.1	0.0	0.0
13	0.7	0.6	0.3	0.5	0.2
14	0.0	0.2	0.0	0.0	0.0
19	0.3	0.2	0.0	0.0	0.0
20	0.1	0.2	0.1	0.0	0.0
21	0.3	0.2	0.2	0.0	0.0
22	0.5	0.4	0.2	0.0	0.2
24	0.3	0.3	0.2	0.0	0.1
25	0.0	0.5	0.3	0.6	0.4
26	0.4	0.1	0.0	0.0	0.0
27	0.3	1.2	0.1	0.0	0.1

APPENDIX 3

Estimated daily energy intake (kJ day^{-1}) for each skua pair on H and EF Blocks

Appendix 3.1. *From penguin eggs during the second half of the penguin incubation stage (21 November to 8 December)*

Pair	1967–68	1968–69	1969–70
EF Block			
144	480	1366	0
145	0	2041	1128
146	1922	—	—
148	1442	—	—
146 new pair	—	0	1128
149	0	680	0
152	961	3412	2252
153	1922	4093	5632
H Block			
8	0	0	—
9	0	2160	—
11	2808	1620	—
12	0	0	—
13	0	1080	—
14	702	0	—
19	0	0	—
20	0	0	—
21	0	0	—
22	0	0	—
24	0	540	—
25	702	2160	—
26	0	0	—
27	0	1080	—

Compiled from numbers of eggs taken on average over this period by each pair converted to energy content in kJ using the value 540 kJ egg^{-1}.

Appendix 3.2. *From penguin eggs and chicks during the chick guard stage*

Pair	8–14 Dec.	15–19 Dec.	20–24 Dec.	25–29 Dec.	30–3 Jan.
EF Block					
1966–67					
144	1570	2155	3728	9903	19 318
145	620	686	862	1554	2542
146	3072	3592	4988	10 447	18 594
148	2189	2893	4583	11 441	21 828
149	766	895	1243	2612	4635
152	1302	1563	2263	5015	8520
153	2940	3210	3900	6662	10 521
1967–68					
144	741	866	1203	2523	4486
145	0	0	0	0	0
146	2746	3746	6444	17 049	33 187
148	2392	2642	3316	5955	9745
149	573	823	1497	4136	8199
152	1937	2187	2861	5500	9358
153	1600	2102	3449	8747	16 785
1968–69					
144	1065	1177	1477	2656	4340
145	1725	1948	2548	4908	8335
146 new pair	0	0	0	0	0
149	405	405	405	405	344
152	1320	1544	2143	4503	7992
153	3090	3202	3502	4681	6060
1969–70					
144	209	301	547	1514	2995
145	209	301	547	1514	2995
146 new pair	0	0	0	0	0
149	332	332	332	332	282
152	1084	1267	1759	3696	6558
153	1293	1568	2306	5210	9553
H Block					
1966–67					
8	0	0	0	0	0
9	960	1380	2505	6945	13 732
11	1520	1520	1520	1520	1292
12	0	0	0	0	0
13	960	1380	2505	6945	13 732
14	0	0	0	0	0
19	480	690	1255	3470	6064
20	0	0	0	0	0
21	0	0	0	0	0
22	0	0	0	0	0
24	760	760	760	760	645
25	480	690	1255	3470	6064
26	0	0	0	0	0
27	1720	2140	3265	7705	14 377

Appendix 3.2. (*cont.*)

Pair	8–14 Dec.	15–19 Dec.	20–24 Dec.	25–29 Dec.	30–3 Jan.
H Block (*cont.*)					
1967–68					
8	0	0	0	0	0
9	0	0	0	0	0
11	1165	1675	3050	8445	16 700
12	24	0	0	0	0
25	1258	1472	1992	4292	8425
26	0	0	0	0	0
27	486	700	1270	3520	6960
1968–69					
8	0	0	0	0	0
9	465	670	1220	3375	6672
11	2440	2440	2440	2440	2070
12	0	0	0	0	0
13	465	670	1220	3375	5572
14	24	0	0	0	0
25	4905	5080	5660	7815	10 412
26, 27	0	0	0	0	0

Appendix 3.3. *During the post-guard stage from the penguin colony determined from the prey remains in territories*

Pair	3–9 Jan.	10–14 Jan.	15–19 Jan.	20–24 Jan.	25–29 Jan.
EF Block					
1965–66					
144	2182	5807	3698	1813	0
145	1316	3748	2251	0	0
146	3328	7636	9847	2264	0
148	1754	4780	2972	0	4406
149	3026	2112	1254	0	0
151	2174	5027	3306	1905	0
152	5524	4863	3148	4471	0
153	8301	10 435	7110	7493	1538
1966–67					
144	4220	2956	949	1950	0
145	1237	1906	4396	1694	0
146	1942	5033	4162	4060	2470
148	3018	2692	7151	1825	3800
149	2921	5355	4314	1175	0
151	3857	2986	2725	1847	1889
152	2384	12 999	6928	2212	1539
153	4758	4792	3710	2161	3539

continued

Appendix 3.3. (*cont.*)

Pair	3–9 Jan.	10–14 Jan.	15–19 Jan.	20–24 Jan.	25–29 Jan.
EF Block (*cont.*)					
1967–68					
144	3153	1478	0	—	—
145	874	0	1653	—	—
146	2580	3460	0	—	—
148	1863	2246	1652	—	—
149	0	0	0	—	—
151	4235	4235	0	—	—
152	4566	3282	3282	—	—
153	4570	8394	1846	—	—
1968–69					
144	1404	0	0	—	—
145	1951	0	0	—	—
146 new pair	2244	2639	0	—	—
149	0	0	0	—	—
151	3245	2118	0	—	—
152	2432	5080	1125	—	—
153	1924	7215	3276	—	—
1969–70					
144	2482	4399	0	—	—
145	2676	2352	0	—	—
146 new pair	0	0	0	—	—
149	2567	2035	0	—	—
151	4317	0	0	—	—
152	4129	0	2544	—	—
153	5064	6200	2227	—	—
H Block					
1965–66					
8	2830	5854	0	0	—
9	5745	1290	0	1847	—
11	3678	3552	5831	0	—
12	4464	1243	1653	0	—
13	5429	3629	4137	0	—
14	0	0	0	0	—
19	6674	1217	2527	0	—
20	1226	0	1553	0	—
21	0	3584	1851	2476	—
22	2826	1959	0	0	—
24	2231	5508	3042	0	—
25	2206	1746	5291	2108	
26	868	4367	2646	1825	—
27	1464	3048	0	4524	—
1966–67					
8	1349	0	0	0	0
9	0	7264	4944	0	0
11	1961	4322	5849	2207	2385
12	3250	5291	2432	0	0

Appendix 3.3. (*cont.*)

Pair	3–9 Jan.	10–14 Jan.	15–19 Jan.	20–24 Jan.	25–29 Jan.
H Block (*cont.*)					
1966–67 (*cont.*)					
13	3916	3600	4260	1954	1749
14	556	0	1745	2555	0
19	2068	3428	0	0	0
20	0	0	2646	0	0
21	2658	3462	0	0	0
22	3667	5028	3492	6142	0
24	2418	5480	0	4216	2985
25	5559	5329	5599	6654	4767
26	0	0	0	1849	0
27	1579	8294	6135	0	0
1967–68					
8	1324	4436	0	0	0
9	849	3854	0	0	0
11	2803	1719	1454	4550	1589
12	0	921	0	0	0
13	1686	4080	2057	2187	0
14	0	0	0	0	0
19	2330	0	0	0	2655
20	2115	840	2150	0	0
21	0	3726	0	0	2455
22	3733	2567	2745	2092	1858
24	806	4882	0	0	0
25	2681	6681	4138	1990	2353
26	1249	0	0	0	0
27	434	1000	0	0	0
1968–69					
8	1126	0	0	—	—
9	2327	0	0	—	—
11	1618	0	0	—	—
12	0	0	0	—	—
13	1399	6147	2157	—	—
14	0	0	0	—	—
19	634	0	0	—	—
20	1771	0	0	—	—
21	0	0	0	—	—
22	0	0	0	—	—
24	1565	0	0	—	—
25	1176	6522	0	—	—
26	0	0	0	—	—
27	400	0	0	—	—
1969–70					
8	0	0	0	—	—
9	1827	0	0	—	—
11	2397	0	3142	—	—
12	0	0	0	—	—

continued

Appendix 3.3. (*cont.*)

Pair	3–9 Jan.	10–14 Jan.	15–19 Jan.	20–24 Jan.	25–29 Jan.
H Block (*cont.*)					
1966–67 (*cont.*)					
13	4420	1884	2954	—	—
14	2344	0	0	—	—
19	2796	0	0	—	—
20	2962	0	0	—	—
21	0	0	0	—	—
22	1413	2332	2646	—	—
24	2490	2520	0	—	—
25	1930	5499	2448	—	—
26	0	0	0	—	—
27	2551	0	0	—	—

APPENDIX 4

The average amount of food received by each skua pair from the colony if shared equally among the skuas of the local area. Net energy intake in kJ day^{-1}

	1965–66	1966–67	1967–68	1968–69	1969–70
EF Block					
Incubation stage (8 pairs)					
	—	—	961	1944	1690
Chick guard stage					
10–14 Dec.	—	1760	1400	1350	520
15–19 Dec.	—	2100	1720	1500	625
20–24 Dec.	—	3025	2605	1900	915
25–29 Dec.	—	6200	6065	3470	2050
30 Dec.–2 Jan.	—	12 070	11 256	5726	3739
Post-guard stage					
3–9 Jan.	3450	2583	2422	1492	2654
10–14 Jan.	5550	4839	2946	2130	1873
15–19 Jan.	4198	4292	1054	550	596
20–24 Jan.	2029	2115	—	—	—
25–30 Jan.	743	1227	—	—	—
H Block					
Incubation stage (14 pairs)					
	—	—	300	615	—
Chick guard stage (14 pairs)					
10–14 Dec.	—	480	185	555	—
15–19 Dec.	—	570	245	615	—
20–24 Dec.	—	810	410	770	—
25–29 Dec.	—	1760	1050	1385	—
30 Dec.–2 Jan.	—	3174	2023	2263	—

continued

Appendix 4 (*cont.*)

	1965–66	1966–67	1967–68	1968–69	1969–70
H Block					
Post-guard stage (17 pairs)					
3–9 Jan.	2331	1704	1135	706	1478
10–14 Jan.	2176	3030	2043	745	720
15–19 Jan.	1678	2280	738	127	658
20–24 Jan.	1000	1504	636	—	—
25–30 Jan.	333	699	642	—	—
31 Jan.–3 Feb.	376	—	508	—	—
5–9 Feb.	248	—	271	—	—
10–14 Feb.	716	—	276	—	—

Note: The nett food intake in grams can be estimated from these figures of energy content by dividing the totals given by the appropriate energy density for each date shown in Table 8.8.

APPENDIX 5

The survival of penguin eggs and chicks in breeding groups protected from skua in exclosures compared with survival in comparable unprotected groups

Breeding group	Total eggs laid	Viable eggs	Egg and chick survival		
			to 15 Dec.	to 1 Jan.	to 20 Jan.
1966–67					
Protected groups					
A17	74	70	69	69	63
C2	105	95	93	83	73
D4	121	114	110	94	84
E4	86	81	81	72	70
E5	76	74	70	64	60
Controls (unprotected groups)					
Groups in the centre of the colony					
A16	57	53	51	51	47
A23	55	50	44	36	34
A31	129	126	120	93	84
D32	77	71	67	66	61
E13	43	41	40	27	24
Groups on the edge of the colony					
C3	71	63	55	19	13
H1	51	46	40	17	11
F9	71	66	57	33	26
D1	43	43	39	38	34
1968–69					
Protected groups					
A17	42	37	32	30	27
C2	48	43	29	29	29
C3	32	29	20	19	17
E4	44	42	39	35	33
E5	54	54	40	37	35

continued

Appendix 5 (*cont.*)

Breeding group	Total eggs laid	Viable eggs	Egg and chick survival		
			to 15 Dec.	to 1 Jan.	to 20 Jan.
1968–69					
Controls (unprotected groups)					
Groups in the centre of the colony					
A4	71	64	56	40	38
A16	31	30	26	25	24
A23	23	22	20	19	15
D32	41	38	31	25	24
E11	47	43	38	28	27
F12	54	50	46	30	25
Groups on the edge of the colony					
B9	103	95	77	60	41
C1	102	96	77	64	63
D1	25	24	17	11	8
D4	31	30	30	25	20
F9	36	35	32	27	25

References

Ainley, D. G. (1975). Displays of Adélie penguins: a reinterpretation. In *The Biology of Penguins*, ed. Stonehouse, B., pp. 449–501. London: MacMillan Press.

Ainley, D. G. (1985). Biomass of birds and mammals in the Ross Sea. In *Antarctic Nutrient Cycles and Food Webs*, ed. Siegfried, W. R., Condy, P. R. and Laws, R. M., pp. 498–515. Heidelberg and Berlin: Springer-Verlag.

Ainley, D. G. and LeResche, R. E. (1973). The effects of weather and ice conditions on breeding in Adélie penguins. *Condor*, **75**, 235–255.

Ainley, D. G., LeResche, R. E. and Sladen, W. J. L. (1983). *Breeding Biology of the Adélie Penguin*. 239 pp. Berkeley, Los Angeles and London: University of California Press.

Ainley, D. G., Morrell, S. H. and Wood R. H. (1986). South Polar Skua breeding colonies in the Ross Sea Region, Antarctica. *Notornis*, **33**, 155–163.

Ainley, D. G., O'Connor, E. F. and Boekelheide, R. J. (1984). The Marine ecology of birds in the Ross Sea, Antarctica. *Ornithological Monographs*, **32**, 97 pp. Washington: American Ornithologists Union.

Ainley, D. G., Ribic, C. A. and Wood, R. C. (1990). A demographic study of the south polar skua *Catharacta maccormicki* at Cape Crozier. *Journal of Animal Ecology*, **59**, 1–20.

Ainley, D. G., Wood, R. C. and Sladen W. J. L. (1978). Bird Life at Cape Crozier, Ross Island. *Wilson Bulletin*, **90**, 492–510.

Andersson, M. (1973). Behaviour of the pomarine skua *Stercorarius pomarinus* Temm. with comparative remarks on Stercorariinae. *Ornis Scandinavica*, **4**, 1–16.

Andersson, M. (1976). Social behaviour and communication in the great skua. *Behaviour*, **58**, 40–77.

Ashmole, N. P. (1971). Sea bird ecology and the marine environment. In *Avian Biology*, ed. Farner, D. S. and King, J. R., volume 1, pp. 223–286. New York and London: Academic Press.

Astheimer, L. B. and Grau, C. R. (1985). The timing and energetic consequences of egg formation in the Adélie penguin. *The Condor*, **87**, 256–268.

Baker, J. R. (1938). The relation between latitude and breeding season in birds. *Proceedings Zoological Society of London*, **108**, 557–582.

Baudinette, R. V. and Schmidt-Nielsen, K. (1974). Energy cost of gliding flight in Herring Gulls. *Nature*, **248**, 83–84.

Birt-Friesen, V. L., Montevecchi, W. A., Cairns, D. K. and Macko, S. A. (1989). Activity specific metabolic rates of free-living northern gannets and other seabirds. *Ecology*, **70**, 357–367.

Bozinovic, F. and Medel, R. G. (1988). Body size, energetic and foraging mode of raptors in central Chile. *Oecologia*, (Berlin), **75**, 456–458.

Broady, P. A., Adams, C. J., Cleary, P. J. and Weaver, S. D. (1989). Ornithological observations at Edward VII Peninsula, Antarctica, in 1987–88. *Notornis*, **36**, 53–61.

Brook, D. and Beck, J. R. (1972). Antarctic petrels, snow petrels and south polar skuas breeding in the Theron Mountains. *British Antarctic Survey Bulletin*, **27**, 131–137.

Burton, R. W. (1968). Agonistic behaviour of the brown skua *Catharacta skua lonnbergi* (Matthews). *British Antarctic Survey Bulletin*, **16**, 15–39.

Caughley, G. (1960). The Adélie penguin of Ross and Beaufort Islands. *Records Dominion Museum New Zealand*, **3**, 263–282.

Clarke, A. and Prince, P. A. (1980). Chemical composition and calorific value of food fed to mollymauk chicks *Diomedea melanophris* and *D. chrysostoma* at Bird Island, South Georgia. *Ibis*, **122**, 488–494.

Costa, D. P. and Prince, P. A. (1987). Foraging energetics of grey-headed albatrosses *Diomedea chrysostoma* at Bird Island, South Georgia. *Ibis*, **129**, 149–158.

Culik, B., Adelung, D., Heise, M., Wilson, R. P., Coria, N. R. and Spairani, H. J. (1989). In situ heart rate and activity of incubating Adélie penguins (*Pygoscelis adeliae*). *Polar Biology*, **9**, 365–370.

Culik, B., Adelung, D. and Woakes, A. J. (1990). The effect of disturbance on the heart rate and behaviour of Adélie penguins (*Pygoscelis adeliae*) during the breeding season. In *Antarctic Ecosystems: Ecological Change and Conservation*, ed. Kerry, K. R. and Hempel, G., pp. 177–182. Heidelberg, Berlin and New York: Springer-Verlag.

Curio, E. (1976). *The Ethology of Predation*. Heidelberg, Berlin and New York: Springer-Verlag.

Daniels, R. A. (1982). Feeding ecology of some fishes of the Antarctic Peninsula. *Fishery Bulletin*, **80**, 575–588.

Darwin, C. (1859). *Origin of Species*. Penguin (1968 edition).

Davis, L. S. (1982). Creching behaviour of Adélie penguin chicks *Pygoscelis adeliae*. *New Zealand Journal of Zoology*, **9**, 279–286.

Davis, L. S. (1988). Coordination of incubation routines and mate choice in Adélie penguins (*Pygoscelis adeliae*). *The Auk*, **105**, 428–432.

Davis, L. S. and McCaffrey, F. T. (1986). Survival analysis of eggs and chicks of Adélie penguins *Pygoscelis adeliae*. *The Auk*, **103**, 379–388.

Davis, L. S., Ward, G. D. and Sadleir, R. M. F. S. (1988). Foraging by Adélie penguins during the incubation period. *Notornis*, **35**, 15–23.

Davis, R. W., Croxall, J. P. and O'Connell, M. J. (1989). The reproductive energetics of gentoo (*Pygoscelis papua*) and macaroni (*Eudyptes chrysolophus*) penguins at South Georgia. *Journal of Animal Ecology*, **58**, 59–74.

Devillers, P. (1978). Distribution and relationships of South American skuas. *Le Gerfaut*, **68**, 374–417.

DeVries, A. L. and Eastman, J. T. (1978). Lipid sacs as a buoyancy adaptation in an Antarctic fish. *Nature*, **271**, 352–353.

DeWitt, H. H. (1970). The character of the midwater fish fauna of the Ross Sea, Antarctica. In *Antarctic Ecology*, ed. Holdgate, M. W., volume 1, pp. 305–314. London: Academic Press.

Duke, G. E., Jegers, A. A., Loff, G. and Evanson, O. A. (1975). Gastric digestion in some raptors. *Comparative Biochemistry and Physiology*, **50A**, 649–656.

Eastman, J. T. (1985). *Pleuragramma antarcticum* (Pisces Nototheniidae) as food for other fishes in McMurdo Sound, Antarctica. *Polar Biology*, **4**, 155–160.

Eastman, J. T. and DeVries, A. L. (1982). Buoyancy studies of notothenioid fishes in McMurdo Sound, Antarctica. *Copeia*, **1982**, 385–393.

Eklund, C. R. (1961). Distribution and life history studies of the south polar skua. *Bird-Banding*, **32**, 187–223.

Emison, W. B. (1968). Feeding preferences of the Adélie penguin at Cape Crozier, Ross Island. In *Antarctic Bird Studies*, ed. Austin, O. L., Antarctic Research Series, volume 12, pp. 191–212. Washington: American Geophysical Union.

Ensor, P. H. (1979). The effect of storms on the breeding success of south polar skuas at Cape Bird, Antarctica. *Notornis*, **26**, 349–353.

Flint, E. N. and Nagy, K. A. (1984). Flight energetics of free-living sooty terns. *The Auk*, **101**, 288–294.

Furness, R. W. (1978). Energy requirements of seabird communities: a bioenergetics model. *Journal of Animal Ecology*, **47**, 39–53.

Furness, R. W. (1984). Influences of adult age and experience, nest location, clutch size and laying sequence on the breeding success of the great skua *Catharacta skua*. *Journal of Zoology*, **202**, 565–576.

Furness, R. W. (1987). *The Skuas*. Calton, UK: Poyser T. and A. D.

Furness, R. W. (1990). A preliminary assessment of the quantities of Shetland sandeels taken by seabirds, seals, predatory fish and the industrial fishery in 1981–83. *Ibis*, **132**, 205–217.

Furness, R. W. and Hislop, J. R. G. (1981). Diets and feeding ecology of great skuas *Catharacta skua* during the breeding season in Shetland. *Journal of Zoology*, **195**, 1–23.

Gabrielsen, G. W., Mehlum, F. and Nagy, K. A. (1987). Daily energy expenditure and energy utilization of free-ranging black-legged kittiwakes. *The Condor*, **89**, 126–132.

Green, K. (1986). Observations on the food of the south polar skua, *Catharacta maccormicki* near Davis, Antarctica. *Polar Biology*, **6**, 185–186.

Gwinner, E. (1986). *Circannual rhythms. Endogenous Annual Clocks in the Organisation of Seasonal Processes*. Berlin: Springer-Verlag.

Hamilton, W. D. (1971). Geometry for the selfish herd. *Journal of Theoretical Biology*, **31**, 295–311.

Harper, P. C., Knox, G. A., Spurr, E. B., Taylor, R. H., Wilson, G. J. and Young, E. C. (1984). The status and conservation of birds in the Ross Sea sector of Antarctica. In *Status and Conservation of the World's Seabirds*, ed. Croxall, J. P., Evans P. G. H., and Schreiber, R. W., pp. 593–608. Cambridge, UK: I.C.B.P.

Heine, J .C. and Speir, T. W. (1989). Ornithogenic soils of the Cape Bird Adélie penguin rookeries, Antarctica. *Polar Biology*, **10**, 89–99.

Hemmings, A. D. (1984). Aspects of the breeding biology of McCormick's skua *Catharacta maccormicki* at Signy Island, South Orkney Islands. *British Antarctic Survey Bulletin*, **65**, 65–79.

Hempel, G. (1985). Antarctic marine food webs. In *Antarctic Nutrient Cycles and Food Webs*, ed. Siegfried, W. R., Condy, P. R. and Laws, R. M. pp. 266–270. Berlin and Heidelberg: Springer-Verlag.

Herbert, C. (1967). A timed series of embryonic developmental stages of the Adélie penguin (*Pygoscelis adeliae*) from Signy Island, South Orkney Islands. *British Antarctic Survey Bulletin*, **14**, 45–67.

Hiller, A., Wand, U., Kampf, H. and Stackebrandt, W. (1988). Occupation of the Antarctic continent during the past 35,000 years: inferences from a 14C study of stomach oil deposits. *Polar Biology*, **9**, 69–77.

Hinde, R. A. (1956). The biological significance of territories of birds. *Ibis*, **98**, 340–369.

Hopkins, T. L. (1987). Midwater food web in McMurdo Sound, Ross Sea, Antarctica. *Marine Biology*, **96**, 93–106.

Howard, T. (1991). Seabirds that live in the mountains. *Anare News*, **March 1991**, 11–12.

Hubold, G. (1984). Spatial distribution of *Pleuragramma antarcticum* (Pisces: Nototheniidae) near the Filchner and Larsen Ice Shelves (Weddell Sea/ Antarctica). *Polar Biology*, **3**, 231–236.

Hubold, G. and Ekau, W. (1990). Feeding patterns of post-larval and juvenile notothenioids in the Southern Weddell Sea (Antarctica). *Polar Biology*, **10**, 255–260.

Hubold, G. and Tomo, A. P. (1989). Age and growth of Antarctic silverfish *Pleuragramma antarcticum* Boulenger, 1902, from the Southern Weddell Sea and the Antarctic Peninsula. *Polar Biology*, **9**, 205–212.

Jouventin, P. (1975). Mortality parameters in Emperor penguins *Aptenodytes forsteri*. In *The Biology of Penguins*, ed. Stonehouse B., pp. 435–446. London and Basingstoke: MacMillan.

Jouventin, P. (1982). Visual and vocal signals in penguins, their evolution and adaptive characters. *Advances in Ethology*, **24**, 1–149.

Kellogg, T. B., Kellogg, D. E. and Stuiver, M. (1990). Late quaternary history of the Southwestern Ross Sea: evidence from debris bands on the McMurdo Ice Shelf Antarctica. *Antarctic Research Series*, **50**, 25–56.

Kendeigh, S. C., Dol'nik, V. R. and Gavrilov, V. M. (1977). Avian energetics. In *Granivorous Birds in Ecosystems*, ed. Pinowski, J. and Kendeigh, S. C., pp. 127–204. Cambridge, UK: Cambridge University Press.

Klopfer, P. H. (1962). *Behavioural Aspects of Ecology*. Englewood Cliffs, New Jersey: Prentice Hall.

Kooyman, G. L. and Mullins, J. L. (1990). Ross Sea Emperor penguin breeding populations estimated by aerial photography. In *Antarctic Ecosystems: Ecological Change and Conservation*, ed. Kerry, K. R. and Hempel, G., pp. 169–176. Berlin and Heidelberg: Springer-Verlag.

Kruuk, H. (1972). *The Spotted Hyena*. Chicago: University of Chicago Press.

Lack, D. (1968). *Ecological Adaptations for Breeding in Birds*. London: Methuen.

Lasiewski, R. C. and Dawson, W. R. (1967). A re-examination of the relation between standard metabolic rate and body weight in birds. *The Condor*, **69**, 13–23.

Laws, R. M. (1984). Seals. In *Antarctic Ecology*, ed. Laws, R. M., volume 2, pp. 621–715. London: Academic Press.

Le Morvan, P., Mougin, J. L. and Prevost, J. (1967). Ecologie du skua antarctique (*Stercorarius skua maccormicki*) dans L'Archipel de Pointe Geologie (Terre Adélie). *L'Oiseau et la Revue Française d'Ornithologie*, **37**, 193–220.

Levick, G. M. (1914). *Antarctic Penguins*. London: William Heinemann.

Lishman, G. S. (1985a). The food and feeding ecology of Adélie penguins (*Pygoscelis adeliae*) and chinstrap penguins (*P. antarctica*) at Signy Island, South Orkney Islands. *Journal of Zoology* (A), **205**, 245–263.

Lishman, G. S. (1985b). The comparative breeding biology of Adélie and Chinstrap penguins (*Pygoscelis adeliae* and *P. antarctica*) at Signy Island, South Orkney Islands. *Ibis*, **127**, 84–99.

Longton, R. E. (1988). *The Biology of Polar Bryophytes and Lichens*. Cambridge, UK: Cambridge University Press.

MacDonald, J. A., Montgomery, J. C. and Wells, R. M. G. (1987). Comparative physiology of Antarctic fishes. *Advances in Marine Biology*, **34**, 322–388.

McGarry, R. (1988). Survey of McCormick skua colonies in Southern McMurdo Sound. *Antarctic Record*, **8**, 5–10.

Martin, P. and Bateson, P. (1986). *Measuring Behaviour. An Introductory Guide*. Cambridge, UK: Cambridge University Press.

Masman, D., Gordijn, M., Daan, S. and Dijkestra, C. (1986). Ecological energetics of the kestrel: field estimates of energy intake throughout the year. *Ardea*, **74**, 24–39.

Moynihan, M. (1962). Hostile and sexual behaviour patterns of South American and Pacific Laridae. *Behaviour*, (Supplement 8), 365 pp.

Müller-Schwarze, D. (1968). Circadian rhythms of activity in the Adléie penguin (*Pygoscelis adeliae*) during the austral summer. In *Antarctic Bird Studies*, ed. Austin, O. L., volume 12, pp. 133–149. Antarctic Research Series, Washington: American Geophysical Union.

Müller-Schwarze, D., Butler, R., Belanger, P., Bekoff, M. and Bekoff, A. (1975). Feeding territories of south polar skuas at Cape Crozier. *Antarctic Journal of the United States*, **10**, 121–122.

Müller-Schwarze, D. and Müller-Schwarze, C. (1973). Differential predation by south polar skuas in an Adélie penguin rookery. *The Condor*, **75**, 127–131.

Müller-Schwarze, D. and Müller-Schwarze, C. (1975). Relations between leopard seals and Adélie penguins. *Rapports et Proces-Verbaux des*

Reunions Conseil Internationat pour l' Exploration de la Mer, **169**, 394–404.

Müller-Schwarze, D. and Müller-Schwarze, C. (1977). Interactions between south polar skuas and Adélie penguins. In *Adaptations within Antarctic Ecosystems*, ed. Llano, G. A., pp. 619–646. Washington: Smithsonian.

Murton, R. K. and Westwood, N. J. (1977). *Avian Breeding Cycles*. Oxford: Clarendon Press.

Myrcha, A. and Kaminski, P. (1982). Changes in body calorific values during nestling development of penguins of the genus *Pygoscelis*. *Polish Polar Research*, **3**, 81–88.

Nagy, K. A. (1987). Food metabolic rate and food requirement scaling in mammals and birds. *Ecological Monographs*, **57**, 111–128.

Neilson, D. R. (1983). Ecological and behavioural aspects of the sympatric breeding of the South Polar Skua (*Catharacta maccormicki*) and the Brown Skua (*Catharacta lonnbergi*) near the Antarctic Peninsula. University of Minnesota: Unpublished thesis.

Norman, F. I. and Ward, S. J. (1990). Foods of the South Polar Skua at Hop Island, Rauer Group, East Antarctica. *Polar Biology*, **10**, 489–493.

Oelke, H. (1975). Breeding behaviour and success in a colony of Adélie penguins, *Pygoscelis adeliae* at Cape Crozier, Antarctica. In *The Biology of Penguins*, ed. Stonehouse, B., pp. 363–395. London: Macmillan.

Parmelee, D. F. (1988). The hybrid skua: a Southern Ocean enigma. *The Wilson Bulletin*, **100**, 345–346.

Parmelee, D. F., Bernstein, N. and Neilson, D. R. (1978). Impact of unfavourable ice conditions on bird productivity at Palmer Station during the 1977–1978 field season. *Antarctic Journal of the United States*, **13**, 146–147.

Parmelee, D. F., Fraser, W. R. and Neilson, D. R. (1977). Birds of the Palmer Station area. *Antarctic Journal of the United States*, **12**, 14–21.

Paulin, C. D. (1975). Feeding of the Adélie Penguin *Pygoscelis adeliae*. *Mauri Ora* (New Zealand) **3**, 27–30.

Penny, R. L. (1967). Molt in the Adélie Penguin. *The Auk*, **84**, 61–71.

Penny, R. L. (1968). Territorial and social behaviour in the Adélie penguin. In *Antarctic Bird Studies*, ed. Austin, O. L., Antarctic Research Series, volume 12, pp. 83–131. Washington: American Geophysical Union.

Penny, R. L. and Lowry, G. (1967). Leopard seal predation on Adélie penguins. *Ecology*, **48**, 878–882.

Pennycuick, C. J. (1987). Flight of Auks (Alcidae) and other northern seabirds compared with southern Procellariiformes: ornithodolite observations. *Journal of Experimental Biology*, **128**, 335–347.

Pennycuick, C. J. (1989). *Bird Flight Performance: A Practical Calculator Manual*. Oxford: Oxford University Press.

Perdeck, A. C. (1960). Observations on the reproductive behaviour of the great skua or bonxie, *Stecorarius skua skua* (Brün) in Shetland. *Ardea*, **48**, 111–136.

Peter, H-U., Kaiser, M. and Gebauer, A. (1990). Ecological and morphological investigation on south polar skuas (*Catharacta maccormicki*) and brown skuas (*Catharacta skua lonnbergi*) on Fildes Peninsula, King George Island, South Shetland Islands. *Zoologische Jahrbüecher Abteilung fuer Systematik Oekologie und Geographie der Tiere*, **117**, 201–218.

Pettit, T. M., Nagy, K. A., Ellis, H. I. and Whittow, G. C. (1988). Incubation energetics of the laysan albatross. *Oecologia* (Berlin), **74**, 546–550.

Pietz, P. J. (1984). Aspects of the Behavioural Ecology of Sympatric South Polar and Brown Skuas near Palmer Station, Antarctica. University of Minnesota: unpublished PhD thesis.

Pietz, P. J. (1985). Long call displays of sympatric South Polar skuas (*Catharacta maccormicki*) and brown skuas (*Catharacta lonnbergi*). *The Condor*, **87**, 316–326.

Pietz, P. J. (1987). Feeding and nesting ecology of sympatric south polar and brown skuas. *The Auk*, **104**, 617–627.

Poncet, S. and Poncet, J. (1987). Census of penguin populations of the Antarctic Peninsula 1983–1987. *British Antarctic Survey Bulletin*, **77**, 109–129.

Ponting, H. G. (1921, reprinted 1950). *The Great White South: Or with Scott in the Antarctic*. London: Duckworth.

Procter, D. L. C. (1975). The problem of chick loss in the south polar skua *Catharacta maccormicki*. *Ibis*, **117**, 452–459.

Pryor, M. E. (1968). The avifauna of Haswell Island, Antarctica. In *Antarctic Bird Studies*, ed. Austin, O. L., pp. 57–82. Antarctic Research Series, volume 12. Washington: American Geophysical Union.

Reid, B. E. (1964). The Cape Hallett Adélie penguin rookery – its size, composition, and structure. *Records Dominion Museum New Zealand*, **5**, 11–37.

Reid, B. E. (1965). The Adélie penguin (*Pygoscelis adeliae*) egg. *New Zealand Journal of Science*, **8**, 503–514.

Richdale, L. E. (1951). *Sexual Behavior in Penguins*. Lawrence, Kansas: University of Kansas Press.

Ricklefs, R. E. and Mathew, K. K. (1983). Rate of oxygen consumption in four species of seabird at Palmer Station, Antarctic Peninsula. *Comparative Biochemistry and Physiology*, **74A**, 885–888.

Roudybush, T. E., Grau, C. R., Petersen, M. R., Ainley, D. G., Hirsch, K. V., Gilman, A. P. and Patten, S. M. (1979). Yolk formation in some charadriform birds. *The Condor*, **81**, 293–298.

Sadleir, R. M. F. S. and Lay, K. M. (1990). Foraging movements of Adélie Penguins (*Pygoscelis adeliae*) in McMurdo Sound. In *Penguin Biology*, ed. Davis, L. S. and Darby, J. T., pp. 157–179. London and San Diego: Academic Press and Harcourt, Brace and Jovanovich.

Sapin-Jaloustre, J. and Bourliére, F. (1952). Parades et attitudes caracteristique de *Pygoscelis adeliae*. *Alauda*, **20**, 39–53.

Sibley, C. G. and Monroe, B. L. (1990). *Distribution and Taxonomy of the Birds of the World*. New Haven, Connecticut: Yale University Press.

Sinclair, M. R. (1982). Weather observations in the Ross Island area, Antarctica. *Technical note*, number 253, *New Zealand Meteorological Service*.

Sinif, D. B., DeMaster, D. P. and Hofman, R. J. (1977). An analysis of the dynamics of a Weddell Seal population. *Ecological Monographs*, **47**, 319–335.

Siple, P. A. and Lindsey, A. A. (1937). Ornithology of the Second Byrd Antarctic Expedition. *The Auk*, **54**, 147–159.

Sladen, W. J. L. (1958). The Pygoscelid Penguins. *Scientific Reports Falkland Islands Dependencies Survey*, **17**. London.

Smith, M. and Rigby, B. (1981). Distribution of polynyas in the Canadian Arctic. In *Polynyas in the Canadian Arctic*, ed. Stirling, I. and Cleator, H., pp. 7–28. Occasional Paper number 45, Canadian Wildlife Service.

Smith, W. J. (1969). Messages of vertebrate communication. *Science*, **165**, 145–150.

Spellerberg, I. F. (1966). Ecology of the McCormick Skua, *Catharacta maccormicki* (Saunders), in Southern McMurdo Sound, Antarctica. University of Canterbury: unpublished thesis.

Spellerberg, I. F. (1967). Distribution of the McCormick Skua (*Catharacta maccormicki*). *Notornis*, **14**, 201–207.

Spellerberg, I. F. (1969). Incubation temperatures and thermoregulation in the McCormick Skua. *The Condor*, **71**, 59–67.

Spellerberg, I. F. (1970a). Abandoned penguin rookeries near Cape Royds, Ross Island, Antarctica and 14C dating of penguin remains. *New Zealand Journal of Science*, **13**, 380–385.

Spellerberg, I. F. (1970b). Body measurements and colour phases of the McCormick skua. *Notornis*, **17**, 280–285.

Spellerberg, I. F. (1971a). Breeding behaviour of the McCormick skua *Catharacta maccormicki* in Antarctica. *Ardea*, **59**, 189–229.

Spellerberg, I. F. (1971b). Aspects of McCormick skua breeding biology. *Ibis*, **113**, 357–363.

Spurr, E. B. (1972). Social organisation of the Adélie penguin. University of Canterbury: unpublished PhD thesis.

Spurr, E. B. (1974). Individual differences in aggressiveness of Adélie penguins. *Animal Behaviour*, **22**, 611–616.

Spurr, E. B. (1975a). Orientation of Adélie penguins on their territories. *The Condor*, **77**, 335–337.

Spurr, E. B. (1975b). Communication in the Adélie penguin. In *The Biology of Penguins*, ed. Stonehouse, B., pp. 449–501. London: MacMillan Press.

Spurr, E. B. (1975c). Breeding of the Adélie penguin *Pygoscelis adeliae* at Cape Bird. *Ibis*, **117**, 324–338.

Spurr, E. B. (1975d). Behavior of the Adélie penguin chick. *The Condor*, **77**, 272–280.

Spurr, E. B. (1977). Adaptive significance of the reoccupation period of the Adélie penguin. In *Adaptations within Antarctic Ecosystems*, ed. Llano, G. A., pp. 605–618. Washington: Smithsonian.

Spurr, E. B. (1978). Diurnal activity of Adélie penguins *Pygoscelis adeliae* at Cape Bird. *Ibis*, **120**, 147–152.

Stahl, J. C., Jouventin, P., Mougin, J. L., Roux, J. P. and Weimerskirch, H. (1985). The foraging zones of seabirds in the Crozet Islands sector of the Southern Ocean. In *Antarctic Nutrient Cycles and Food Webs*, ed. Siegfried, W. R., Condy, P. R. and Laws, R. M., pp. 478–486. Berlin and Heidelberg: Springer-Verlag.

Stephens, D. W. and Krebs, J. R. (1986). *Foraging Theory*. Princeton, New Jersey: Princeton University Press.

Stirling, I. (1969). Ecology of the Weddell Seal in McMurdo Sound, Antarctica. *Ecology*, **50**, 573–586.

Stirling, I. (1971). Population dynamics of the Weddell Seal (*Leptonychotes weddelli*) in McMurdo Sound, Antarctica 1966–1968. In *Antarctic Pinnipedia*, ed. Burt, W. H., pp. 141–161. Antarctic Research Series 18. Washington: American Geophysical Union.

Stirling, I. (1980). The biological importance of polynyas in the Canadian Arctic. *Arctic*, **33**, 303–315.

Stonehouse, B. (1963). Observations on Adélie penguins (*Pygoscelis adeliae*) at Cape Royds, Antarctica. In *Proceedings. XIII International Ornithology Congress*, pp. 766–779.

Stonehouse, B. (1964). Emperor penguins at Cape Crozier. *Nature*, **203**, 849–851.

Stonehouse, B. (1965). Counting Antarctic animals. *New Scientist*, **29 July**, pp. 273–276.

Stonehouse, B. (1967a). Occurrence and effects of open water in McMurdo Sound, Antarctica, during winter and early spring. *Polar Record*, **13**, 775–778.

Stonehouse, B. (1967b). The general biology and thermal balances of penguins. In *Advances in Ecological Research*, ed. Cragg, J. B., volume 4, pp. 131–196. London and New York: Academic Press.

Stonehouse, B. (1970). Recent climate changes in Antarctica suggested from 14C dating of penguin remains. *Palaeogeography Palaeoclimatology Palaeoecology*, **7**, 341–343.

Sturman, A. P. and Anderson, M. R. (1986). On the sea-ice regime of the Ross Sea, Antarctica. *Journal of Glaciology*, **32**, 54–58.

Takahasi, M. and Nemoto, T. (1984). The food of some Antarctic fish in the Western Ross Sea in summer 1979. *Polar Biology*, **3**, 237–239.

Taylor, R. H. (1962a). The Adélie penguin *Pygoscelis adeliae* at Cape Royds. *Ibis*, **104**, 176–204.

Taylor, R. H. (1962b). Speed of Adélie penguins on ice and snow. *Notornis*, **10**, 111–113.

Taylor, R. H. and Roberts, H. S. (1962). Growth of Adélie penguin (*Pygoscelis adeliae* Hombron and Jacquinot) chicks. *New Zealand Journal of Science*, **5**, 191–197.

Taylor, R. H., Wilson, P. R. and Thomas, B. W. (1990). Status and trends of Adélie penguin populations in the Ross Sea region. *Polar Record*, **26**, 293–304.

Taylor, R. J. (1984). *Predation*. New York and London: Chapman and Hall.

Tenaza, R. (1971). Behaviour and nesting success relative to nest location in Adélie penguins (*Pygoscelis adeliae*). *The Condor*, **73**, 81–92.

Thomas, J. A. and DeMaster, D. P. (1983). Parameters affecting survival of Weddell Seal pups (*Leptonychotes weddelli*) to weaning. *Canadian Journal of Zoology*, **61**, 2078–2083.

Thomas, P. G. and Green, K. (1988). Distribution of *Euphausia crystallorophias* within Prydz Bay and its importance to the inshore marine ecosystem. *Polar Biology*, **8**, 327–331.

Todd, F. S. (1977). Permanent breeding colony of high Antarctic penguins for research and education. *Antarctic Journal of the United States*, **12**, 13–14.

Trillmich, F. (1978). Feeding territories and breeding success of south polar skuas. *The Auk*, **95**, 23–33.

Trivelpiece, W., Butler, R. G. and Volkman, N. J. (1980). Feeding territories of brown skuas (*Catharacta lonnbergi*). *The Auk*, **97**, 669–676.

Trivelpiece, W., Ainley, D. G., Fraser, W. R. and Trivelpiece, S. G. (1990). Skua Survival. *Nature*, **345**, 211–212.

Trivelpiece, W. and Volkman, N. J. (1979). Nest site competition between

Adélie and chinstrap penguins: an ecological interpretation. *The Auk*, **96**, 675–691.

Trivelpiece, W. and Volkman, N. J. (1982). Feeding strategies of sympatric south polar (*Catharacta maccormicki*) and brown (*C. lonnbergi*) skuas. *Ibis*, **124**, 50–54.

van Heezik, Y. (1988). Diet of Adélie penguins during the incubation period at Cape Bird, Ross Island, Antarctica. *Notornis*, **35**, 23–26.

Vermeu, G. J. (1982). Unsuccessful predation and evolution. *American Naturalist*, **120**, 701–720.

Volkman, N. J., Presler, P. and Trivelpiece, W. (1980). Diets of pygoscelid penguins at King George Island, Antarctica. *The Condor*, **82**, 373–378.

Volkman, N. J. and Trivelpiece, W. (1980). Growth in pygoscelid penguin chicks. *Journal of Zoology*, **191**, 521–530.

Volkman, N. J. and Trivelpiece, W. (1981). Nest-site selection among Adélie, chinstrap and gentoo penguins in mixed species rookeries. *Wilson Bulletin*, **93**, 243–248.

Volrath, F. (1984). Kleptobiotic interactions in invertebrates. In *Producers and Scroungers*, ed. Bernard, C. J., pp. 61–94. London: Croom Helm.

Whitehead, M. D., Johnstone, G. W. and Burton, H. R. (1990). Annual fluctuations in productivity and breeding sucesss of Adélie penguins and fulmarine petrels in Prydz Bay, East Antarctica. In *Antarctic Ecosystems: Ecological Change and Conservation*, ed. Kerry, K. R. and Hempel, G., pp. 214–223. Berlin, Heidelberg and New York: Springer-Verlag.

Wijnandts, H. (1984). Ecological energetics of the long-eared owl (*Asio otus*). *Ardea*, **72**, 1–92.

Williams, A. J., Siegfried, W. R. and Cooper, J. (1982). Egg composition and hatching precocity in seabirds. *Ibis*, **124**, 456–470.

Williams, J. M. (1969). Territorial Ecology and Ethology of McCormick's Skua *Catharacta maccormicki* (Saunders) at Cape Bird, Ross Island, Antarctica. University of Canterbury: unpublished thesis.

Wilson, E. A. (1907). National Antarctic Expedition 1901–1904. *Natural History, Zoology*, volume 2, part 2. *Aves*. London: British Museum, Natural History.

Wilson, G. J. (1983). Distribution and abundance of Antarctic and Sub-antarctic penguins: a synthesis of current knowledge. *BIOMASS Scientific Series*, **4**. Cambridge: SCAR, SCOR.

Wilson, K-J. (1990). Fluctuations in populations of Adélie penguins at Cape Bird, Antarctica. *Polar Record*, **26**, 305–308.

Wilson, K-J., Taylor, R. H. and Barton, K. J. (1990). The impact of man on Adélie penguins at Cape Hallett, Antarctica. In *Antarctic Ecosystems: Ecological Change and Conservation*, ed. Kerry, K. R. and Hempel, G., pp. 183–190. Berlin, Heidelberg and New York: Springer-Verlag.

Wilson, R. P., Nagy, K. A. and Obst, B. S. (1989). Foraging ranges of penguins. *Polar Record*, **25**, 303–307.

Wolf, L. L. and Hainsworth, F. R. (1978). Energy: expenditures and intakes. In *Chemical Zoology*, ed. Florkin, M. and Scheer, B. E., pp. 307–358. New York, San Francisco and London: Academic Press.

Wood, R. C. (1971). Population dynamics of breeding south polar skuas of unknown age. *The Auk*, **88**, 805–814.

Yeates, G. W. (1968). Studies on the Adélie penguin at Cape Royds 1964–1965 and 1965–1966. *New Zealand Journal of Marine and Freshwater Research*, **2**, 472–496.

Yeates, G. W. (1971). Observations on orientation of penguins to wind and on colonisation in the Adélie penguin rookery at Cape Royds Antarctica. *New Zealand Journal of Science*, **14**, 901–906.

Yeates, G. W. (1975). Microclimate, climate and breeding success in Antarctic penguins. In *The Biology of Penguins*, ed. Stonehouse, B., pp. 397–410. London: MacMillan.

Young, E. C. (1963a). The breeding behaviour of the South Polar skua *Catharacta maccormicki*. *Ibis*, **105**, 201–233.

Young, E. C. (1963b). Feeding habits of the South Polar skua *Catharacta maccormicki*. *Ibis*, **105**, 301–318.

Young, E. C. (1970). The techniques of a skua–penguin study. In *Antarctic Ecology 1*, ed. Holdgate, M. W., pp. 568–584. London: Academic Press.

Young, E. C. (1972). Territory establishment and stability in McCormick's skua. *Ibis*, **114**, 234–244.

Young, E. C. (1977). Egg laying in relation to latitude in Southern Hemisphere skuas. *Ibis*, **119**, 191–195.

Young, E. C. (1981). The ornithology of the Ross Sea. *Journal of the Royal Society of New Zealand*, **11** (4): 287–315.

Young, E. C. (1990a). Diet of the south polar skua *Catharacta maccormicki* from regurgitated pellets: limitations of a technique. *Polar Record*, **26**, 124–125.

Young, E. C. (1990b). Long-term stability and human impact in Antarctic skuas and Adélie penguins. In *Antarctic Ecosystems: Ecological Change and Conservation*, ed. Kerry, K. R. and Hempel, G., pp. 231–236. Heidelberg, Berlin and New York: Springer-Verlag.

Young, P. M. (1971). *Penguin Summer*. Wellington, New Zealand: Reed.

Index